Biocommunication of Fungi

Guenther Witzany
Editor

Biocommunication of Fungi

Springer

Editor
Guenther Witzany
Telos-Philosophische Praxis
Bürmoos, Austria

ISBN 978-94-007-4263-5   ISBN 978-94-007-4264-2 (eBook)
DOI 10.1007/978-94-007-4264-2
Springer Dordrecht Heidelberg New York London

Library of Congress Control Number: 2012940531

Printed on acid-free paper

Springer is part of Springer Science+Business Media (www.springer.com)

Preface

Why Biocommunication of Fungi?

When we consider the biocommunication of fungi, we first must become familiar with the current terms of communication (and with the signalling system, what we would term language or code, which is used to communicate). Therefore, we should consider the results of the pragmatic turn in the philosophy of science in the 1970s and 1980s, which was the result of discourse from 1920 until 1980, and was intended to clarify the conditions for generating correct sentences in science.

Communication is defined as the sign mediated interaction between at least two living agents, which share a repertoire of signs (which represents a kind of natural language) that are combined (according to syntactic rules) in varying contexts (according to pragmatic rules) to transport content (according to semantic rules).

Contrary to former concepts the importance of this result is that these three levels of semiotic rules are complementary parts of any natural language or code. If one level is missing, according to Charles Morris, we cannot seriously speak of language or signal mediated communication. Therefore, the most recent definition of communication is: sign-mediated and rule-governed interactions, i.e. interactions that depend on a shared repertoire of signs and rules. However, these features are lacking in abiotic interactions; no semiotic rules are necessary if water freezes to ice.

Additionally, we know that mathematical and mechanistic theories of language are less helpful in investigating natural language and real-life communication processes, because such theories cannot explain typical features of communication between living entities, which are not formal (i.e. for which no algorithm is available), such as the *de novo* generation of coherent, sentences or sequences. This means that no natural language or code speaks, or codes, itself but requires living agents that are competent in such languages or codes.

In the biology of the twentieth century, the physiology of all manner of cells, tissues, organs and organisms, was the mainstream direction of biological research and experiments. In the 1970s, an increasing use of "communication" as a metaphor also arose in biology. During the last decade of the twentieth century, interest in communication (no longer being used as a metaphor) within, and between, organisms overtook that of the purely physiological understanding of organisms. This was due to concrete communication processes designating varying contexts in real life circumstances. Cell-to-cell communication now dominates contemporary cell biology, including knowledge of a great variety of signalling pathways, serving for both organization and coordination of production, release, uptake, interpretation and processing of context-dependent "information" (content) within and between cells. Context dependency determines the crucial fact that, it is not the syntax (grammar) of a sequence of signs (information) which determines the meaning (semantics), but the context (pragmatics) in which the concrete use of the sequence occurs.

In parallel, the use of "language" as a metaphor has increased since the middle of the twentieth century with the improved knowledge of the genetic code. Most of the processes that evolve, constitute, conserve and rearrange the genetic storage medium (DNA), are terms that were originally used in linguistics. For example: nucleic acid language, genetic code, "code without commas" (F. Crick), coding, copying, translation, transcription, "genetic text" (F. Jacob), sequence homology, etc. Meanwhile, the linguistic approach also lost its metaphorical character, and the similarity between natural languages and codes, and the genetic storage medium of DNA have not only been accepted, but have been adapted in epigenetics, comparative genomics, bioinformatics, biolinguistics, biocommunication theory and biosemiotics.

The advantage of a methodical adaptation of communication and linguistic terminology is that it provides appropriate tools for differentiation at specific levels, which is otherwise difficult to describe in non-reductive terms by pure physiology. The result of this is that language like structures and communication processes occur at the simplest levels of nature. Language and communication are not the evolutionary inventions of humans, nor are they anthropomorphic adaptations to describe nonhuman entities. It is an empirical fact that all coordination and organisation within and between cells, tissues, organs and organisms needs signs, i.e. molecules that serve as signals or symbols in messages, or serve as vital indicators of environmental conditions. Because no natural code can encode itself, in the way that no natural language can speak itself, these signs must be sensed and interpreted in the correct way by biological agents, i.e. there must be subjects of sign production and sign interpretation. The consequence of this is that sensing, as well as interpretation, may fail. This can result in inappropriate behaviour, or even be of fatal consequence, for cells, tissues, organs, and organisms.

The method of analysing any part of a machine in detail to get a picture of the whole functional blueprint, which can then be used to reproduce or manipulate it, or to produce an even more perfect example, taking artificial genetic engineering as an example, is still useful if we are dealing with machines. In contrast communication between cells, cellular components, tissues, organs and organisms is far from being a procedure that can be reduced to mechanistic input/output or cause/reaction descriptions. It is evident that communication processes between living organisms include a variety of non-mechanistic circumstances and competences that must be fulfilled in parallel if communicative acts are to have successful consequences, for example, to innovate common coordination to adapt to new environmental conditions. Machines cannot create new programs out of a functional blueprint, which is in contrast to the abilities of living organisms that are able to communicate between each other.

Firstly, no single organism is able to communicate as an emerging property; it must be part of a community, society or swarm of organisms that share an identity and have a competence to sense whether others are part of this identity or not (self/non-self differentiation competence), even if this competence is solely shared genetically. In contrast to former opinions, it is now evident that communication is not primarily the exchange of information between a sender and a receiver, but it is actually a variety of social interactions. To communicate it is necessary that organisms have assets that serve as signs, signals or symbols, such as chemical molecules, either produced directly by the organism, or as secondary metabolites, or even molecules in the surroundings, but which can be manipulated according to the organisms needs.

Secondly, organisms must also share a competence to use these signs in a coherent manner, which means using these signs in a strict temporal and spatial context. In most cases it is not just one signalling molecule, but several, that are combined in a specific manner to transport messages or information. This represents a common feature of sign use in communication processes, and is termed the correct combination or syntax.

Thirdly, organisms are part of the natural habitat in which they live, together with similar organisms of the same or related species, but usually also with an abundance of unrelated organisms. This historically developed context exactly represents the natural history of the swarms or communities in the way that they have evolved certain abilities and are able to mount appropriate response behaviours to enable their survival. These competencies, which include sensing, monitoring, learning and memory, are preconditions for faster adaptation.

Finally, the signalling molecules, which serve as signs, transport messages with meanings (semantics). The informational (semantic) content which is transported, triggers certain response behaviours in the same, related, or even unrelated, organisms. Interestingly, the signal sequence or content does not necessarily depict a strict meaning, i.e. a function, but can vary according to different situational contexts. This means that identical signs can transport a variety of different messages according to different contextual needs. This is important in very dense ecological habitats, for example in mycelial differentiation, or root ecospheres. The different uses of identical signs, or sequences, enable the generation of dialects within the same species that can transport messages which are micro-ecosphere specific. This includes a very sensitive self/non-self recognition between slightly differently adapted populations of the same species in the same ecological habitat.

Although sign-mediated interactions (i.e. communication processes) are very reliable in most cases, they do not function mechanistically in a strict sense. Syntax (combination), pragmatics (context) and semantics (content), must function in parallel to ensure and optimize coordination, and thus survival of group members.

These three levels of semiotic rules (syntax, pragmatics and semantics), do not function mechanistically but can be varied, deleted, or, in certain circumstances, and in contrast to the capabilities of machines, generated *de novo*. Additionally, semiotic rules do not function by themselves but need semiotic subjects, i.e. living organisms that utilise such rules. If no living organism is present, semiotic rules, signs and communication are absent. Although highly conserved semiotic rules are modifiable, environmental circumstances, such as stress, can trigger adaptive responses. In such cases, signals may transport new messages which previously did not exist, broadening the communicative competences of organisms, i.e. broadening their evolutionary capabilities.

To answer the preface question, we can see that biocommunication in fungi integrates the biology of fungi with their communicative competencies, and gives a more coherent explanation and description of the full range of fungi capabilities than would be possible by mechanistic or even reductionist approaches. Natural communication assembles the full range of signal mediated interactions that are necessary in order to organise all evolutionary, and developmental coordination within, and between, cells, tissues, organs and organisms.

Contributions to the Biocommunication of Fungi

After the introduction in which a general overview on the key levels of communication of fungi is given in the first section on intraorganismic biocommunication of fungi, Jaqueline Servin, Asharie Campbell and Katherine Borkovich begin with the G protein signaling components in filamentous fungal genomes. Maria Bertolini, Fernanda Freitas, Renato de Paula, Fernanda Cupertina and Rodrigo Goncalves describe metabolic regulation processes in *Neurospora crassa*. Robert Cichewicz reports in the important role of epigenetic regulation processes. Jeremy Bruen investigates the role of double stranded RNA Viruses in nuclear genomes of fungi. David Soll reports on signal transduction pathways used by *Candida albicans* and related species. Tatiana Potapova investigates cell- cell communication in the tip growth of mycelial fungi. Elizabeth Hutchison and Louise Glass report on programmed cell death and heterokaryon incompatibility in filamentous fungi.

In the second section on interorganismic biocommunication of fungi Zdena Palková and Libuse Váchová investigate communication and differentiation in the development of yeast colonies. Philippe Silare focus on self versus non-self fungal recognition. Silvia Polaino and Alexander Idnurm describe the important role of pheromone signalling. Kenneth Nickerson, Audrey Atkin, Jessica Hargarten, Ruvini Pathirana and Sahar Hasim.

The third section in transorganismic biocommunication of fungi starts with Danielle Troppens and John Morrissey report on metabolite-mediated interactions between bacteria and fungi. Rusty Rodriguez and Marilyn Roossinck describe cross-kingdom communication and mutualism between viruses, fungi and plants. Aurélie Deveau, Jonathan Plett, Valérie Legué, Pascale Frey-Klett and Francis

Martin describe communication processes between plant, fungi and bacteria whereas Eli Borrego and Michael Kolomiets focus on signalling processes between fungi and plants. Andrey Averyanov, Tatiana Belozerskaya and Natalia Gessler focus on phytopathological aspects of the reactive oxygen species in fungal development. Massimo Reverberi, Anna Fabbri and Corrado Fanelli report on the role of oxidative stress and oxylipins in interactions between plants and fungi. Katharyn Affeldt and Nancy Keller investigate the role of oxylipins in fungal-mammalian interactions. Drion Boucias, Verena Lietze and Peter Teal investigate chemical signals that mediate insect-fungal interactions.

Contents

Part II Interorganismic Communication

Part III Transorganismic Communication

Introduction: Keylevels of Biocommunication in Fungi

Guenther Witzany

Abstract In contrast to former opinions, communicative acts are not restricted only to information exchange between a sender and a receiver, but designate a variety of social interactions mediated by signals according to syntactic (combinatorial), pragmatic (context dependent) and semantic (content-specific) rules. On one side, development and growth of fungal organisms depends upon successful communication processes within, and between, cells of fungal organisms. However, on the other side, sign mediated interactions are necessary to coordinate behaviour with the same, or related, fungal species and with non-related organisms such as bacteria, plants and animals. In order to generate an appropriate response behaviour, fungal organisms must be able to sense, interpret and memorise important indices from the abiotic environment and adapt to them appropriately. However, these communication and interpretation processes can also fail. In such cases the overall consequences could be disease-causing, or even lethal. Interestingly, certain rules of fungal communication are very similar to those of animals, while others more closely resemble those of plants.

Keywords Semiochemicals • Sign-mediated interactions • Context dependence • Symbiotic signalling

1 Introduction

Currently it is estimated that there are at least 1.5 million fungal species, out of which about 300,000 are described in the scientific literature. It is estimated that fungi account for at least one-fourth of the global biomass. On an evolutionary time scale, the kingdom of fungi emerged approx. 300 million years after the

G. Witzany (✉)
Telos-Philosophische Praxis, Vogelsangstraße 18c, A-5111 Buermoos, Austria
e-mail: witzany@sbg.at

G. Witzany (ed.), *Biocommunication of Fungi*, DOI 10.1007/978-94-007-4264-2_1,

appearance of the first animal species, although they descended from a common ancestor (Lang et al. 2002). In contrast to animal- and higher plant-life, monocellular representatives are fairly common among fungi, i.e., fungi are by no means mere multicellular organisms. This fact can be easily and coherently reconstructed through the lineages of Protoctista – in that coordinated behavioural patterns are found among single-celled eukaryotes, which closely resemble those of single-celled fungi. However, there are significant differences in protoctist structure (i.e. flagellated) and those of fungi (i.e. non-flagellated). Obviously, fungi have evolved out of protoctists, such as red and joch-algae (Margulis and Schwartz 1988).

As with plants, fungi are sessile organisms that can live for extremely long periods or extend over large areas: one example (*Armillaria gallica*) has been found which covers as much as 15 ha with an age of approx. 1,500 years (Casselman 2007). Endolithic fungi from the Antarctic are known to be among the most long-lived organisms on this planet (Villarreal 2005). Another example (*Armillaria ostoyae*) covers 900 ha with an estimated age of 2,400 years (Casselman 2007).

Higher fungi are modular hyphal organisms in that they reproduce by clonation or are parasexual. They establish interlocking networks. Like red algae they merge their cytoplasms to form a multi-nucleated cell. This holistically-merged body is also found among some plants. A spore germinates under appropriate environmental conditions and is followed by the formation of filaments called hyphae. The latter are characterised by nuclear division and spore formation, which develop into monokaryotic filaments (tip-growing). As in other tip-growing organisms, this process requires a high density of tip-localised mitochondria (Levina and Lew 2006).

The embryological stage, a characteristic of higher plants and animals, is completely absent in fungi (Margulis and Schwartz 1988; Ingold and Hudson 1993). Hyphae formation is also found among certain bacteria, such as Streptomyces and Actinomyces. When hyphae are tightly packed together into a mat we term this a mycelium. Each filament of hyphae has tubular side walls made mostly of chitin, a feature that is also common in Arthropoda. The cell walls that seem to separate adjacent cells in a filament are called septa – however, their porous nature does not really give them separating properties. The merger of filamental tips of the same or different species triggers a self/non-self identification process. This process is sign-mediated and results either in repulsion or attraction. If the latter occurs, merger into a dikaryotic mycelium takes place and initiates the formation of a fruiting body.

Most fungi are saprobes and decompose and feed on non-living organic matter. They secrete powerful enzymes that enable the cells to digest organic matter from the nearby environment outside their body, in turn breaking this down to smaller molecules that can be absorbed in a dissolved form (Jennings 1995). To deter potential predators, a number of complex and highly efficient deterring substances are produced by fungi (Margulis and Schwartz 1988).

Fungi are known to utilise a broad variety of symbiotic interactions with animals, plants and eukaryotic protozoa for both mutual benefit as well as parasitic and even lethal associations (Lott et al. 2006). They also settle on specific types of tissue. Fungal diseases are known to affect both plant and animal life, in some cases inducing devastating effects on and also in organisms (Humber 2008),

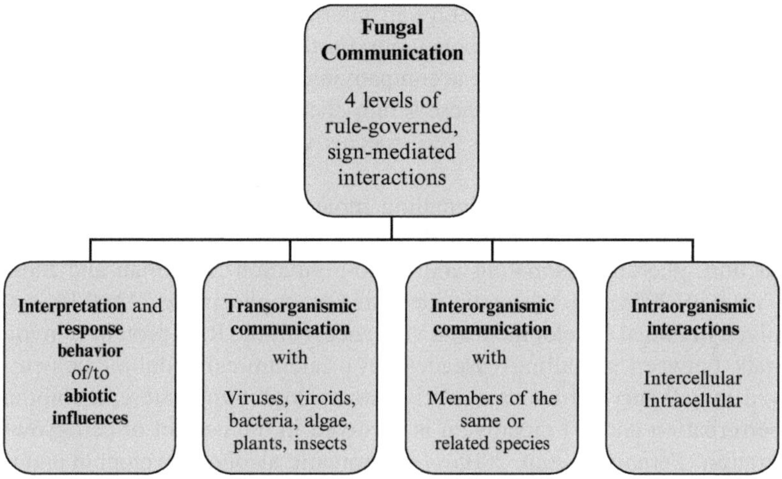

General levels of fungal communication

General levels of fungal communication

e.g. in agriculture. On the other hand, fungal activity has been actively exploited by humans for thousands of years in beer, wine, cheese and bread production. Since fungi are very simple organisms, sequencing and technical manipulation are relatively easy, making them ideal organisms for laboratory experiments (Weld et al. 2006), e.g. *Neurospora grassa* (Dunlap et al. 2004; FGSC 2011). Their application in the medical field as producers of various antibiotics is comprehensive and well-appreciated (e.g. penicillin).

Fungi communicate and therefore are able to organize and coordinate their behaviour. Coordination and organisation processes occur in all organismic kingdoms, and in fungi these are seen at the intraorganismic level e.g., during the formation of fruiting-bodies, between species of the same kind (inter-organismic) and between fungal and non-fungal organisms (trans-organismic). Without sign-mediated interactions, i.e. biocommunicative processes no coordination would occur (Witzany 2010). The semio-chemicals used are of biotic origin, in contrast to abiotic indicators that trigger the fungal organism to react in a specific manner.

2 Semiochemical Vocabulary

A variety of different primary signalling molecules have been shown to be involved in very different behavioural patterns such as filamentation, mating, growth, and pathogenicity (Adachi and Hamer 1998; Wang and Heitman 1999; Hemenway and Heitman 1999; Alspaugh et al. 2000; Borges-Walmsley and Walmsley 2000; Leeder et al. 2011). Behavioural coordination and the production of such substances can only be achieved through interpretation of incoming information followed by appropriate response behaviour. Furthermore, there are numerous, less-investigated subunits that play an accompanying role as they are weaker in effect (Lengeler et al. 2000). Globally, these semio-chemicals serve to coordinate similar goals in different fungal species, yet species-variation among them cannot be ignored.

The roles of some of these signalling molecules are as follows: (i) Mitogen activated protein kinase signalling (MAPK) is involved in cell integrity, cell wall construction, pheromones/mating and osmo-regulation (Dohlman and Slessareva 2006; Yu et al. 2008), (ii) the cyclic adenosine monophosphate cAMP/PKA system is involved in fungal development and virulence, (iii) the RAS protein is involved in cross-talk between signalling cascades, (iv) calcium-calmodulin-calcineurin are involved in cell survival under oxidative stress, high temperature, membrane/cell wall perturbation and (v) rapamycin is involved in the control of cell growth and proliferation (Fernandes et al. 2005), (vi) aromatic alcohols tryptophol and phenylathylalcohol are used as quorum sensing molecules, (vii) a variety of volatiles (alcohols, esters, ketones, acids, lipids) and non-volatile inhibitory compounds (farnesol, H_2O_2) (Leeder et al. 2011).

To date, 400 different secondary metabolites have been documented. These are known to contain mycotoxins and are used both for defensive and aggressive behaviours.

3 Sensing, Interpretation, Memory and Coordination of Response to Abiotic Indices

Fungi differentiate varying contexts such as nutrient availability and nutrient fluxes, and react by organising appropriate intra-organismic communication. For example, in the case of carbon or nitrogen insufficiencies, the internal communication system of the organism responds adequately and is phenotypically expressed by a change in hyphal growth. To date, two specific signalling pathways have been found that coordinate such behaviour. These diverging pathways have also been documented among other fungi, including those that are pathogenic to plants and animals (Lengeler et al. 2000).

An abundance of nutrients for example, results in increased production of cAMP (Lee et al. 2003), which suppresses that of Ste11, itself a mating inhibitor. Such processes are sign-mediated processes that originate from cyclase-associated proteins (CAP) (Lengeler et al. 2000). Nutrient availability has to be communicated to induce cell growth (Dechant and Peter 2008).

As with animals and plants, seasonality is found in fungi as a part of the circadian rhythm (Dunlap et al. 2004; De Paula et al. 2008), e.g. light-regulated physiological processes that coordinate the internal fungal clock (Bell-Pederson et al. 1996). *Neurospora crassa*, for example, responds according to irradiance patterns arising from the diurnal cycle. In such cases we can speak of a *Neurospora* chronotype (Dunlap et al. 2004; Tan et al. 2004). However, this responsiveness is not connected to photosynthetic activity, as this is only found in green algae, i.e. in the symbiotic association of lichens.

Fungi are capable to generate adaptive response behavioural patterns against environmental stress situations (Alonso-Monge et al. 2009) such as ROS reactive oxygen species (Moradas-Ferreira and Costa 2000) or in the case of varying osmotic conditions (Hohmann 2002; Santos and Shiozaki 2004; Krantz et al. 2006) which includes constant monitoring of cell wall integrity (Levin 2005).

Almost all fungi are saprobes and feed on dead organic matter. The excretion of extracellular digestive enzymes fragments larger biomolecules and makes them soluble, and more readily accessible for the fungal organism. This is particularly important for the digestion of cellulose, which is broken down by the enzymatic activity of exocellulase and endocellulase, and lignin (lignin peroxidase and manganese peroxidase). Enzymatic breakdown of organic matter yields simple sugars, amino acids, fatty acids and other smaller molecular components (Margulis and Schwartz 1988).

4 Remarkable Genomic Interactions Between Fungi and Viruses

Although prokaryotic ancestors of fungi were colonised mainly by large DNA viruses, in fungal populations they are missing but are commonly colonised by persistent double-stranded RNA viruses, single-stranded RNA viruses, double-stranded DNA viruses, retroviruses and even prions. They are the simplest organisms to be colonised by retroviruses, in the form of fungal retroposons. In contrast to all other eukaryotes, fungal mitochondria are infected by either double-stranded RNA or DNA viruses. Most fungus-infecting viruses are not pathogenic although they are associated with toxin genes that are pathologically dangerous, not for the infected host, but for non-infected fungal relatives and non-fungal organisms. This phenomenon has been used quite successfully by brewing industries, to protect useful yeasts from exogenous yeast genetic parasites by colonising the industrial yeast colonies with protective versions of killer viruses (Villarreal 2005). An example for mutually beneficial symbiosis is that of the black aspergillus species and dsRNA viruses, as in the case of mycovirus (van Diepeningen et al. 2006).

The interconnected filamentous fungi are a very attractive habitat for viruses because they allow rapid motility through the whole cellular network. Fungal networks also allow replacement of nuclei that are left behind in mitochondria. This may be an advantage if mitochondria are colonised by parasites, which is common (Villarreal 2005). It has been suggested that non-pathogenic viral colonisation of fungal mitochondria protects mtDNA from age-dependent degradation and is beneficial for the longevity of the fungal host. As a consequence, the growth radius of the fungal mats may increase up to tenfold (Villarreal 2005).

Interestingly, higher fungi lack repeat elements in DNA, whilst lower fungi can have up to 50% of their genomes colonised by repeat elements. Recent research shows that these repeat elements represent remnants of early retroviral genomic colonisation events. Fungi were the first organisms in evolutionary history to be colonised by viral RNA. This assumption is underlined by the presence of SINEs and LINEs. So far up to 15 SINE-families have been recognised (Whisson et al. 2005), a fact that supports the assumption that very early viral colonisation of the fungal genome must have taken place (Villarreal 2005; Rooney and Ward 2005).

Although plants and animals seem to be descendants of fungi, the relationship between fungi and viruses is different from virus–animal or virus–plant relationships (Villarreal 2005). The latter two lack linear plasmids, killer phenotypes, mitochondrial infections, distorted senescence and ubiquitous double-stranded RNA colonisation, which are all characteristics of the fungal virus–host relationship. This may be an indication of co-evolutionary interactions between fungi and viruses (Villarreal 2005).

Genomic studies show that most methylated regions of *Neurospora crassa* are derived from repeat-induced point mutations (RIP), which comprise a pre-meiotic homology-based genome defense system. This early immune system consists of a variety of inactivated transposons that include DNA modifications and chromatin modifications (Selker et al. 2003; Galagan and Selker 2004). As shown by comparative genomics, such inactivated transposons descended from viral infection events that reached a persistent status, providing the host with a new phenotype of an innate immune system in order to prevent similar infections (Villarreal 2005).

In yeast species this transposon is derived from an endogenous retrovirus, i.e. the characteristic Ty1 element, which encodes a functional reverse transcriptase, and a GAG gene, which encodes structural proteins. It is inserted into the silent regions of chromatin, which means that this is a good example of a beneficial situation for both the colonising virus and the host. Interestingly, yeasts also persistently harbour killer viruses that are lethal to related yeast species that are not infected persistently. Generally it is assumed that the high rate of different toxins in the organismic kingdom of fungi represents such toxin/antitoxin modules that are persistently integrated into the fungal host genomes (Villarreal 2009).

Interestingly RNAi that is lost in some yeast can be restored by introducing Argonaute and Dicer from close relatives. The reconstitutives silence endogenous transposons. Restoring RNAi affected maintenance of killer. This endemic viral system is inherited as a dsRNA virus (L-A) and its satellite dsRNA, M that encodes a toxin, that kills nearby cells and confers immunity to cells making the toxin. This clearly resembles a toxin/antitoxin addiction module provided by viral infection events to yeast hosts (Drinnenberg et al. 2011).

In comparison to animal or plant genomes fungal genomes are not highly populated by genetic settlers. Only one family of endogenous retroviruses is present in several full-length copies, some of them completely deactivated in specific fungal lineages. Such chromovirus sequences sometimes match with fungal host sequences. In Ascomycetes and Basidiomycetes these sequences are distinct but inactivated and not transcribed, which could be an indicator of several massive viral colonisation events during which the previously integrated viruses were displaced. Interestingly, a small number of viral copies are highly conserved and transcribed, so we can assume that they are closely connected with fungal evolution (Villarreal 2009). The high infection rate by dsRNA viruses and endogenous retroviruses is coherent with the presence of a small interfering RNA system with the ability for post-transcriptional silencing. This new system for identifying RNA and repeated DNA sequences by double-stranded RNA sequences is able to silence RNA expression, its main purpose being to silence endogenous retroviruses (chromovirus) and retroposons (Villarreal 2009).

Interestingly, the mitochondria of fungi are highly colonised by persistent double-stranded RNA viruses, each of them from another lineage but present in the host genome as multiple interconnected settlers. Especially the oxidative competence of mitochondrial respiration which is essential for life span of fungal host share some antagonist which regulates its function. If regulation of this process is disturbed, the life span of the host can be interrupted abruptly (Villarreal 2009).

Generally the remarkable genomic interactions between fungi and viruses seems to be an indicator of the virus-driven evolutionary history of fungi in that the virus–host relationship, especially in the endogenous persistent status (represented by dsRNA and small linear dsDNA viruses), is linked to all immunity mechanisms in fungi, such as killer toxins in mitochondria or siRNA, i.e. dsRNA-based identification competence, with an important role in silencing endogenous retroviruses and limiting retroviral infections (Villarreal 2009). This would be coherent with the fact that most fungal genomes are small in comparison to other eukaryotic genomes that lack the ability to ward off RNA viruses and retroviruses.

5 Intraorganismic Communication of Fungi

The countless variety of fungal organisms represents a major challenge when establishing a homogeneous designation of the signalling processes employed. Research activities so far have predominantly focused on those fungal species that pose a serious threat to agriculture, are pathogenic to humans or possess antibacterial properties. Species of this kind are relatively well investigated, whereas species with obviously insignificant properties are little known.

Growth in *Saccharomyces cerevisiae*, for example, is entirely dependent on the final purpose of growth, i.e. its growth and behavioural patterns are correlated. Thus mating of haploid fungal cells is a completely different behavioural cycle than filamentous growth of diploid cells. The signals involved can be differentiated by four different signal-related production patterns (Lengeler et al. 2000).

There is new evidence that certain proteins that have been restricted to signalling pathways in animals so far are also common in fungal signalling pathways (Herranz et al. 2005). In addition, apoptotic processes, which have been considered to play important roles in the development of multicellular organisms of animals and plants, have now been found to be important for regulating development, growth and aging processes in single-celled fungi (Hamann et al. 2008).

A very large family of transmembrane receptors is responsible for transducing extracellular signals into intracellular responses such as G protein-coupled receptors (Xue et al. 2008).

5.1 Intercellular Communication

Hyphal growth is a totally different behavioural pattern of conduct than normal cell growth: such cells change shape, become elongated, and re-orientate themselves into specific directions to physically come into contact or even merge with each other, only to colonise a potential growth resource.

Fungal hyphae simultaneously extend in a given direction only once nutritional resources are ideally distributed. However, this is a rare event. Usually the fungus propagates in the direction that is enriched with organic matter (carbon and its derivatives) just to halt growth once little or no resources are available; here the fungus coordinates its growth by establishing certain priorities. In order to do so, the fungus employs intercellular signals that enable it to comprehend the overall state of the organism. This includes an integrated ability to discriminate between self and non-self. Once the fungus encounters a resource-depleted substrate or even poisonous compounds, it responds by halting its growth cycle or by propagating in another direction. The protein signals involved in such processes are quite complex: the apex of hyphae houses specialised receptors that are able to respond to any environmental condition.

Any carbon-enriched substrate causes these receptors to become active, which in turn results in the production and release of protein signals into the hyphal cytosol where the corresponding signalling cascade is triggered. In turn, the mycelium responds with the mobilisation and translocation of resources into the activated area. In the absence of carbon-rich substrates, or at increasingly acidic pH levels, the hyphae respond by activating yet other receptors that slow down growth and eventually cause the organism to withdraw resources from the affected area. Since septa within the hyphae are perforated, they perform similar functions as the gap junctions in higher animals, micro-plasmodesmata in cyanobacteria and plasmodesmata in higher plants (Belozerskaya 1998; Gessler et al. 2007).

An interesting example of intercellular communication in the sexual cycle of Saccharomyces cerevisiae. The sexual cycle of Saccharomyces cerevisiae begins by cells of the opposite type in producing a and alpha pheromones. The haploid cells possess receptors for pheromenes of the opposite mating type. They sense them and respond by stopping normal growth (G1) phase and develop cell outgrowth in the direction of the opposite cell. These outgrowths may fuse into a joint cell with two haploid nuclei that can fuse to produce a nucleus of an a/alpha diploid cell. This means that intercellular communication process initiates three subsequent steps of internal coordination as appropriate response, G1 phase stop of the cell cycle, cell und nuclear fusion, and cell morphogenesis into direction of pheromone emitting partner (Shapiro 2011). This is clearly a concrete example of how cells communicate and trigger a genomic expression event to coordinate an appropriate response. Complex sensing and signalling are step by step coordination. Are these signalling pathways disturbed or damaged communication will not function appropriately.

5.2 Intracellular Communication

By investigating a great number of signal transduction events from the outside through the cell membrane into the cytoplasm it has been possible to decode some important intracellular communication processes. Thus, it was found that signal processes coordinate cell polarity, mating, pheromone control and cellular morphology. Some of these processes even adjust the cell cycle, cause polarised growth activity and modify the transcription profiles of fungal cells (Lengeler et al. 2000; Bardwell 2004; Fernandes et al. 2005).

By examining the fungal pathogen *Paracoccidioides brasilienus* it was realised that some signalling pathways are identical to those of other species such as *Saccharomyces cerevisiae*, *Cryptococcus neoformans*, *Candida albicans* and *Aspergillus fumigatus* (Fernandes et al. 2005).

The "protein cascades" that characterise production pathways of appropriate chemicals and messenger signals reflect the behavioural contexts, which are to some extent completely different (Fernandes et al. 2005):

- Cell integrity, cell wall construction, pheromone/mating, and osmo-regulation by mitogen-activated protein kinase signalling (MAPK)
- Fungal development and virulence by the cAMP/PKA system
- Cross-talk between cascades by the RAS protein
- Cell survival under oxidative stress, high temperature, membrane/cell wall perturbation by calcium-calmodulin-calcineurin
- Control of cell growth and proliferation by rapamycin

Combinatorial communication procedures such as those involving the MAPK and cAMP pathways are also part of the behavioural contexts (Lengeler et al. 2000). These in turn serve to multiply the semantic contents of the encoded messages.

Several signalling pathways have been found that sense extracellular stimuli and convert them into intracellular signals that regulate developmental and growth processes signalling. The guanine nucleotide-binding protein (G-Protein) is essential for extracellular detection of nutrients and sexual partners (Dohlman 2002; Kays and Borkovich 2004; Hoffman 2005). The TOR protein kinases, which are bound and inhibited by rapamycin, function as nutrient-sensing signals and regulate cellular responses like proliferation, transcription, translation, autophagy and ribosome biogenesis (Beck and Hall 1999; Cutler et al. 2001).

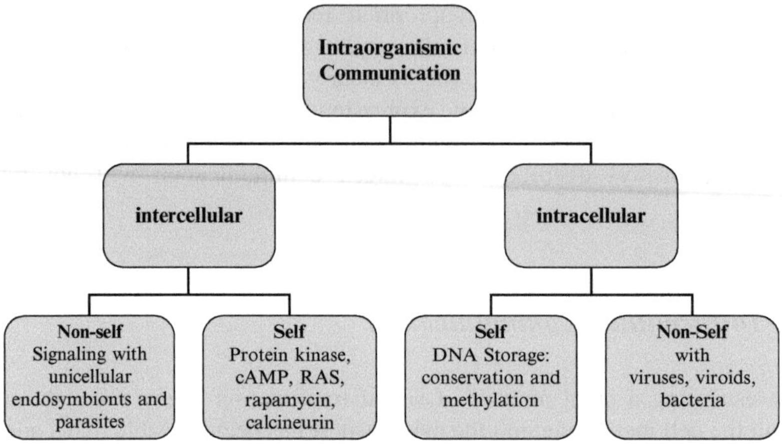

Levels of intraorganismic fungal communication

As with any signal-mediated interaction that can be achieved with molecules, the same components are employed for differing behavioural contexts and in varying messages. That is, different modes of behaviour can be achieved by syntactically identical signalling. The signalling pathways use identical proteins to coordinate different response patterns. Even if they are syntactically identical they semantically contain a completely different meaning: for example, activated cAMP

- triggers filamentous growth in *Saccharomyces cerevisiae*,
- regulates positive virulence in *Cryptococcus neoformans*,
- suppresses mating in *Schizosaccharomyces pombe* and
- inhibits filamentous growth in *Ustilago maydis* (Lengeler et al. 2000; D'Souza and Heitman 2001) or
- activates protein kinase for directly or indirectly induced developmental changes in *Magnaporthe grisea* during infection of rice (Mitchel and Dean 1995).

Another example is the *Ustilago maydis* pheromone response, which regulates both cell fusion and the pathogenicity programme for plant infection (Krüger et al. 1998; Hartmann et al. 1996, 1999).

These examples show that different behavioural contexts determine different meanings of identical signalling molecules. In such cases, biosemiotically identical signals can induce opposite responses in different organisms. It is interesting to note that fungi are not just capable of differentiating among varying messages and responding appropriately, but moreover are able to differentiate among molecules that are chemically identical to signalling molecules and obviously contain no relevant meaning ("noise"), i.e. are not parts of biotic messages.

Recent genome comparisons have provided new insights into the evolutionary aspects of fungi. The hypothesis that evolution happened by whole genome duplication events followed by selective gene-loss and stabilisation is strengthened by analysis on *Saccharomyces cerevisiae* (Kellis et al. 2004). Interestingly the signal-to-noise ratio in yeasts is approximately 70% protein coding regions and 15% regulatory elements in the non-protein coding regions, in comparison to humans with ratios of 3% and 97% respectively (Kellis et al. 2003).

6 Interorganismic Communication of Fungi

Since there are both single and multi-cellular species, the communication processes between the same species and related fungal species cannot be distinguished unambiguously from intercellular communication (intra-organismic). Thus communication processes in the mono-cellular yeast (Banuett 1998), which resemble that of amoeba like *Dictyostelium*, must be considered as inter-organismic communication, whereas those within multi-cellular fungal species are truly intercellular. To verify whether cell-to-cell communication is inter- or intra-organismic, we have to consider intercellular processes on a case-by-case basis.

Herein we find another fundamental characteristic of biota, concerning the competence for identifying "self" and "non-self" (Muirhead et al. 2002). This competence was successfully demonstrated in *Neurospora crassa* (Glass et al. 2000; Glass and Saupe 2002; Glass and Kaneko 2003). It is obvious that this capacity to distinguish between oneself and others is vital for fungi, since the encounter of mycelia from the same species results in the merger of their fungal hyphae. However, such dikaryotic mycelia can also result from the merger of different fungal species.

While peripheral hyphae tend to avoid merging with hyphae of other species, the opposite is the case with those at the centre of the mycelium (Glass et al. 2000; Glass and Saupe 2002; Glass and Kaneko 2003). If one assigns mycelia the role of a wrapper within which the fungus is enveloped, so to speak, into a fluid-like continuum, then the nuclei of compatible but different species are "flowing" through the same mycelium. The overall result is an organism, which houses nuclei of different genetic origin in its cytoplasm (Wu and Glass 2001). However, if specific genetic sequences are incompatible, then repulsion sets in, forcing the approaching hyphae into an immune-like response, much like those found in plants.

Resource competition in fungi occurs directly, indirectly and via mechanical interaction. Indirect competition involves absorption of all available resources within the reach of the mycelium, thereby starving potential competitors by maintaining a nutritional deficiency gradient. Direct interaction on the other hand, involves secondary metabolites, which suppress growth or even induce death of the competing fungi. The antibiotics employed in such cases can be either volatile or non-volatile (Kettering et al. 2004). Mechanical interaction simply requires overgrowth of one fungal species by the other, in which the overgrowing species exerts its lytic action on the other. In some cases lysis is induced via antibiotic agents (Dix and Webster 1995; Griffin 1994).

Like bacteria, single-celled fungi also use quorum sensing to regulate and affect biofilm formation and pathogenesis (Reynolds and Fink 2001). This is mediated by small molecules that accumulate in the extra-cellular environment. If these reach a critical concentration a response regulator is activated within the local population of cells, leading to the coordination of gene expression (Hogan 2006).

In parasitic interactions between fungal organisms cytoplasmic fusion was found during infection processes, which indicated genetic transfer in the host–parasite relationship. The recognition pattern in this predator-prey relationship is mediated by trisporoids, which are also involved in a non-parasitic behavioural pattern, being responsible for sexual communication (Schultze et al. 2005).

Some fungi have drawn attention only as saprobes, ectomycorrhizal symbionts or parasites of plants, while their role as parasites of other fungi has not yet been mentioned e.g. basidiomycetes. Their role in a great variety of interactions is still being studied (Bauer and Oberwinkler 2008).

7 Transorganismic Communication of Fungi

Although viruses can be considered as essential agents of life they are lacking metabolism, that means they cannot be considered as organisms. In addition to the above outlined relationship to viruses, fungi are competent to generate a great variety of signalling processes with non-fungal organisms such as plants, animals (insects, nematodes) and bacteria. A typical example of this kind is the mutually beneficial symbiosis between bark beetles and quite a few different fungi (Sullivan and Berisford 2004). The fungal spores benefit from the locomotion provided by the beetles in several ways e.g. access to new hosts, while the beetle benefits from the availability of fungal nutrients and pheromones. Some fungi provide nitrogen, amino acids and sterols that are crucial for the development of beetle larvae – however, this takes effect only once the adult beetle has colonised a host. Many bark beetles have even evolved transportation pockets for fungal hyphae, a fact that points to their common evolutionary history (Kopper et al. 2004).

Another trans-organismic symbiotic signalling process happens between fungi and ants, and is derived from a co-evolutionary relationship that lasted over millions of years (Poulsen and Boomsma 2005). Interestingly some lignin-degrading fungi also produce semiochemicals that have effects on the feeding and foraging behaviour of a Formosan subterranean termite (Cornelius et al. 2004).

Some fungi live as both soprophytes and parasites. Nematode trapping fungi we can find in terrestrial and aquatic ecosystems. They play important symbiotic roles in maintaining nematode population density and therefore can control parasitic nematodes of plants and animals. They are capable of developing trapping devices such as adhesive networks and knobs, constricting rings to capture nematodes and then extract nutrients out of them. To organize this nematode trapping these fungi have to coordinate various steps intracellularly (Yang et al. 2011).

There is much evidence to suggest that the fungal and animal kingdom share common ancestors, such as protoctists, which have a true nucleus like choanoflagellates (Villarreal 2005). In this sense, fungi and animals are more related to each other than is the case with the plant kingdom. This is further strengthened by the sign-mediated processes, which regulate cellular functions. Yet another indicator of their common ancestry is found in a particular signalling pathway, termed the mitogen-activated protein kinase cascade (MAPK). This plays a crucial role in cell wall stabilisation of fungi and pheromone/mating interactions among mammalian cells. On top of that, MAPK is highly conserved (Lengeler et al. 2000).

Then there are fungi which parasitise plants. For example, they may colonise host tissues with an intercellular mycelium that forms haustoria, i.e. fungal mats within plant cells (Jakupovic et al. 2006) that penetrate the cell to utilise the nutrients of the plant. Studies on hazardous fungal infections on plants have revealed the crucial role of enzymes such as cutinase, pisatin, demethylase and HC-toxins (NIAID 1993).

Today, several hundred species of fungi colonise more than 100,000 different plant species. This type of cohabitation relies on and requires symbiotic signalling (Lammers 2004), e.g. the initiation of filamentous growth of fungi through plant hormones (Prusty et al. 2004).

Roots of plants provide better conditions for mycorrhizal fungi, which in turn supply plants with better nutrients (Witzany 2006). For the fungus, such a relationship is either balanced or predatory. Endophytic fungi, however, live in plants without triggering symptoms of disease (Brundrett 2002). Today, scientists consider the origin of the plant cell to be the result of the terrestrial activity of mycorrhiza, i.e. settlement on land was a co-evolutionary event that is comparable to that between flowering plants and insects (Villarreal 2005). The mutually beneficial relationship between subterranean fungi and plant roots is a fine-tuned network of sign-mediated interactions developed over millions of years (Besserer et al. 2006), whereby fungi excrete digestive enzymes into the surrounding soil, and convert nutrients into aqueous solutions that in turn can be readily absorbed by the plant. A staggering 80% of all terrestrial plants rely on the activity of mycorrhiza, especially trees (Schwarze et al. 2004).

Fungi that affect animals are usually dependent on the host's body temperature, i.e. host colonisation by the fungi is only possible if the body temperature is sufficiently high. *Aspergillus fumigatus* in particular colonises animal hosts if they are under thermal stress (Bhabhra et al. 2006). Although fungal disease is common in birds, the relative resistance of endothermic vertebrates to fungal diseases may be a result of immune responses connected with higher body temperature (Casadevall 2006).

A typical fungus cultivated in research laboratories is *Neurospora*. Several subspecies of this genus contain toxins that are pathogenic for animals and plants. However, they are essential components for the large-scale industrial production of antibiotics, chemicals, enzymes and pharmaceuticals (Dunlap et al. 2004).

One of the most striking trans-organismic communication processes between fungi and non-fungal species can be found in lichens (Sanders 2001, 2006). Lichens are polyphyletic. They have been derived many times independently from different kinds of ascomycetes, so their nature of symbiosis doubtless varies (Raven 2001). In lichens, the algae provide the fungi with photosynthate, while the fungi provide the algae with nutrients. Lichens constitute one of the oldest known fungal members and are capable of resisting quite adverse environmental conditions. The symbiosis between fungi and algae or fungi and bacteria results in a mutual supply of nutrients and their associated competences. Through quorum sensing, the fungi benefit from the bacterial association (Hogan 2006). In turn, the bacteria utilise dissolved fungal metabolites to satisfy their nutritional requirements.

A similar co-dependence is observed with algae as symbiosis partners. But there is also a great variety of other sign-mediated interactions between fungi and bacteria that are beneficial for both and are based on reciprocal signalling (McAlester et al. 2008). Trans-organismic communication can be found in any interaction that involves fungi and viruses, bacteria, protoctista (algae), animals (insects) and plants.

8 Summary

An overview of all significant levels of fungal communication shows that identification of signal-mediated processes in signalling pathways are context dependent – both within and among fungal cells as well as between fungi and other organisms.

Depending on the context, semio-chemicals (molecular components) are integrated into unique signalling pathways where they are used to transport certain meanings. Such meanings are subject to change, i.e. they rely on various behavioural contexts, which differ under altered conditions. These contexts concern cell adhesion, pheromone response, calcium/calmodulin, cell integrity, osmotic growth, stress response or cell growth. The interactional context determines the semantic relationship, i.e., its meaning and the function of the chemical components, and forms a signal-mediated communication pattern in fungi. This is a common feature in all eukaryotic kingdoms: the context determines the meaning of trans-, inter- and intra-organismic (inter- and intracellular) communication, while differences in abiotic and biotic signal perception determine the content arrangement of response behaviour.

After recognising how versatile fungal communication competences really are we can see that one main principle is followed throughout all these signalling processes: fungal organisms coordinate all their behavioural patterns with a core set of chemical molecules. The interactional context and the different modes of coordinating appropriate response behaviour in e.g. development, growth, mating, attack, defense, virulence, etc. determines the combinations of signals that generate the appropriate meaning-function, i.e. informational content of messages. These generating processes normally function in a very conservative way but under certain circumstances may fail, or selective pressure may lead to changes that can be a driving force in fungal evolution.

Additionally it can be recognised that the persistent lifestyle of viruses is a driving force in fungal evolution in that they are the main resource for immunity, group identity and a large number of important secondary metabolites.

References

Adachi K, Hamer JE (1998) Divergent cAMP signaling pathways regulate growth and pathogenesis in the rice blast fungus *Magnaporthe grisea*. Plant Cell 10:1361–1374

Alonso-Monge R, Román E, Arana DM, Pla J, Nombela C (2009) Fungi sensing environmental stress. Clin Microbiol Infect 15:17–19

Alspaugh JA, Cavallo LM, Perfect JR, Heitman J (2000) Ras1 regulates filamentation, mating and growth at high temperature of *Cryptococcus neoformans*. Mol Microbiol 36:352–365

Banuett F (1998) Signalling in the yeasts: an informational cascade with links to the filamentous fungi. Microbiol Mol Biol Rev 62:249–274

Bardwell L (2004) A walk-through of the yeast mating pheromone response pathway. Peptides 25:1465–1476

Bauer R, Oberwinkler F (2008) Cellular basidiomycete – fungus interactions. In: Varma A, Abbott L, Werner D, Hampp R (eds) Plant surface microbiology. Springer, Berlin/Heidelberg, pp 267–279

Beck T, Hall MN (1999) The TOR signalling pathway controls nuclear localization of nutrient-regulated transcription factors. Nature 402:689–692

Bell-Pederson D, Dunlap JC, Loros JJ (1996) Distinct cis-acting elements mediate clock, light, and developmental regulation of the Neurospora crassa eas (ccg2) gene. Mol Cell Biol 16:513–521

Belozerskaya TA (1998) Cell-to-cell communication in differentiation of mycelial fungi. Membr Cell Biol 11(6):R831–R840

Besserer A, Puech-Pages V, Kiefer P, Gomez-Roldan V, Jauneau A, Roy S, Portais JC, Roux C, Becard G, Sejalon-Delmas N (2006) Strigolactones stimulate arbuscular mycorrhizal fungi by activating mitochondria. PLoS Biol 4(7):e226. doi:10.1371/journal.pbio.0040226

Bhabhra R, Zhao W, Rhodes JC, Askew DS (2006) Nucleolar localization of aspergillus fumigatus CgrA is temperature-dependent. Fungal Genet Biol 43:1–64

Borges-Walmsley MI, Walmsley AR (2000) cAMP signalling in pathogenic fungi: control of dimorphic switching and pathogenicity. Trends Microbiol 8:133–141

Brundrett MC (2002) Coevolution of roots and mycorrhizas of land plants. New Phytol 154:275–304

Casadevall A (2006) Fungal virulence, vertebrate endothermy, and dinosaur extinction: Is there a connection? Fungal Genet Biol 42:98–106

Casselman A (2007) Strange but true: the largest organism on earth is a fungus. Sci Am 4 Oct

Cornelius ML, Bland JM, Daigle DJ, Williams KS, Lovisa MP, Connick WJ, Lax AR (2004) Effect of a lignin-degrading fungus on feeding preferences of formosan subterranean termite (Isoptera: Rhinotermitidae) for different commercial lumber. J Econ Entomol 97:1025–1035

Cutler NS, Pan X, Heitman J, Cardenas MA (2001) The TOR signal transduction cascade controls cellular differentiation in response to nutrients. Mol Biol Cell 12:4103–4113

Dechant R, Peter M (2008) Nutrient signals driving cell growth. Curr Opin Cell Biol 20:678–687

De Paula RM, Lamb TM, Bennett L, Bell-Pedersen D (2008) A connection between MAPK pathways and circadian clocks. Cell Cycle 7:2630–2634

Dix NJ, Webster J (1995) Fungal ecology. Chapman & Hall, London

Dohlman HG (2002) G proteins and pheromone signalling. Annu Rev Physiol 64:129–152

Dohlman HG, Slessareva JE (2006) Pheromone signaling pathways in yeast. Sci Signal 364:cm6

Drinnenberg IA, Fink GR, Bartel DP (2011) Compatibility with killer explains the rise of RNAi-deficient fungi. Science 333:1592

D'Souza CA, Heitman J (2001) Conserved cAMP signaling cascades regulate fungal development and virulence. FEMS Microbiol Rev 25:349–364

Dunlap JC, Loros JJ, Denault D, Lee K, Froelich A, Colot H, Shi M, Pregueiro A (2004) Genetics and molecular biology of circadian rhythms. In: Brambl R, Marzluf GA (eds) The mycota III. Biochemistry and molecular biology, 2nd edn. Springer, Berlin/Heidelberg, pp 209–229

Fernandes L, Araujo MAM, Amaral A, Reis VCB, Martins NF, Felipe MS (2005) Cell signaling pathways in *Paracoccidioides brasiliensis* – inferred from comparisons with other fungi. Genet Mol Res 4:216–231

FGSC – Fungal Genetics Stock Center (2011) The Neurospora homepage. http://www.fgsc.net. Accessed 12 Sept 2011

Galagan JE, Selker EU (2004) RIP: the evolutionary cost of genome defense. Trends Genet 20:417–423

Gessler NN, Aver'yanov AA, Belozerskaya TA (2007) Reactive oxygen species in regulation of fungal development. Biochemistry (Mosc) 72:1091–1109

Glass NL, Kaneko I (2003) Fatal attraction: nonself recognition and heterokaryon incompatibility in filamentous fungi. Eukaryot Cell 2:1–8

Glass NL, Saupe SJ (2002) Vegetative incompatibility in filamentous ascomycetes. In: Osiewacz HD (ed) Molecular biology of fungal development. Marcel Dekker, New York, pp 109–131

Glass NL, Jacobson DJ, Shiu PKT (2000) The genetics of hyphal fusion and vegetative incompatibility in filamentous ascomycete fungi. Annu Rev Genet 34:165–186

Griffin DH (1994) Fungal physiology, 2nd edn. Wiley-Liss, New York

Hamann A, Brust D, Osiewacz HD (2008) Apoptosis pathways in fungal growth, development and aging. Trends Microbiol 16:276–283

Hartmann HA, Kahmann R, Bölker M (1996) The pheromone response factor coordinates filamentous growth and pathogenicity in *Ustilago maydis*. EMBO J 15:1632–1641

Hartmann HA, Krüger J, Lottspeich F, Kahmann R (1999) Environmental signals controlling sexual development of the corn smut fungus *Ustilago maydis* through the transcriptional regulator Prf1. Plant Cell 11:1293–1305

Hemenway CS, Heitman J (1999) Calcineurin: structure, function, and inhibition. Cell Biochem Biophys 30:115–151

Herranz S, Rodriguez JM, Bussink HJ, Sanchez-Ferrero JC, Arst HN Jr, Penalva MA, Vincent O (2005) Arrestin-related proteins mediate pH signalling in fungi. Proc Natl Acad Sci USA 102:12141–12146

Hoffman CS (2005) Except in every detail: comparing and contrasting G-protein signaling in *Saccharomyces cerevisiae* and *Schizosaccharomyces pombe*. Eukaryot Cell 4:495–503

Hogan DA (2006) Talking to themselves: autoregulation and quorum sensing in fungi. Eukaryot Cell 584:613–619

Hohmann S (2002) Osmotic stress signalling and osmoadaptation in yeasts. Microbiol Mol Biol Rev 66:300–372

Humber RA (2008) Evolution of entomopathogenicity in fungi. J Invertebr Pathol 98:262–266

Ingold CT, Hudson HJ (1993) The biology of fungi. Chapman & Hall, London

Jakupovic M, Heintz M, Reichmann P, Mendgen K, Hahn M (2006) Microarray analysis of expressed sequence tags from haustoria of the rust fungus *Uromyces fabae*. Fungal Genet Biol 43:8–19

Jennings DH (1995) Physiology of fungal nutrition. Cambridge University Press, New York

Kays AM, Borkovich KA (2004) Signal transduction pathways mediated by heterotrimeric G proteins. In: Brambl R, Marzluf GA (eds) The mycota III. Springer, Berlin, pp 175–207

Kellis M, Patterson N, Endrizzi M, Birren B, Lander ES (2003) Sequencing and comparison of yeast species to identify genes and regulatory elements. Nature 423:241–254

Kellis M, Birren BW, Lander ES (2004) Proof and evolutionary analysis of ancient genome duplication in the yeast *Saccharomyces cerevisiae*. Nature 428:617–624

Kettering M, Sterner O, Anke T (2004) Antibiotics in the chemical communication of fungi. Z Naturforsch 59:816–823

Kopper BJ, Klepzig KD, Raffa KF (2004) Components of antagonism and mutualism in Ips pini – fungal interactions: relationship to a life history of colonizing highly stressed and dead trees. Environ Entomol 33:28–34

Krantz M, Becit E, Hohmann S (2006) Comparative genomics of the HOG-signalling system in fungi. Curr Genet 49:137–151

Krüger J, Loubradou G, Regenfelder E, Hartmann A, Kahmann R (1998) Crosstalk between cAMP and pheromone signalling pathways in *Ustilago maydis*. Mol Gen Genet 260:193–198

Lammers PJ (2004) Symbiotic signaling: new functions for familiar proteins. New Phytol 161:324–326

Lang BF, O'Kelly C, Nerad T, Gray MW, Burger G (2002) The closest unicellular relatives of animals. Curr Biol 12:1773–1778

Lee N, D'Souza CA, Kronstad JW (2003) Of smuts, blasts, mildews, and blights: cAMP signaling in phytopathogenic fungi. Annu Rev Phytopathol 41:399–427

Leeder AC, Palma-Guerrero J, Glass NL (2011) The social network: deciphering fungal language. Nat Rev Microbiol 9:440–451

Lengeler KB, Davidson RC, D'Souza C, Harashima T, Shen WC, Wang P, Pan X, Waugh M, Heitman J (2000) Signal transduction cascades regulating fungal development and virulence. Microbiol Mol Biol Rev 64:746–785

Levin D (2005) Cell wall integrity signalling in *Saccharomyces cerevisiae*. Microbiol Mol Biol Rev 69:262–291

Levina NN, Lew RR (2006) The role of tip-localized mitochondria in hyphal growth. Fungal Genet Biol 43:65–134

Lott TJ, Fundyga RE, Kuykendall RJ, Arnold J (2006) The human commensal yeast, *Candida albicans*, has an ancient origin. Fungal Genet Biol 42:444–451

Margulis L, Schwartz KV (1988) Five kingdoms. W. H. Freeman and Company, New York

McAlester G, O'Gara F, Morrissey JP (2008) Signal-mediated interactions between *Pseudomonas aeruginosa* and *Candida albicans*. J Med Microbiol 57:563–569

Mitchell TK, Dean RA (1995) The cAMP-dependent protein kinase catalytic subunit is required for appressorium formation and pathogenesis by the rice blast pathogen *Magnaporthe grisea*. Plant Cell 7:1869–1878

Moradas-Ferreira P, Costa V (2000) Adaptive response of the yeast *Saccharomyces cerevisiae* to reactive oxygen species: defenses, damage and death. Redox Rep 5:277–285

Muirhead CA, Glass NL, Slatkin M (2002) Multilocus self-recognition systems in fungi as a cause of trans-species polymorphism. Genetics 161:633–641

National Institute of Allergy and Infectious Diseases (NIAID) (1993) Molecular medical mycology SECRETION. In: Proceedings of the NIAID workshop in medical mycology, University of Minnesota, MN, http://www.niaid.nih.gov/dmid/meetings/mycology/secretion.htm. Accessed 4 Apr 2012

Poulsen M, Boomsma JJ (2005) Mutualistic fungi control crop diversity in fungus-growing ants. Science 307:741–744

Prusty R, Grisafi P, Fink GR (2004) The plant hormone indoleacetic acid induces invasive growth in *Saccharomyces cerevisiae*. Proc Natl Acad Sci USA 101:4153–4157

Raven JA (2001) Selection pressures on stomatal evolution. New Phytol 153:371–386

Reynolds TB, Fink GR (2001) Bakers' yeast, a model for fungal biofilm formation. Science 291:806–807

Rooney AP, Ward TJ (2005) Evolution of a large ribosomal RNA multigene family in filamentous fungi: birth and dead of a concerted evolution paradigm. Proc Natl Acad Sci USA 102:5084–5089

Sanders WB (2001) Lichens: interface between mycology and plant morphology. Bioscience 51:1025–1035

Sanders WB (2006) A feeling for the superorganism: expression of plant form in the lichen thallus. Bot J Linn Soc 150:89–99

Santos JL, Shiozaki K (2004) Phosphorelay signaling in yeast in response to changes in osmolarity. Sci Signal 262. doi:10.1126/stke.2622004tr12

Schultze K, Schimek C, Wostemeyer J, Burmester A (2005) Sexuality and parasitism share common regulatory pathways in the fungus *Parasitella parasitica*. Gene 348:33–44

Schwarze FWMR, Engels J, Mattheck C (2004) Fungal strategies of wood decay in trees, 2nd edn. Springer, Heidelberg

Selker EU, Tountas NA, Cross SH, Margolin BS, Murphy JG, Bird AP, Freitag M (2003) The methylated component of the *Neurospora crassa* genome. Nature 422:893–897

Shapiro JA (2011) Evolution: a view from the 21st century. FT Press Science, New Jersey

Sullivan BT, Berisford CW (2004) Semiochemicals from fungal associates of bark beetles may mediate host location behavior of parasitoids. J Chem Ecol 30:703–717

Tan Y, Merrow M, Roenneberg T (2004) Photoperiodism in *Neurospora crassa*. J Biol Rhythms 19:135–143

Van Diepeningen AD, Debets AJM, Hoekstra RF (2006) Dynamics of dsRNA mycoviruses in black aspergillus populations. Fungal Genet Biol. doi:10.1016/j.fgb.2006.01.014

Villarreal LP (2005) Viruses and the evolution of life. ASM Press, Washington, DC

Villarreal LP (2009) Origin of group identity. Viruses, addiction and cooperation. Springer, New York

Wang P, Heitman J (1999) Signal transduction cascades regulating mating, filamentation and virulence in *Cryptococcus neoformans*. Curr Opin Microbiol 2:358–362

Weld RJ, Plummer KM, Carpenter MA, Ridgway HJ (2006) Approaches to functional genomics in filamentous fungi. Cell Res 16:31–44

Whisson SC, Avrova AO, Lavrova O, Pritchard L (2005) Families of short interspersed elements in the genome of the oomycete plant pathogen, *Phytophtora infestans*. Fungal Genet Biol 42:351–365

Witzany G (2006) Plant communication from biosemiotic perspective. Plant Signal Behav 1:169–178

Witzany G (2010) Biocommunication and natural genome editing. Springer, Dordrecht

Wu J, Glass NL (2001) Identification of specificity determinants and generation of alleles with novel specificity at the het-c heterokaryon incompatibility locus of *Neurospora crassa*. Mol Cell Biol 21:1045–1057

Xue C, Hsueh YP, Heitman J (2008) Magnificent seven: roles of G protein-coupled receptors in extracellular sensing in fungi. FEMS Microbiol Rev 32:1010–1032

Yang J, Wang L, Ji X, Feng Y, Li X et al (2011) Genomic and proteomic analyses of the fungus *Arthrobotrys oligospora* provide insights into nematode-trap formation. PLoS Pathog 7(9): e1002179. doi:10.1371/journal.ppat.1002179

Yu RC, Pesce CG, Colman-Lerner A, Lok L, Pincus D, Serra E, Holl M, Benjamin K, Gordon A, Brent R (2008) Negative feedback that improves information transmission in yeast signalling. Nature 456:755–761

Part I
Intraorganismic Communication

G Protein Signaling Components in Filamentous Fungal Genomes

Jacqueline A. Servin, Asharie J. Campbell, and Katherine A. Borkovich

Abstract In fungi, heterotrimeric G proteins regulate a number of critical developmental processes including growth, mating and pathogenesis. Signals may originate from extracellular ligands or from internal sources. These signals are conveyed from G protein coupled receptors (GPCRs) or nonreceptor guanine nucleotide exchange factors (GEFs) to a heterotrimeric G protein composed of α, β and γ subunits. GPCRs and nonreceptor GEFs activate the Gα subunit, causing the disassociation of the heterotrimer. Both the Gα and Gβγ heterodimer are free to act upon downstream effectors. Two prominent output pathways are the cyclic adenosine monophosphate (cAMP) and mitogen-activated protein kinase (MAPK) pathways. Changes in growth, mating and pathogenesis can be initiated through G protein signals and executed downstream by these or as yet uncharacterized signaling pathways.

1 Introduction

In fungi, heterotrimeric G proteins function in integral signal transduction pathways that regulate growth, mating and pathogenesis. In canonical G protein signaling, an extracellular ligand binds to a G protein coupled receptor (GPCR) located at the plasma membrane (Fig. 1). The GPCR conveys the extracellular message to the G protein heterotrimer bound to the intracellular side of the plasma membrane.

J.A. Servin • A.J. Campbell • K.A. Borkovich (✉)
Department of Plant Pathology and Microbiology, Institute of Integrative Genome Biology, University of California, Riverside, 900 University Avenue, Riverside, CA 92521, USA
e-mail: jacqueline.servin@ucr.edu; asharie.campbell@ucr.edu; Katherine.Borkovich@ucr.edu

G. Witzany (ed.), *Biocommunication of Fungi*, DOI 10.1007/978-94-007-4264-2_2,

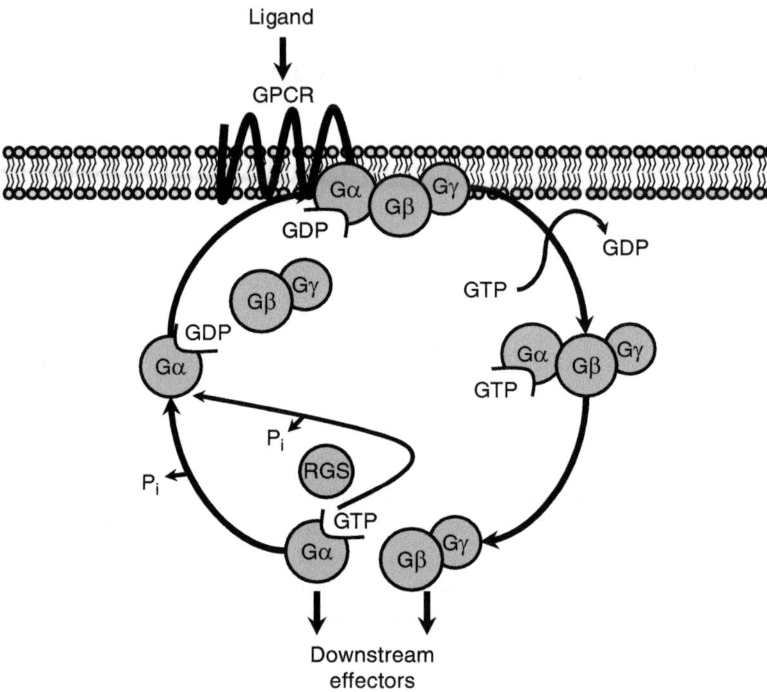

Fig. 1 The G protein cycle. Ligand binding to the G protein–coupled receptor (GPCR) leads to exchange of GDP for GTP on the Gα subunit and dissociation of the Gα and Gβγ heterodimer. Both Gα-GTP and Gβγ may regulate downstream effectors. GTP hydrolysis by the Gα subunit results in reassociation of GDP-bound Gα with the Gβγ heterodimer and the GPCR, thus completing the cycle. RGS (regulator of G protein signaling) proteins accelerate the rate of GTP hydrolysis by the Gα subunit

The heterotrimer is composed of a Gα subunit and a Gβ-Gγ heterodimer (Neves et al. 2002). Upon binding of an extracellular ligand, the GPCR acts as a guanine nucleotide exchange factor (GEF), thereby activating the Gα by causing exchange of guanosine diphosphate (GDP) for guanosine triphosphate (GTP). Alternatively, activation can be achieved through GTP exchange facilitated by nonreceptor cytoplasmic GEFs. The GTP-bound Gα dissociates from the Gβγ moiety and both are free to activate downstream effectors. GTP hydrolysis by the Gα subunit can be accelerated by a regulator of G protein signaling (RGS) protein. The GDP-bound Gα reassociates with the Gβγ heterodimer and awaits the next cycle of activation.

This chapter will provide a brief introduction to the signaling components that function during heterotrimeric G protein signaling in filamentous fungi. Due to space limitations we are unable to reference all the work in this field. Two output pathways, cAMP (cyclic adenosine monophosphate) and MAPK (mitogen-activated protein kinase) signaling, will be discussed. Finally, examples will be given for two major processes regulated via G protein signaling: mating and pathogenesis.

2 G-Protein Signaling Components

2.1 *Gα Subunits*

Most characterized filamentous fungi contain three Gα proteins which belong to three distinct groups: Group I, Group II and Group III (Bölker 1998; Li et al. 2007a). This is in contrast to the budding and fission yeasts which contain only two Gα proteins. Heterotrimeric G proteins are members of the GTPase superfamily (Coleman and Sprang 1996; Hamm and Gilchrist 1996; Oldham and Hamm 2008). Gα subunits are composed of an α-helical and a GTPase domain (Bohm et al. 1997). The GTPase domain consists of five α helices surrounding a β sheet. The guanine nucleotide-binding pocket is located within the GTPase domain and includes the GXGXXGKS consensus sequence for GTP binding (Hamm and Gilchrist 1996). Most Gα subunits undergo post-translational lipid modifications which aid in tethering the protein to the membrane, thereby facilitating the interaction with the Gβγ heterodimer (Resh 1996). Of the three fungal G protein groups, Group I Gα proteins share the most sequence similarity to those in the mammalian $G\alpha_i$ superfamily and typically contain a myristolyation sequence at their amino terminus (Li et al. 2007a; Turner and Borkovich 1993). Group I proteins are highly conserved in filamentous fungi and regulate multiple pathways, including pheromone response and mating, nutrient sensing, vegetative growth and pathogenesis.

An analogous function for Group III Gα proteins and mammalian $G\alpha_s$ subunits has been inferred, given their shared influence on cAMP levels through the regulation of adenylyl cyclases (Bölker 1998). However, Group III proteins are not homologous to mammalian $G\alpha_s$ proteins (Li et al. 2007a). Like Group I, Group III Gα proteins possess a myristolyation sequence at the amino terminus. In addition to positively influencing cAMP levels, evidence suggests that Group III Gα proteins function during sexual spore development and nutrient sensing (Kamerewerd et al. 2008; Kays et al. 2000).

Group II Gα proteins are not as well conserved as Group I and Group III proteins and the function of these proteins is less obvious (Li et al. 2007a). Evidence from *Neurospora crassa* and *Sordaria macrospora* suggests that Group II Gα proteins play a compensatory role to Group I and Group III Gα subunits; when comparing phenotypes of strains depleted of either a Group I or III Gα, additional loss of a Group II Gα intensifies the phenotype (Baasiri et al. 1997; Kamerewerd et al. 2008). More recent work in *N. crassa* suggests that the Group II Gα, GNA-2, plays an independent role in regulating optimal growth on poor carbon sources (Li and Borkovich 2006).

2.2 *Gβ and Gγ Subunits*

Multiple isoforms of Gβ and Gγ subunits are common in higher eukaryotes (Hamm 1998; Li et al. 2007a, b). At present, only a single Gβ has been predicted in most fungi (Wang et al. 2000; Whiteway et al. 1989). Gβ proteins are composed of an amino terminal region followed by seven repeating units of a common WD motif, that consists of 36–46 amino acids with a tryptophan (W) or aspartate (D) amino acid at the end (WD-40) (Fong et al. 1986; Neer et al. 1994; Simon et al. 1991). In *Saccharomyces cerevisiae*, *Schizosaccharomyces pombe* and *Cryptococcus neoformans*, alternate Gβ proteins with this characteristic WD-40 repeat have been shown to function as Gβ subunits (Goddard et al. 2006; Zeller et al. 2007; Palmer et al. 2006; see below).

The gene for the Gβ subunit in *N. crassa*, *gnb-1*, is required for normal asexual sporulation and female fertility (Krystofova and Borkovich 2005; Yang et al. 2002). In *Cryptococcus neoformans*, deletion of the *GPB-1* Gβ gene causes defects in monokaryotic fruiting and sterility (Wang et al. 2000). The *sfaD* Gβ gene in *Aspergillus nidulans* is required for normal conidiation and vegetative growth (Rosén et al. 1999). In *Magnaporthe oryzae*, the Gβ subunit MGB1 is required for growth and conidiation, appressorium formation, and maintenance of intracellular cAMP levels (Nishimura et al. 2003).

Most sequenced fungi have one predicted Gγ subunit, while *C. neoformans*, *Coprinus cinereus*, and *Podospora anserina* possess more than one Gγ in their genomes (Palmer et al. 2006). Gγ subunits are associated with the Gβ and are equally important for G-protein signaling. Gγ subunits belong to a large family of small proteins that are typically characterized by a CaaX box motif at the carboxy terminus for post translational isoprenylation and subsequent methylation (Gautam et al. 1998; Zhang and Casey 1996). The lipid modification targets the Gβγ complex to the plasma membrane and mediates the effective interaction of Gβγ with other proteins, including Gα, downstream effectors and receptors (Cabrera-Vera et al. 2003; Simon et al. 1991).

2.3 Alternative Gβ Subunits

General control of amino acid biosynthesis is a central regulatory mechanism in fungi. This phenomenon was first described in *N. crassa* (Carsiotis and Jones 1974; Carsiotis et al. 1974) and has been extensively studied in several species, including the yeast *S. cerevisiae* (Delforge et al. 1975; Hinnebusch 1988). One of the proteins involved in this pathway is encoded by the cross-pathway control *cpc-2* gene (Krüger et al. 1990), a homolog of the RACK1 (receptor for activated C-kinase) gene in humans (Ron et al. 1994), Interestingly, the amino acid sequence of CPC-2 is made up of WD-40 tandem repeats and shares sequence similarity with Gβ subunits (Müller et al. 1995). This highlighted the possibility that alternate Gβ subunits may interact with Gα proteins to regulate signal transduction in fungi.

Studies in *C. neoformans* revealed that Gib2, the RACK1 homolog, complexes with the Group III Gα Gpa1 and functions as an atypical Gβ in G-protein cAMP signaling (Palmer et al. 2006). Similarly, the *S. pombe* CPC-2/RACK1 homolog Gnr1 interacts with the Gpa1 Gα protein and functions as a negative regulator of the pheromone response pathway (Goddard et al. 2006). In *S. cerevisiae* the CPC-2/RACK1 homolog Asc1p is the Gβ for the Group III Gα Gpa2. Asc1p plays a critical role in the glucose signaling pathway mediated by adenylyl cyclase (Zeller et al. 2007) and is required for integrity of the cell wall near the bud site (Melamed et al. 2010). Deletion of *cpc-2* in *N. crassa* leads to female sterility (Müller et al. 1995), a phenotype characteristic of the canonical Gβ mutant Δ*gnb-1*, and the Gα mutant Δ*gna-1* (Kays and Borkovich 2004; Krystofova and Borkovich 2005). Further work is necessary to determine whether CPC-2 is an alternate Gβ in this organism.

2.4 G-Protein Coupled Receptors (GPCRs)

G-protein coupled receptors (GPCRs) represent the largest family of transmembrane receptors, with well over 600 members in the human genome (Venter et al. 2001). GPCRs contain seven transmembrane helices connected by intracellular and extracellular loops. Several GPCR classification schemes have been proposed for fungal genomes (Brunner et al. 2008; Galagan et al. 2003; Kulkarni et al. 2005; Lafon et al. 2006; Li et al. 2007a; Xue et al. 2008; Zheng et al. 2010). GPCRs common to all classification systems include pheromone receptors similar to *S. cerevisiae* Ste2p and Ste3p, receptors involved in nutrient, carbon, amino acid, and nitrogen sensing, receptors with similarity to cAMP receptors and receptors with similarity to microbial opsins. Additional classes exist, but do not appear to be as widespread across the fungal kingdom. Three classes were first identified in *Magnaporthe grisea* (Kulkarni et al. 2005). The PTH11-like class gains its name from the seven transmembrane protein PTH11, required for pathogenicity in *M. grisea* (Kulkarni et al. 2005). PTH11-like receptors are present in a subset of ascomycetes, but are absent from basidiomycetes and both budding and fission yeasts. The steroid receptor mPR class was also found in *M. grisea* and homologs are known to exist in basidiomycetes and budding and fission yeasts (Kulkarni et al. 2005). The third class, MG00532, displays weak homology to rat growth hormone releasing factor. This class is present in ascomycetes, but apparently not in basidiomycetes (Li et al. 2007a). Another class of GPCRs containing a RGS domain was identified in the *Aspergilli* and additional ascomycetes (Lafon et al. 2006). Analysis of this group for RGS characteristics has been conducted in the model plant *Arabidopsis thaliana*, where loss of AtRGS1 results in increased activity of the Gα subunit (Chen et al. 2003). This class of receptors is intriguing, suggesting that GPCRs may both positively and negatively regulate Gα subunits. Most recently, an additional three classes of GPCRs have been identified in *Verticillum spp.* and homologs are present in ascomycetes (Zheng et al. 2010). The first class, PTM1-like, is also found in the budding and fission yeasts. The GPR89A like and Family C-like classes have been identified in other ascomycetes but are absent from yeasts (Zheng et al. 2010).

2.5 Cytosolic Non-receptor GEFs

2.5.1 RIC8

Studies in the invertebrate animal systems *Caenorhabditis elegans* and *Drosophila melanogaster* have revealed that the cytoplasmic protein RIC8 serves as a positive regulator of Gα proteins, affecting protein localization, stability and downstream processes (Afshar et al. 2004, 2005; Hampoelz et al. 2005). Phylogenetic analysis of genes encoding G-protein signaling components revealed that GPCRs, G proteins and RGS proteins are found in animals as well as plants and fungi (Wilkie and Kinch 2005). Interestingly, this same study showed that RIC8 is present in animals and certain fungi, but is absent from plants and baker's yeast (Wilkie and Kinch 2005).

RIC8 was first discovered in *C. elegans* during screens for mutants that were resistant to inhibitors of cholinesterase. RIC8 localizes at the cortex, mitotic spindle, nuclear envelope and around chromatin (Couwenbergs et al. 2004) and was found to be important for G protein-mediated asymmetric cell division in zygotes and priming of synaptic vesicles (Miller and Rand 2000; Wilkie and Kinch 2005). In *D. melanogaster*, RIC8 accumulates at high concentrations around the mitotic spindle where it maintains polarity during asymmetric cell division (Hampoelz et al. 2005).

The GEF activity of RIC8 toward Gα proteins was first demonstrated in mammalian (rat) cells (Tall et al. 2003). The Ric8A homolog binds to the GDP-Gα in the absence of the Gβγ and facilitates exchange of GDP for GTP. The Ric8B homolog also has GEF activity. In addition, it stabilize the Gα subunit by inhibiting its ubiquitination (Nagai et al. 2010).

To date, RIC8 has been implicated as a positive regulator of Gα proteins in two fungal species: the rice pathogen *M. oryzae* (Li et al. 2010) and the saprobe and model system *N. crassa* (Wright et al. 2011). Studies of the *M. oryzae* homolog MoRic8 in Li et al. (2010) showed that RIC8 is found in the cytoplasm of vegetative hyphae and conidia and is highly expressed in appressoria, the specialized structures required for infection of plant tissue. RIC8 interacts with the Gα subunit MagB and appears to act upstream of cAMP production during the G-protein mediated regulation of infection-related morphogenesis (Li et al. 2010). Similar to MoRIC8, *N. crassa* RIC8 is localized to the cytoplasm of vegetative hyphae and mature conidia. Loss of *ric8* leads to the same extreme phenotypes observed for mutants lacking both a Group I and Group III Gα gene: i.e., dramatically impaired growth, short aerial hyphae, inappropriate conidiation in submerged culture, and female sterility (Li et al. 2010; Wright et al. 2011). RIC8 was also shown to be essential for the stability of all three Gα proteins, the Gβ subunit and the adenylyl cyclase protein. Importantly, similar to the mammalian homologs, *N. crassa* RIC8 demonstrated GEF activity towards Group I and Group III Gα proteins, with the greatest effect on the Group III Gα GNA-3 (Wright et al. 2011).

2.5.2 Arr4/GET3

Recent studies in *S. cerevisiae* have revealed a cytoplasmic protein, Arr4p, that possesses GEF activity (Lee and Dohlman 2008). Arr4p binds to the Gpa1p Gα protein, accelerates the exchange of GDP for GTP while stabilizing the nucleotide-free transient state, a similar activity to that observed for Ric8A in human cells. Arr4p also plays a major role in G protein-dependent MAPK phosphorylation, transcriptional regulation and mating. Arr4p was originally named because the protein sequence is homologous to the catalytic subunit of a bacterial ATP-dependent arsenite extrusion pump, ArsA, with roles in metal stress tolerance (Shen et al. 2003). Since its discovery, the *ARR4* gene has also been investigated for its involvement in Golgi/ER trafficking, and has been given the alternative name *GET3* (Schuldiner et al. 2005).

2.6 RGS Proteins

As mentioned above, Gα proteins have intrinsic GTPase activity, however, the process is greatly accelerated by RGS proteins that act as GTPase activating proteins (GAPs) (Ross and Wilkie 2000). RGS proteins comprise a large and diverse family found in animals, plants and fungi, as well as *Dictyostelium* and *Entamoeba* (Wilkie and Kinch 2005). The characteristic feature of all RGS proteins is a ~130 residue RGS box that serves as a site for interaction with Gα proteins. Comparative structural analysis of human RGS proteins revealed variable modulatory residues on the periphery of the RGS domain-G protein interaction site that aid in G-protein recognition and that optimize G-protein binding and inactivation (Kosloff et al. 2011).

The first RGS protein characterized in filamentous fungi was *A. nidulans* FlbA (Lee and Adams 1994). It modulates asexual sporulation by regulating the Group I Gα protein FadA (Yu et al. 1996). Since then, additional RGS proteins have been identified in this organism. RgsA regulates colony growth, aerial hyphae and pigment formation by regulating Group III Gα proteins (Han et al. 2008), while RgsB and RgsC have yet to be characterized.

In the budding yeast *S. cerevisiae*, the RGS proteins with the greatest contributions to G protein regulation are "*supersensitivity" to pheromone-2* (Sst2p) and Rgs2p, a homolog of an RGS protein in Drosophila, Loco (Yu et al. 2005). Sst2p and Rgs2p regulate the Gα subunits, Gpa1p and Gpa2p, respectively (Hill et al. 2006). Deletion of *sst2* leads to loss of pheromone adaptation and mating defects (Dohlman et al. 1996; Jackson and Hartwell 1990), while Rgs2p acts through Gpa2 to negatively regulate glucose-induced cAMP signaling (Versele et al. 1999).

As mentioned above, some potential GPCRs contain both a 7-TM region and a RGS domain (Lafon et al. 2006). One such protein (GprK) was found in *Aspergillus sp*. (Lafon et al. 2006). This protein is similar to AtRGS1, which negatively regulates the Gpa1 Gα subunit during cellular proliferation in *A. thaliana* (Chen et al. 2003).

3 G-Protein Signaling Output Pathways

3.1 cAMP (Cyclic Adenosine Monophosphate) Pathway

In fungi, G protein signaling has been shown to influence effector pathways including phosphatidylinositol-3-kinase dependent responses to mating pheromones (Koelle 2006), as well as cAMP-dependent and mitogen-activated protein kinase (MAPK) pathways (discussed below). The central component of the cAMP pathway is the enzyme adenylyl cyclase, which converts ATP to cAMP and pyrophosphate (Fig. 2). Fungal adenylyl cyclases are cytoplasmic peripheral membrane proteins (Adachi and Hamer 1998; Gold et al. 1994; Matsumoto et al. 1984). Changes in the level of cAMP are sensed by the cell through binding of cAMP to the regulatory subunit of the cAMP-dependent protein kinase A (PKA). This triggers dissociation of the catalytic and regulatory subunits, thus activating the catalytic subunit to phosphorylate downstream targets (Fig. 2).

cAMP has been implicated in regulation of many cellular processes in fungi, including nutrient sensing, asexual and sexual development, pathogenesis and secondary metabolism (Kozubowski et al. 2009; Li et al. 2007a, b; Shimizu and Keller 2001). Nutrient sensing is perhaps the most extensively studied function of the cAMP signaling pathway in yeast and filamentous fungi (Li et al. 2007a, b; Xue et al. 2006). In *S. pombe*, the Gβγ (Git5/Git11) dimer is required for cAMP-mediated glucose sensing (Landry and Hoffman 2001). In *S. cerevisiae,* the GPCR Gpr1 and the Gα protein Gpa2 regulate cAMP levels during glucose and sucrose (agonists) and mannose (antagonist) sensing (Lemaire et al. 2004; Li et al. 2007a, b; Lorenz et al. 2000). In *C. neoformans*, the GPCR Gpr4 acts upstream of the Group III Gpa1 Gα protein to regulate cAMP levels during amino acid sensing (Xue et al. 2006). In *N. crassa*, the GPCR GPR-4 signals through the Group I Gα subunit and adenylyl cyclase to regulate cAMP levels during growth on poor carbon sources (Li and Borkovich 2006). The Gβγ dimer has also been implicated in modulating cAMP levels (Krystofova and Borkovich 2005), however this may be an indirect effect resulting from decreased levels of the Gα protein in the absence of the Gβγ dimer. Finally, recent studies in *N. crassa* have revealed that the non-receptor GEF RIC8 is a positive regulator of cAMP signaling (Wright et al. 2011).

3.2 MAPK (Mitogen-Activated Protein Kinase) Pathways

In fungi, Gα and Gβγ subunits have been shown to signal to MAPK cascades, thus regulating a variety of cellular processes, including mating, morphogenesis, and pathogenesis (Fig. 2) (Lopez-Ilasaca 1998; Nishimura et al. 2003; Raudaskoski

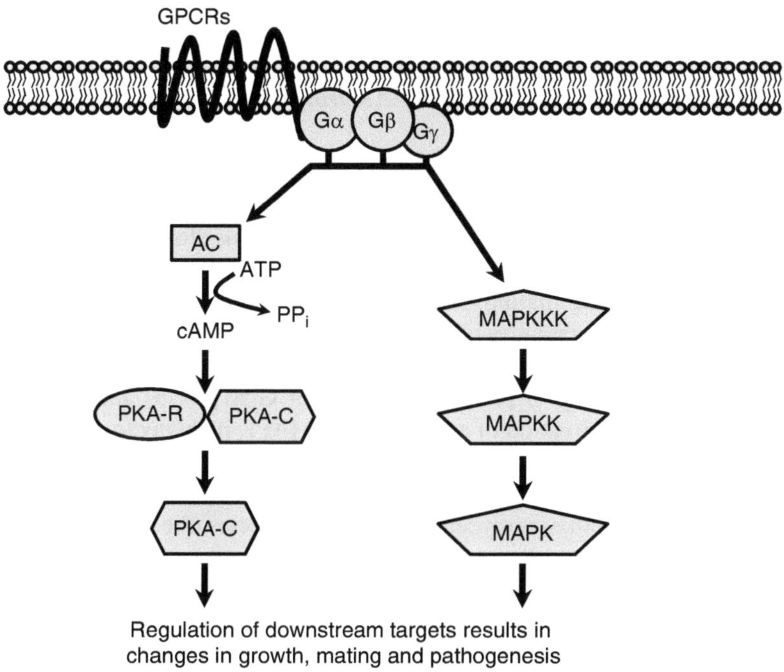

Fig. 2 Overview of signal transduction pathways downstream of heterotrimeric G proteins. Activation of heterotrimeric G proteins can result in modulation of adenylyl cyclase activity/ cAMP levels and MAPK pathway signaling. This results in changes in growth, mating and pathogenesis

and Kothe 2010; Wang et al. 2000). MAPKs are serine/threonine protein kinases that operate in a three-tiered cascade of phosphorylation to regulate cellular processes. Within each MAPK unit, a MAPK kinase kinase (MAPKKK) is first activated by phosphorylation. The MAPKKK then phosphorylates and activates the downstream MAPK kinase (MAPKK) which in turn phosphorylates and activates the terminal MAPK. The terminal MAPK phosphorylates transcription factors and regulatory proteins to execute the cellular changes described above (Bardwell 2005). Filamentous fungi possess three major MAPK cascades homologous to those for the pheromone response, osmoregulation, and cell wall integrity pathways present in *S. cerevisiae* (Borkovich et al. 2004). The roles of heterotrimeric G proteins in regulating downstream MAPK pathways have been extensively studied in *S. cerevisiae* (Jones and Bennett 2011). In response to pheromones, the β subunit, GPB1, signals via a MAPK cascade to regulate mating and haploid fruiting in C. *neoformans* (Wang et al. 2000). In *M. grisea*, the β subunit, MGB1, has been postulated to signal to downstream MAPK pathways regulating the pathogenic invasion of rice plants (Nishimura et al. 2003; see below).

4 Communication

4.1 Mating in the Filamentous Ascomycete *N. crassa*

Pheromone receptors are GPCRs that relay an extracellular pheromone signal to an intracellular G-protein signaling pathway, resulting in mating. In filamentous fungi, mating can take place between homothallic organisms displaying self-fertility, heterothallic organisms that are outcrossing, or in organisms capable of both homothallic and heterothallic mating (Li et al. 2007a). Studies of fungal pheromone signaling are best understood yeast *S. cerevisiae* (Bardwell 2005; Kurjan 1992). In this system, the pheromone receptors Ste2p and Ste3p transduce a pheromone signal to a heterotrimeric G protein composed of Gpa1p (Gα) and Ste4p/Ste18p (Gβγ). This leads to activation of a downstream MAPK pathway to achieve mating between two cell types, *MATa* and *MATα* (Bardwell 2005). Mating in filamentous fungi is similar, but more complex, than in *S. cerevisiae*. For a comparison of mating programs in both ascomycetes and basidiomycetes, the reader is directed to a recent review (Jones and Bennett 2011).

Mating in heterothallic *N. crassa* occurs between cells of two mating types, *mat a* and *mat A*. In contrast to yeast, mating is achieved through specialized interaction of male and female cells. Under nitrogen starvation, vegetative hyphae form into a spherical female reproductive structure known as a protoperithecium (Raju 1992). A specialized hypha called the trichogyne emanates from the protoperithecium and then grows chemotropically towards a male cell (macroconidium, microconidium, hyphal segment) of opposite mating type. Upon contact, the trichogyne and male cell fuse, bringing the haploid nuclei from two opposite mating type cells into a single cellular compartment. Following this chemotropic stage of mating, *mat a* and *mat A* haploid nuclei undergo multiple rounds of mitosis, developing a mass of ascogenous hyphae within the enlarging fruiting body (Raju 1992). Nuclei of opposite mating type then fuse to form *mat a mat A* diploids, and undergo meiosis to produce four meiotic products. Each product divides in a final post-meiotic round of mitosis, resulting in eight haploid sexual spores (ascospores). The fruiting body is now fully mature and contains a pore from which the ascospores will be ejected (Raju 1992).

The involvement of G-protein signaling components in mating has been extensively studied in *N. crassa*. The *N. crassa* Gα subunit GNA-1 was the first Gα subunit to be characterized in filamentous fungi (Turner and Borkovich 1993). In addition to defects in vegetative and asexual growth, loss of this Group I Gα subunit leads to female sterility (Ivey et al. 1996). Examination of Δ*gna-1* mutants reveal that while protoperithecia are formed, the emanating trichogynes are unable to grow chemotropically towards and fuse with male cells of opposite mating type (Kim and Borkovich 2004). Loss of the Gβ subunit GNB-1 or the Gγ subunit GNG-1 also leads to blind trichogynes in protoperithecia (Krystofova and Borkovich 2005; Yang et al. 2002). Group II (GNA-2) and Group III (GNA-3) Gα subunits also influence sexual development, albeit to a lesser extent than GNA-1 (Baasiri et al. 1997; Kays et al. 2000; Kays and Borkovich 2004).

Upstream of the heterotrimer, examination of the pheromone receptors PRE-1 and PRE-2 has provided further insight into the roles of G-protein signaling components in mating. Expression of pheromone receptor in female cells and pheromone ligand in male cells is required for successful mating (Kim and Borkovich 2004, 2006; Kim et al. 2012). Females lacking the proper pheromone receptor experience an arrest in sexual development. Protoperithecia and trichogynes are formed, but further examination exposes the inability of trichogynes to chemotropically interact with males in a manner reminiscent of Δ*gna-1* females (Kim and Borkovich 2004). The chemotropic requirement for mating can be bypassed through forcing fusion of two opposite mating type cells in heterokaryons, but the presence of at least one pheromone-receptor/ligand pair and *gna-1* are required for meiosis and sexual spore production (Kim et al. 2012). These data suggest that pheromone receptors are not only required for early steps in mating but are also important for post-fertilization events. While this phenomenon has not been observed in yeast, it has been noted in basidiomycete mushroom fungi (Casselton and Feldbrügge 2010). Studying the functions of G protein signaling components in post-fertilization processes will open new avenues for discovery.

4.2 Pathogenesis

Fungal pathogens are responsible for severe economic and ecological damage world-wide. Their prevalence stimulates millions in spending to control agricultural losses, increases morbidity in humans (especially among the immunocompromised), and significantly depletes the numbers in many animal and insect populations (Blaustein et al. 1994; Wilson and Talbot 2009a).

The pathogenic morphological process is fairly conserved among the disease-causing fungi, including secretion, penetration, osmoregulation and sporulation (Liu et al. 2011; Wilson and Talbot 2009a). It is apparent that G proteins play a significant role in the signal-initiated processes in pathogenesis. In human fungal pathogens, morphogenesis, toxin secretion (mycotoxins) and resistance to host defenses are the key factors that determine pathogenicity and are processes that are regulated by G protein signaling (Li et al. 2007a, b). For example, the human pathogen *C. neoformans* produces a polysaccharide capsule that prevents phagocytosis, while melanin production protects the fungi from anti-fungal oxidants. G protein signaling regulates both of these processes (Kozel and Gotschlich 1982). In the pneumonia-inducing *Aspergilli*, Group III Gα proteins mediate survival mechanisms for conidia in macrophages (Liebmann et al. 2003).

Similar to animal fungal pathogens, morphogenesis plays a critical role in pathogenicity in many plant fungal pathogens. The interplay of signaling systems between plants and fungi determines the pathogenic efficiency of fungi. Signaling begins as early as the initial contact between the fungus and the plant surface. If conditions are favorable, the fungus initiates host penetration and invasion. Some plant pathogens produce cell-wall degrading enzymes and specialized structures to enhance the efficiency of host penetration (Liu et al. 2011). Plants initiate defense-based signaling mechanisms to counteract the presence of fungi. In turn, the invading fungi respond by implementing mechanisms to thwart the anti-fungal agents produced (Kempken 2011).

The impact of G protein signaling in these fungal responses and plant invasion mechanisms has been well studied (Kempken 2011). In *Cryphonectria parasitica*, deletion of the Group I Gα Cpg-1 severely reduces pathogenicity by affecting vegetative growth and conidiation (Gao and Nuss 1996). In *Ustilago maydis*, the Gα Gpa3 (Group III) is required for mating, a precursor to invasion of maize tissue (Krüger et al. 1998).

In the rice blast pathogen, *M. oryzae*, G protein signaling is triggered in response to the hardness and hydrophobicity of the leaf surface (Wilson and Talbot 2009b). *M. oryzae* begins infection when conidia attach to the leaf tissue and produce a germ tube which differentiates to form the appressorium, a structure that facilitates the penetration and invasion of leaves of rice, millets and other grasses (Wilson and Talbot 2009b). G protein signaling regulates the MAPK cascade for appressorium formation, invasive growth and disease development, as well as the cAMP pathway for control of conidiation and appressorium maturation (DeZwaan et al. 1999). During appressorium development, the Mst11/Mst7 MAPKK kinase/MAPK kinase and subsequently the Pmk1 MAPK become activated (Nishimura et al. 2003; Zhao and Xu 2007; Zhao et al. 2005). The interaction and function of these kinases hinge on the presence of the scaffolding protein Mst50. The Gβ subunit physically interacts with Mst50 and is essential for appressorium development (Nishimura et al. 2003). This phosphorelay leads to the activation of Mst12 and other transcription factors (Park et al. 2002). The cAMP pathway is regulated by all three of the Gα proteins, MagA, MagB and MagC, in *M. oryzae* (Liu et al. 2007). Activation of MagA induces appressorium development, while MagB may negatively regulate appressorium development on non-inductive surfaces. Both responses occur via the upregulation or downregulation of the adenylyl cyclase, Mac1 (Choi and Dean 1997). cAMP generated by Mac1 acts on the regulatory subunit of cAMP-dependent protein kinase A Sum1, which results in the detachment and activation of the catalytic subunit CpkA (Choi and Dean 1997; Wilson and Talbot 2009b). A link between the MAPK and cAMP cascades has been established, but questions still remain about the exact physical interaction between the proteins involved (Wilson and Talbot 2009b).

Recently, study of the role of G protein signaling in pathogenesis has been extended to the wheat fungal pathogen *Stagonospora nodorum* (Solomon et al. 2004). In *S. nodorum,* the Group I Gna1 is expressed during infection and the Δ*gna1* mutant produces considerably smaller lesions, is deficient in the ability to sporulate, has reduced abundance of putative cell wall degrading enzymes and cannot penetrate the leaf surface except through natural openings (Solomon et al. 2004; Tan et al. 2009). Δ*gna1* mutants exhibit differential expression of the transcripts and proteins of several genes throughout infection, including those associated with asexual development, stress and pathogen responses. Among the regulated proteins is Sch1, a short chain dehydrogenase required for sporulation (Casey et al. 2010; Tan et al. 2008).

5 Conclusion

In fungi, heterotrimeric G proteins play instrumental roles in signaling pathways that regulate a number of critical developmental processes. These signals are conveyed from GPCRs or nonreceptor GEFs to the heterotrimer composed of the α, β and γ subunits. Upon activation, the Gα disassociates from the βγ dimer and both are free to act upon downstream effectors. Two notable output pathways downstream of G proteins are the cAMP and MAPK pathways. Signals for many developmental processes including growth, mating and pathogenesis are initiated by G proteins and implemented by downstream cAMP and/or MAPK cascades. Future studies will aim to further establish the connections between heterotrimeric G proteins and control of downstream responses in fungi.

References

Adachi K, Hamer JE (1998) Divergent cAMP signaling pathways regulate growth and pathogenesis in the rice blast fungus *Magnaporthe grisea*. Plant Cell 10:1361–1374

Afshar K, Willard FS, Colombo K, Johnston CA, McCudden CR, Siderovski DP, Gönczy P (2004) RIC-8 is required for GPR-1/2-dependent Galpha function during asymmetric division of *C. elegans* embryos. Cell 119:219–230

Afshar K, Willard FS, Colombo K, Siderovski DP, Gönczy P (2005) Cortical localization of the Galpha protein GPA-16 requires RIC-8 function during *C. elegans* asymmetric cell division. Development 132:4449–4459

Baasiri RA, Lu X, Rowley PS, Turner GE, Borkovich KA (1997) Overlapping functions for two G protein alpha subunits in *Neurospora crassa*. Genetics 147:137–145

Bardwell L (2005) A walk-through of the yeast mating pheromone response pathway. Peptides 26:339–350

Blaustein AR, Hokit DG, O'Hara RK, Holt RA (1994) Pathogenic fungus contributes to amphibian losses in the Pacific northwest. Biol Conserv 67:251–254

Bohm A, Gaudet R, Sigler PB (1997) Structural aspects of heterotrimeric G-protein signaling. Curr Opin Biotechnol 8:480–487

Bölker M (1998) Sex and crime: heterotrimeric G proteins in fungal mating and pathogenesis. Fungal Genet Biol 25:143–156

Borkovich KA, Alex LA, Yarden O, Freitag M, Turner GE, Read ND, Seiler S et al (2004) Lessons from the genome sequence of *Neurospora crassa*: tracing the path from genomic blueprint to multicellular organism. Microbiol Mol Biol Rev 68:1–108

Brunner K, Omann M, Pucher ME, Delic M, Lehner SM, Domnanich P, Kratochwill K et al (2008) Trichoderma G protein-coupled receptors: functional characterisation of a cAMP receptor-like protein from *Trichoderma atroviride*. Curr Genet 54:283–299

Cabrera-Vera TM, Vanhauwe J, Thomas TO, Medkova M, Preininger A, Mazzoni MR, Hamm HE (2003) Insights into G protein structure, function, and regulation. Endocr Rev 24:765–781

Carsiotis M, Jones RF (1974) Cross-pathway regulation: tryptophan-mediated control of histidine and arginine biosynthetic enzymes in *Neurospora crassa*. J Bacteriol 119:889–892

Carsiotis M, Jones RF, Wesseling AC (1974) Cross-pathway regulation: histidine-mediated control of histidine, tryptophan, and arginine biosynthetic enzymes in *Neurospora crassa*. J Bacteriol 119:893–898

Casey T, Solomon PS, Bringans S, Tan K-C, Oliver RP, Lipscombe R (2010) Quantitative proteomic analysis of G-protein signalling in *Stagonospora nodorum* using isobaric tags for relative and absolute quantification. Proteomics 10:38–47

Casselton L, Feldbrügge, M (2010) Mating and Sexual Morphogenesis in Basidiomycete Fungi. In: Borkovich, KA, Ebbole, DJ (eds) Cellular and Molecular Biology of Filamentous Fungi. ASM Press, Washington, DC, pp 536–555

Chen J-G, Willard FS, Huang J, Liang J, Chasse SA, Jones AM, Siderovski DP (2003) A seven-transmembrane RGS protein that modulates plant cell proliferation. Science 301:1728–1731

Choi W, Dean RA (1997) The adenylate cyclase gene MAC1 of *Magnaporthe grisea* controls appressorium formation and other aspects of growth and development. Plant Cell 9:1973–1983

Coleman DE, Sprang SR (1996) How G proteins work: a continuing story. Trends Biochem Sci 21:41–44

Couwenbergs C, Spilker AC, Gotta M (2004) Control of embryonic spindle positioning and Galpha activity by *C. elegans* RIC-8. Curr Biol 14:1871–1876

Delforge J, Messenguy F, Wiame JM (1975) The regulation of arginine biosynthesis in *Saccharomyces cerevisiae*. The specificity of argR- mutations and the general control of amino-acid biosynthesis. Eur J Biochem 57:231–239

DeZwaan TM, Carroll AM, Valent B, Sweigard JA (1999) *Magnaporthe grisea* Pth11p is a novel plasma membrane protein that mediates appressorium differentiation in response to inductive substrate cues. Plant Cell 11:2013–2030

Dohlman HG, Song J, Ma D, Courchesne WE, Thorner J (1996) Sst2, a negative regulator of pheromone signaling in the yeast *Saccharomyces cerevisiae*: expression, localization, and genetic interaction and physical association with Gpa1 (the G-protein alpha subunit). Mol Cell Biol 16:5194–5209

Fong HK, Hurley JB, Hopkins RS, Miake-Lye R, Johnson MS, Doolittle RF, Simon MI (1986) Repetitive segmental structure of the transducin beta subunit: homology with the *CDC4* gene and identification of related mRNAs. Proc Natl Acad Sci USA 83:2162–2166

Galagan JE, Calvo SE, Borkovich KA, Selker EU, Read ND, Jaffe D, FitzHugh W et al (2003) The genome sequence of the filamentous fungus *Neurospora crassa*. Nature 422:859–868

Gao S, Nuss DL (1996) Distinct roles for two G protein alpha subunits in fungal virulence, morphology, and reproduction revealed by targeted gene disruption. Proc Natl Acad Sci USA 93:14122–14127

Gautam N, Downes GB, Yan K, Kisselev O (1998) The G-protein betagamma complex. Cell Signal 10:447–455

Goddard A, Ladds G, Forfar R, Davey J (2006) Identification of Gnr1p, a negative regulator of G alpha signalling in *Schizosaccharomyces pombe*, and its complementation by human G beta subunits. Fungal Genet Biol 43:840–851

Gold S, Duncan G, Barrett K, Kronstad J (1994) cAMP regulates morphogenesis in the fungal pathogen *Ustilago maydis*. Genes Dev 8:2805–2816

Hamm HE (1998) The many faces of G protein signaling. J Biol Chem 273:669–672

Hamm HE, Gilchrist A (1996) Heterotrimeric G proteins. Curr Opin Cell Biol 8:189–196

Hampoelz B, Hoeller O, Bowman SK, Dunican D, Knoblich JA (2005) *Drosophila* Ric-8 is essential for plasma-membrane localization of heterotrimeric G proteins. Nat Cell Biol 7:1099–1105

Han K-H, Kim JH, Moon H, Kim S, Lee S-S, Han D-M, Jahng K-Y et al (2008) The *Aspergillus nidulans esdC* (early sexual development) gene is necessary for sexual development and is controlled by *veA* and a heterotrimeric G protein. Fungal Genet Biol 45:310–318

Hill C, Goddard A, Davey J, Ladds G (2006) Investigating RGS proteins in yeast. Semin Cell Dev Biol 17:352–362

Hinnebusch AG (1988) Mechanisms of gene regulation in the general control of amino acid biosynthesis in *Saccharomyces cerevisiae*. Microbiol Rev 52:248–273

Ivey FD, Hodge PN, Turner GE, Borkovich KA (1996) The G alpha i homologue Gna-1 controls multiple differentiation pathways in *Neurospora crassa*. Mol Biol Cell 7:1283–1297

Jackson CL, Hartwell LH (1990) Courtship in *S. cerevisiae*: both cell types choose mating partners by responding to the strongest pheromone signal. Cell 63:1039–1051

Jones SK, Bennett RJ (2011) Fungal mating pheromones: choreographing the dating game. Fungal Genet Biol 48:668–676

Kamerewerd J, Jansson M, Nowrousian M, Pöggeler S, Kück U (2008) Three alpha-subunits of heterotrimeric G proteins and an adenylyl cyclase have distinct roles in fruiting body development in the homothallic fungus *Sordaria macrospora*. Genetics 180:191–206

Kays AM, Borkovich KA (2004) Severe impairment of growth and differentiation in a *Neurospora crassa* mutant lacking all heterotrimeric G alpha proteins. Genetics 166:1229–1240

Kays AM, Rowley PS, Baasiri RA, Borkovich KA (2000) Regulation of conidiation and adenylyl cyclase levels by the Galpha protein GNA-3 in *Neurospora crassa*. Mol Cell Biol 20:7693–7705

Kempken F (2011) Fungal defences against animal antagonists – lectins and more. Mol Ecol 20:2876–2877

Kim H, Borkovich KA (2004) A pheromone receptor gene, *pre-1*, is essential for mating type-specific directional growth and fusion of trichogynes and female fertility in *Neurospora crassa*. Mol Microbiol 52:1781–1798

Kim H, Borkovich KA (2006) Pheromones are essential for male fertility and sufficient to direct chemotropic polarized growth of trichogynes during mating in *Neurospora crassa*. Eukaryot Cell 5:544–554

Kim H, Wright SJ, Park G, Ouyang S, Krystofova S, Borkovich KA (2012) Roles for Receptors, Pheromones, G proteins and Mating Type Genes During Sexual Reproduction in Neurospora crassa. Genetics (in press). doi: 10.1534/genetics.111.136358

Koelle MR (2006) Heterotrimeric G protein signaling: getting inside the cell. Cell 126:25–27

Kosloff M, Travis AM, Bosch DE, Siderovski DP, Arshavsky VY (2011) Integrating energy calculations with functional assays to decipher the specificity of G protein-RGS protein interactions. Nat Struct Mol Biol 18:846–853

Kozel TR, Gotschlich EC (1982) The capsule of *Cryptococcus neoformans* passively inhibits phagocytosis of the yeast by macrophages. J Immunol 129:1675–1680

Kozubowski L, Lee SC, Heitman J (2009) Signalling pathways in the pathogenesis of *Cryptococcus*. Cell Microbiol 11:370–380

Krüger D, Koch J, Barthelmess IB (1990) *cpc-2*, a new locus involved in general control of amino acid synthetic enzymes in *Neurospora crassa*. Curr Genet 18:211–215

Krüger J, Loubradou G, Regenfelder E, Hartmann A, Kahmann R (1998) Crosstalk between cAMP and pheromone signalling pathways in *Ustilago maydis*. Mol Gen Genet 260:193–198

Krystofova S, Borkovich KA (2005) The heterotrimeric G-protein subunits GNG-1 and GNB-1 form a Gbetagamma dimer required for normal female fertility, asexual development, and Galpha protein levels in *Neurospora crassa*. Eukaryot Cell 4:365–378

Kulkarni RD, Thon MR, Pan H, Dean RA (2005) Novel G-protein-coupled receptor-like proteins in the plant pathogenic fungus *Magnaporthe grisea*. Genome Biol 6:R24

Kurjan J (1992) Pheromone response in yeast. Annu Rev Biochem 61:1097–1129

Lafon A, Han K-H, Seo J-A, Yu J-H, d' Enfert C (2006) G-protein and cAMP-mediated signaling in *Aspergilli*: a genomic perspective. Fungal Genet Biol 43:490–502

Landry S, Hoffman CS (2001) The Git5 Gbeta and Git11 Ggamma form an atypical Gbetagamma dimer acting in the fission yeast glucose/cAMP pathway. Genetics 157:1159–1168

Lee BN, Adams TH (1994) Overexpression of *flbA*, an early regulator of *Aspergillus* asexual sporulation, leads to activation of *brlA* and premature initiation of development. Mol Microbiol 14:323–334

Lee MJ, Dohlman HG (2008) Coactivation of G protein signaling by cell-surface receptors and an intracellular exchange factor. Curr Biol 18:211–215

Lemaire K, Van de Velde S, Van Dijck P, Thevelein JM (2004) Glucose and sucrose act as agonist and mannose as antagonist ligands of the G protein-coupled receptor Gpr1 in the yeast *Saccharomyces cerevisiae*. Mol Cell 16:293–299

Li L, Borkovich KA (2006) GPR-4 is a predicted G-protein-coupled receptor required for carbon source-dependent asexual growth and development in *Neurospora crassa*. Eukaryot Cell 5:1287–1300

Li L, Wright SJ, Krystofova S, Park G, Borkovich KA (2007a) Heterotrimeric G protein signaling in filamentous fungi. Annu Rev Microbiol 61:423–452

Li L, Shen G, Zhang Z-G, Wang Y-L, Thompson JK, Wang P (2007b) Canonical heterotrimeric G proteins regulating mating and virulence of *Cryptococcus neoformans*. Mol Biol Cell 18:4201–4209

Li Y, Yan X, Wang H, Liang S, Ma W-B, Fang M-Y, Talbot NJ et al (2010) MoRic8 is a novel component of G-protein signaling during plant infection by the rice blast fungus *Magnaporthe oryzae*. Mol Plant Microbe Interact 23:317–331

Liebmann B, Gattung S, Jahn B, Brakhage AA (2003) cAMP signaling in *Aspergillus fumigatus* is involved in the regulation of the virulence gene *pksP* and in defense against killing by macrophages. Mol Genet Genomics 269:420–435

Liu H, Suresh A, Willard FS, Siderovski DP, Lu S, Naqvi NI (2007) Rgs1 regulates multiple Galpha subunits in *Magnaporthe* pathogenesis, asexual growth and thigmotropism. EMBO J 26:690–700

Liu W, Zhou X, Li G, Li L, Kong L, Wang C, Zhang H et al (2011) Multiple plant surface signals are sensed by different mechanisms in the rice blast fungus for appressorium formation. PLoS Pathog 7:e1001261

Lopez-Ilasaca M (1998) Signaling from G-protein-coupled receptors to mitogen-activated protein (MAP)-kinase cascades. Biochem Pharmacol 56:269–277

Lorenz MC, Pan X, Harashima T, Cardenas ME, Xue Y, Hirsch JP, Heitman J (2000) The G protein-coupled receptor Gpr1 is a nutrient sensor that regulates pseudohyphal differentiation in *Saccharomyces cerevisiae*. Genetics 154:609–622

Matsumoto K, Uno I, Ishikawa T (1984) Identification of the structural gene and nonsense alleles for adenylate cyclase in *Saccharomyces cerevisiae*. J Bacteriol 157:277–282

Melamed D, Bar-Ziv L, Truzman Y, Arava Y (2010) Asc1 supports cell-wall integrity near bud sites by a Pkc1 independent mechanism. PLoS One 5:e11389

Miller KG, Rand JB (2000) A role for RIC-8 (Synembryn) and GOA-1 (G(o)alpha) in regulating a subset of centrosome movements during early embryogenesis in *Caenorhabditis elegans*. Genetics 156:1649–1660

Müller F, Krüger D, Sattlegger E, Hoffmann B, Ballario P, Kanaan M, Barthelmess IB (1995) The *cpc-2* gene of *Neurospora crassa* encodes a protein entirely composed of WD-repeat segments that is involved in general amino acid control and female fertility. Mol Gen Genet 248:162–173

Nagai Y, Nishimura A, Tago K, Mizuno N, Itoh H (2010) Ric-8B stabilizes the alpha subunit of stimulatory G protein by inhibiting its ubiquitination. J Biol Chem 285:11114–11120

Neer EJ, Schmidt CJ, Nambudripad R, Smith TF (1994) The ancient regulatory-protein family of WD-repeat proteins. Nature 371:297–300

Neves SR, Ram PT, Iyengar R (2002) G protein pathways. Science 296:1636–1639

Nishimura M, Park G, Xu J-R (2003) The G-beta subunit MGB1 is involved in regulating multiple steps of infection-related morphogenesis in *Magnaporthe grisea*. Mol Microbiol 50:231–243

Oldham WM, Hamm HE (2008) Heterotrimeric G protein activation by G-protein-coupled receptors. Nat Rev Mol Cell Biol 9:60–71

Palmer DA, Thompson JK, Li L, Prat A, Wang P (2006) Gib2, a novel Gbeta-like/RACK1 homolog, functions as a Gbeta subunit in cAMP signaling and is essential in *Cryptococcus neoformans*. J Biol Chem 281:32596–32605

Park G, Xue C, Zheng L, Lam S, Xu J-R (2002) MST12 regulates infectious growth but not appressorium formation in the rice blast fungus *Magnaporthe grisea*. Mol Plant Microbe Interact 15:183–192

Raju NB (1992) Genetic control of the sexual cycle in *Neurospora*. Mycol Res 96:241–262

Raudaskoski M, Kothe E (2010) Basidiomycete mating type genes and pheromone signaling. Eukaryot Cell 9:847–859

Resh MD (1996) Regulation of cellular signalling by fatty acid acylation and prenylation of signal transduction proteins. Cell Signal 8:403–412

Ron D, Chen CH, Caldwell J, Jamieson L, Orr E, Mochly-Rosen D (1994) Cloning of an intracellular receptor for protein kinase C: a homolog of the beta subunit of G proteins. Proc Natl Acad Sci USA 91:839–843

Rosén S, Yu JH, Adams TH (1999) The *Aspergillus nidulans sfaD* gene encodes a G protein beta subunit that is required for normal growth and repression of sporulation. EMBO J 18:5592–5600

Ross EM, Wilkie TM (2000) GTPase-activating proteins for heterotrimeric G proteins: regulators of G protein signaling (RGS) and RGS-like proteins. Annu Rev Biochem 69:795–827

Schuldiner M, Collins SR, Thompson NJ, Denic V, Bhamidipati A, Punna T, Ihmels J et al (2005) Exploration of the function and organization of the yeast early secretory pathway through an epistatic miniarray profile. Cell 123:507–519

Shen J, Hsu C-M, Kang B-K, Rosen BP, Bhattacharjee H (2003) The *Saccharomyces cerevisiae* Arr4p is involved in metal and heat tolerance. Biometals 16:369–378

Shimizu K, Keller NP (2001) Genetic involvement of a cAMP-dependent protein kinase in a G protein signaling pathway regulating morphological and chemical transitions in *Aspergillus nidulans*. Genetics 157:591–600

Simon MI, Strathmann MP, Gautam N (1991) Diversity of G proteins in signal transduction. Science 252:802–808

Solomon PS, Tan K-C, Sanchez P, Cooper RM, Oliver RP (2004) The disruption of a Galpha subunit sheds new light on the pathogenicity of *Stagonospora nodorum* on wheat. Mol Plant Microbe Interact 17:456–466

Tall GG, Krumins AM, Gilman AG (2003) Mammalian Ric-8A (synembryn) is a heterotrimeric Galpha protein guanine nucleotide exchange factor. J Biol Chem 278:8356–8362

Tan K-C, Heazlewood JL, Millar AH, Thomson G, Oliver RP, Solomon PS (2008) A signaling-regulated, short-chain dehydrogenase of *Stagonospora nodorum* regulates asexual development. Eukaryot Cell 7:1916–1929

Tan K-C, Heazlewood JL, Millar AH, Oliver RP, Solomon PS (2009) Proteomic identification of extracellular proteins regulated by the Gna1 Galpha subunit in *Stagonospora nodorum*. Mycol Res 113:523–531

Turner GE, Borkovich KA (1993) Identification of a G protein alpha subunit from *Neurospora crassa* that is a member of the Gi family. J Biol Chem 268:14805–14811

Venter JC, Adams MD, Myers EW, Li PW, Mural RJ, Sutton GG, Smith HO et al (2001) The sequence of the human genome. Science 291:1304–1351

Versele M, de Winde JH, Thevelein JM (1999) A novel regulator of G protein signalling in yeast, Rgs2, downregulates glucose-activation of the cAMP pathway through direct inhibition of Gpa2. EMBO J 18:5577–5591

Wang P, Perfect JR, Heitman J (2000) The G-protein beta subunit *GPB1* is required for mating and haploid fruiting in *Cryptococcus neoformans*. Mol Cell Biol 20:352–362

Whiteway M, Hougan L, Dignard D, Thomas DY, Bell L, Saari GC, Grant FJ et al (1989) The *STE4* and *STE18* genes of yeast encode potential beta and gamma subunits of the mating factor receptor-coupled G protein. Cell 56:467–477

Wilkie TM, Kinch L (2005) New roles for Galpha and RGS proteins: communication continues despite pulling sisters apart. Curr Biol 15:R843–R854

Wilson RA, Talbot NJ (2009a) Fungal physiology – a future perspective. Microbiology 155 (Pt 12):3810–3815

Wilson RA, Talbot NJ (2009b) Under pressure: investigating the biology of plant infection by *Magnaporthe oryzae*. Nat Rev Microbiol 7:185–195

Wright SJ, Inchausti R, Eaton CJ, Krystofova S, Borkovich KA (2011) RIC8 is a guanine-nucleotide exchange factor for Galpha subunits that regulates growth and development in *Neurospora crassa*. Genetics 189:165–176

Xue C, Bahn Y-S, Cox GM, Heitman J (2006) G protein-coupled receptor Gpr4 senses amino acids and activates the cAMP-PKA pathway in *Cryptococcus neoformans*. Mol Biol Cell 17:667–679

Xue C, Hsueh Y-P, Heitman J (2008) Magnificent seven: roles of G protein-coupled receptors in extracellular sensing in fungi. FEMS Microbiol Rev 32:1010–1032

Yang Q, Poole SI, Borkovich KA (2002) A G-protein beta subunit required for sexual and vegetative development and maintenance of normal G alpha protein levels in *Neurospora crassa*. Eukaryot Cell 1:378–390

Yu JH, Wieser J, Adams TH (1996) The Aspergillus FlbA RGS domain protein antagonizes G protein signaling to block proliferation and allow development. EMBO J 15:5184–5190

Yu F, Wang H, Quan H, Kaushik R, Bownes M, Yang X, Chia W (2005) Locomotion defects, together with Pins, regulates heterotrimeric G-protein signaling during *Drosophila* neuroblast asymmetric divisions. Genes Dev 19:1341–1353

Zeller CE, Parnell SC, Dohlman HG (2007) The RACK1 ortholog Asc1 functions as a G-protein beta subunit coupled to glucose responsiveness in yeast. J Biol Chem 282:25168–25176

Zhang FL, Casey PJ (1996) Protein prenylation: molecular mechanisms and functional consequences. Annu Rev Biochem 65:241–269

Zhao X, Xu J-R (2007) A highly conserved MAPK-docking site in Mst7 is essential for Pmk1 activation in *Magnaporthe grisea*. Mol Microbiol 63:881–894

Zhao X, Kim Y, Park G, Xu J-R (2005) A mitogen-activated protein kinase cascade regulating infection-related morphogenesis in *Magnaporthe grisea*. Plant Cell 17:1317–1329

Zheng H, Zhou L, Dou T, Han X, Cai Y, Zhan X, Tang C et al (2010) Genome-wide prediction of G protein-coupled receptors in *Verticillium* spp. Fungal Biol 114:359–368

Glycogen Metabolism Regulation in *Neurospora crassa*

Maria C. Bertolini, Fernanda Z. Freitas, Renato M. de Paula, Fernanda B. Cupertino, and Rodrigo D. Goncalves

Abstract Microorganisms accumulate glycogen as carbon and energy reserves to face adverse environmental conditions during growth and development. The processes of glycogen synthesis and degradation share similarities among different microorganisms. However, the regulation of the metabolism as a whole shows differences, likely due to the environmental conditions to which they individually respond. This chapter aims to present some molecular mechanisms that regulate glycogen metabolism in the fungus *Neurospora crassa*. The availability of its genome sequence (Galagan et al., Nature 422:859–868, 2003) and a collection of mutant strains, each carrying a deletion in a specific ORF, allowed investigation into the role of specific proteins as regulators of glycogen metabolism to begin. Here we present some biochemical and molecular mechanisms that have already been described for this fungus, and additionally, we focused on more recent findings including the molecular basis underlying the metabolism regulation, mainly at transcriptional level.

1 Introduction

Glycogen is a polymer of glucose which is widely distributed in nature, being found in microorganisms all the way through to higher eukaryotes including plants and animals. The glycogen structure is characterized by glucose units

M.C. Bertolini (✉) • F.Z. Freitas • F.B. Cupertino • R.D. Goncalves
Departamento de Bioquímica e Tecnologia Química, Instituto de Química, UNESP, R. Prof. Francisco Degni, 55, 14800-900 Araraquara, São Paulo, Brazil
e-mail: mcbertol@iq.unesp.br; fzfreitas@iq.unesp.br; fernanda_cupertino@yahoo.com.br; rdgoncalves@gmail.com

R.M. de Paula
Nalco Company, 7705 Hwy 90A, Sugar Land, TX 77478, USA
e-mail: rdepaula@nalco.com

G. Witzany (ed.), *Biocommunication of Fungi*, DOI 10.1007/978-94-007-4264-2_3,

linked by α-1,4 linear glycosidic bonds and α-1,6 linked glucose at the branching points. One great advantage of glycogen being used as a reserve carbohydrate is that this molecule has little effect on the cell's osmotic pressure. Together with starch in plants, glycogen is considered the main intracellular carbon and energy storage molecule. Most of the biochemical and molecular studies of glycogen metabolism regulation were performed in mammalian cells and in the yeast *Saccharomyces cerevisiae*. In mammalian cells, the liver and skeletal muscle cells are the main depository of glycogen. Yeast cells accumulate glycogen in the diauxic phase of growth, or in response to a limitation of carbon, nitrogen, sulfur or phosphorus, and they hydrolyze it under conditions of carbon starvation (Johnston and Carlson 1992; François and Parrou 2001). The fungus *Neurospora crassa*, the focus of this chapter, accumulates glycogen during the exponential growth phase and degrades when the growth rate decreases (de Paula et al. 2002). Although *N. crassa* and *S. cerevisiae* accumulate glycogen, they differ in the way they regulate their intracellular storage under environmental conditions. For example, under stressful environmental conditions, such as heat shock, *Neurospora* degrades glycogen while the yeast accumulates glycogen. The genome sequence of *Neurospora* has been completed (Galagan et al. 2003) and comparing the multiple filamentous fungi genomic sequences available with that of the yeast has revealed how divergent they are at the genomic level. An interesting feature of *N. crassa* is the high number of genes without identifiable homologues to known proteins ("orphan" genes) (Galagan et al. 2005); only approximately 40% of the proteins have been functionally annotated in the *N. crassa* genome database (http://www.broadinstitute.org/annotation/genome/neurospora/MultiHome.html) (Wang et al. 2011). This information indicates how dynamic the fungal genomes are compared to other microorganisms, and this stimulates the investigation of specific aspects of cell biology in a particular organism. Many reviews on glycogen metabolism have been published focusing on the yeast *S. cerevisiae* and mammalian cells (François and Parrou 2001; Roach et al. 2001; Wilson et al. 2010); while this chapter describes the data we have obtained with the fungus *N. crassa* and compares them to the yeast system.

2 Glycogen Synthesis and Degradation

Synthesis of glycogen involves three steps, which are: initiation, elongation, and branching; and requires the activities of glycogenin, glycogen synthase, and the branching enzyme, respectively. Degradation of glycogen requires the activities of glycogen phosphorylase and the debranching enzyme. An overall schematic representation is shown in Fig. 1.

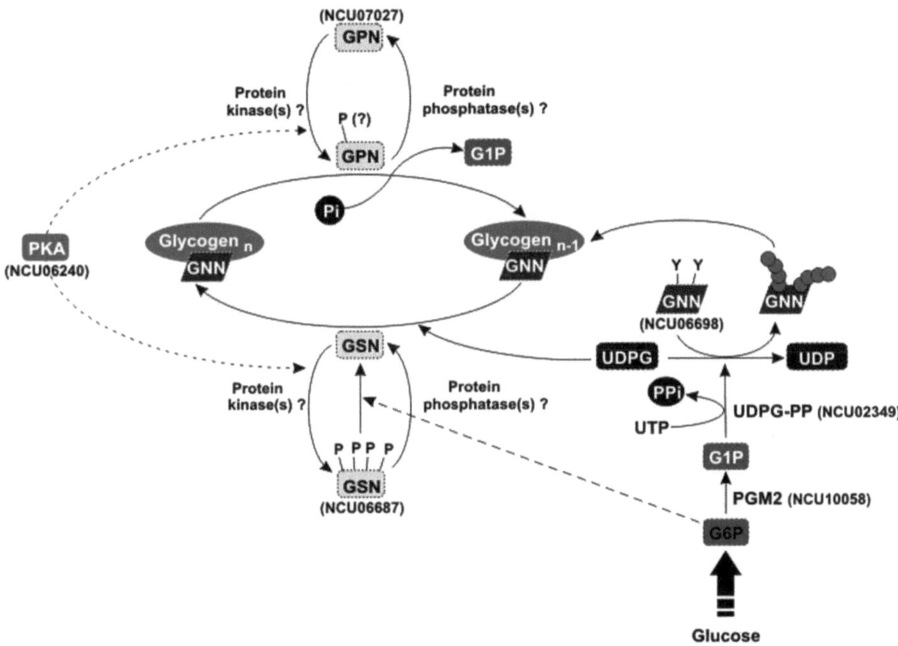

Fig. 1 Schematic representation of glycogen synthesis and degradation. The *N. crassa* orthologs are represented by their respective ORFs (http://www.broadinstitute.org/annotation/genome/neurospora/MultiHome.html). Glucose enters the cell and is converted to UDPG, the glucose donor for synthesis. GNN, the initiator of glycogen synthesis, is self-glucosylated, providing the non-reducing ends that ensure the glycogen elongation by the GSN enzyme through the formation of the α-1,4-glycosidic linkages, using UDPG as glucose donor. GPN enzyme catalyzes the phosphorolysis of glycogen yielding G1P and shortened glycogen as product. GSN and GPN are both controlled by reversible phosphorylation by the action of protein kinases and protein phosphatases. The four putative phosphorylation sites in GSN are shown. *G6P* glucose-6-P, *PGM2* phosphoglucomutase 2, *G1P* glucose-1-P, *UDPG-PP* uridine 5′-diphosphoglucose pyrophosphorylase, *UDPG* uridine 5′-diphosphoglucose, *GNN N. crassa* glycogenin, *GSN N. crassa* glycogen synthase, *GPN N. crassa* glycogen phosphorylase, *PKA* cyclic AMP-dependent protein kinase

2.1 Glycogen Initiation

The de novo synthesis of glycogen has been subject of intense study over decades. The question on how new glycogen molecules start to be synthesized came from a first study reported by Krisman and Barengo (1975). These authors identified a glucan-protein complex composed of a protein moiety attached to an α-1,4 glucan chain and proposed that this protein might function as an initiator molecule. Later, a protein that was covalently attached to the glycogen molecule and co-purified with glycogen synthase was identified (Alonso et al. 1995). The initiator protein was named later as glycogenin (Viskupic et al. 1992). One of the most fascinating characteristics of glycogenin is its ability to self-glucosylate using UDP-glucose (UDPG) as the glucan donor, a reaction highly stimulated by

Mn^{2+} ions (Lomako et al. 1988). A glycosidic bond is formed between the anomeric C1 of the glucose moiety derived from UDPG and the hydroxyl oxygen of a tyrosine side-chain of glycogenin. The glucosylated Tyr in the rabbit muscle glycogenin was identified as Tyr194 (Smythe et al. 1988) and mutations in this residue resulted in a protein unable to self-glucosylate (Cao et al. 1993a). Glycogenin then catalyzes glucosylation at C4 of the attached glucose once again (using UDPG as donor) to yield an O-linked disaccharide with the α-1,4 glycosidic linkage. This is repeated until a short linear glucose polymer (8–12 unities) with α-1,4 glycosidic linkages is built up on the glycogenin. Thus, glycogenin possesses two distinct enzymatic actions: the former C1-O-tyrosil and the subsequent α-1,4 glycosyl activities. Additionally, glycogenin is able to transfer glucose residues to a number of small acceptors in a process named trans-glucosylation (Smythe et al. 1990). However, the physiological significance of this process is not really understood. The three-dimensional structure of the rabbit skeletal muscle glycogenin has provided insight into understanding the catalytic action of this complex enzyme (Gibbons et al. 2002). The model revealed the dimeric nature of the protein but the mechanism of attachment of the first glucose residue is not yet fully understood. From the structure, it was suggested that transfer of the first residue is via an intermolecular reaction whereas the transfer of subsequent residues may be achieved by an intramolecular reaction.

The yeast *S. cerevisiae* has two glycogenin isoforms, which are products of the genes *glg1* and *glg2* (Cheng et al. 1995). The Glg1p isoform carries only one self-glucosylating Tyr residue (Tyr232) while the Glg2p isoform can be modified into two adjacent residues (Tyr230 and Tyr232) (Mu et al. 1996). Cells lacking either one of these genes can accumulate glycogen at levels comparable to wild-type cells, indicating that these genes are redundant in function. The absence of both genes completely abolishes the glycogen accumulation in cells under normal conditions. However, significant amounts of glycogen can be found in cells where glycogen synthase activity has been enhanced, suggesting the existence of alternative primers for the glycogen synthesis in the absence of a functional glycogenin (Torija et al. 2005). An auxiliary protein called GNIP (for glycogenin interacting protein) has been identified in yeast cells (Skurat et al. 2002), which enhances the self-glucosylation reaction upon the binding of GNIP (Zhai et al. 2004).

In the filamentous fungus *N. crassa*, only one isoform of glycogenin (GNN) has been identified (de Paula et al. 2005a) and gene inactivation by RIP (repeat-induced point mutation) completely abolished the accumulation of glycogen in this organism, suggesting that GNN is likely the only glycogen initiator present in *N. crassa* cells. This protein is unusually long (664 amino acids) and all domains required for the self- and trans-glucosylation activities are located within the 300 amino acids in the N-terminal region (de Paula et al. 2005a). However, the long C-terminal extension seems to be important to enhance the interaction with the glycogen synthase (GSN) enzyme as determined by yeast two-hybrid assays (de Paula et al. 2005b). Furthermore, expression of a truncated form of GNN containing only the first 360 amino acids was enough to rescue the glycogen deficient phenotype in yeast cells, indicating that the truncated protein is fully active (de Paula et al. 2005a). GNN has two glucosylation sites – Tyr196 and Tyr198; however, each residue contributes differently to the self-glucosylation process (de Paula et al. 2005b). Tyrosine 196 is the major glucosylation site as mutation of this residue to Phe resulted in the accumulation of only 30% of total glycogen, compared to the wild-type protein. It is unclear whether both sites are glucosylated at the same time or if Tyr198 represents an alternative glucosylation site which can be modified when Tyr196 is unavailable. The regulation of glycogenin expression has been poorly understood and only a few studies on this subject have been reported. In *N. crassa*, GNN expression seems to be regulated by nutrient availability and stress (de Paula et al. 2005a).

Once glucosylated, glycogenin acts as substrate for glycogen synthase (GS), which will elongate the glucose chain. It is speculated that the length of the oligosaccharide chain in glycogenin necessary to allow the proper elongation by GS is determined by the balance between the activities of glycogen phosphorylase and glycogen synthase (Cao et al. 1993b). Direct interactions by the two-hybrid approach between glycogenin and glycogen synthase (Pitcher et al. 1987) confirmed the physical interactions between these two proteins (Skurat et al. 2006; de Paula et al. 2005b).

2.2 Glycogen Maturation

Primed glycogenin will serve as the substrate for the two enzymes involved in the maturation process of glycogen particles: glycogen synthase and branching enzyme, which are the proteins that catalyze the formation of α-1,4 and α-1,6 glycosidic bonds, respectively. Similar to glycogenin, GS uses UDP-glucose as the donor of the glucose residues. Once GS extends the glycosidic chain to a certain length, the branching enzyme will transfer a string of approximately seven glucoses from the nascent chain to the glucose C6 in an adjacent chain, thus creating a ramification point. These ramifications occur, equidistantly, every 10–14 glucose residues apart. Bacterial glycogen synthase differs from its eukaryotic counterparts, due to the fact that they utilize ADP-glucose as the glucan donor (Preiss and Romeo 1994) and they lack regulation.

Overall, all glycogen synthases are conserved proteins among microbes and higher organisms and the differences are located mostly in the N- and C-termini of the protein, where the regulatory phosphorylation sites are located. *S. cerevisiae* contains two GS isoforms, Gsy1p and Gsy2p, in which the latter accounts for 80% of the total GS activity in the cells (Farkas et al. 1990, 1991). Three phosphorylation sites were found in the C-terminus of this enzyme (Hardy and Roach 1993). The kinases PKA, Snf1p, Yak1, Psk2, and Pho85 were described as involved in phosphorylation and regulation of Gsy2p, either directly or indirectly (Cameron et al. 1988; Wang et al. 2001; Huang et al. 1998, reviewed in Wilson et al. 2010).

In *N. crassa*, one GS isoform was identified (GSN), which shared much conservation with the counterparts in yeast and mammals (de Paula et al. 2002). Although the mechanisms of GSN regulation have not been completely unraveled, there is evidence that the PKA signaling pathway exerts a very important role in the control of its activity. Strains defective in the PKA signaling pathway showed impairments in accumulation of glycogen and in *gsn* expression. The in vitro studies showed that the PKA pathway also influences the GSN phosphorylation status (Freitas et al. 2010). However, GS is not an essential enzyme; *gsn* inactivation by RIP (Repeat Induced Point Mutation) does not seem to impair growth and development of the mutant strain (unpublished results). Four putative phosphorylation sites were identified in GSN (Ser632, Ser636, Thr641, and Thr645) based on a sequence alignment of different GS enzymes, all located at the C-terminus. Systematic site-directed mutagenesis of the four sites was carried out, mutating all residues to alanine. In vitro incorporation of radioactive phosphate by the mutant proteins produced in *E. coli* using cellular extract from a non-synthesizing GSN strain confirmed that all residues can be phosphorylated. However, compared to the yeast enzyme Gsy2p, GSN seems to have an additional phosphorylation site, since a truncated protein missing the C-terminal region containing the putative phosphorylation sites was still able to incorporate radioactive phosphate (unpublished results).

Among the phosphatases involved in the dephosphorylation and activation of glycogen synthases in microorganisms, PP1A seems to be the main player in yeast cells. Mutant alleles in *GLC7*, which encodes for the catalytic subunit of PP1A, displayed reduced glycogen accumulation, consistent with its inability to dephosphorylate Gsy2p (Cannon et al. 1994). Four different PP1A regulatory subunits (Gac1p, Pig1p, Pig2p and Gip2p) were found to be associated with Gsy2p, indicating that the specificity of the reaction is dictated by proper association with these subunits (Cheng et al. 1997). Although the protein phosphatases involved in the dephosphorylation of GSN in *Neurospora* have not been identified yet, GLC7 and GAC1 are conserved in this organism.

Only recently, the tri-dimensional structure of a eukaryotic GS was determined and thus provided the first insights into the role of the activator glucose-6-phosphate (Baskaran et al. 2011). Crystal structures of the basal activity state and those in the glucose-6-phosphate-activated form of the yeast protein (Gsy2p) showed an unusual tetramer assembling. Binding of glucose-6-phosphate induced conformational changes of the subunits leading to increased catalytic efficiency. Also, site-directed mutagenesis demonstrated that the residues Arg583 and Arg587 are necessary and sufficient for glucose-6-phosphate activation.

2.3 Glycogen Degradation

The breakdown of glycogen particles is accomplished by the action of glycogen phosphorylase and debranching enzymes. Glycogen phosphorylase releases glucose-1-phosphate from a terminal α-1,4 glycosidic bond, which then is converted to glucose-6-phosphate by the action of phosphoglucomutase before entering the catabolic pathway. The debranching enzyme carries out two distinct enzymatic activities: glucosyltransferase – the transfer of three glucose residues from one branch to another, and glucosidase – the breaking of α-1,6 glycosidic bonds. Similar to glycogen synthase, glycogen phosphorylase is regulated by reversible phosphorylation and allosteric effectors, mainly AMP.

In *N. crassa*, the two forms of glycogen phosphorylase and their dependency on the allosteric modulators have been described many years ago (Téllez-Iñón and Torres 1970). Moreover, the activities of glycogen synthase and glycogen phosphorylase oppose each other during stress conditions, favoring a breakdown of glycogen to allow enough energy availability for survival (Noventa-Jordão et al. 1996).

3 Regulation of Glycogen Metabolism

Regulation of glycogen metabolism has been studied for decades focusing on the activities of the enzymes catalyzing its synthesis and degradation, which are glycogen synthase and glycogen phosphorylase, respectively. The main control point is thought to be at glycogen synthase level, which is subject to multiple forms of regulation. The two enzymes are regulated by reversible covalent modification, in which phosphorylation activates glycogen phosphorylase and inhibits glycogen synthase (Téllez-Iñón et al. 1969; Fletterick and Madsen 1980). In addition, they are also regulated by allosterism, where glucose-6-phosphate and AMP are the allosteric effectors of glycogen synthase and glycogen phosphorylase, respectively. Glucose-6-phosphate reverses the glycogen synthase inactivation by phosphorylation and AMP is the allosteric activator for the dephosphorylated glycogen phosphorylase. The ratio between the activities in the absence and in the presence of the allosteric effectors is an index of the phosphorylation state for both enzymes. Multiple phosphorylation sites were identified in glycogen synthases, which are phosphorylated by different protein kinases, depending on the organism, whereas glycogen phosphorylase is phosphorylated in a single residue, Ser14, which is modified by the phosphorylase kinase protein.

Besides reversible changes in the regulatory enzymes activities, glycogen levels are also correlated with physiological conditions through control of gene expression mainly at transcriptional level of the gene encoding glycogen synthase. Environmental conditions indirectly regulate glycogen metabolism by activating signaling pathways and leading to either activation or repression of gene expression.

3.1 Environmental Conditions Regulating Glycogen Accumulation in N. crassa. Regulation of the gsn Gene Expression

N. crassa accumulated glycogen late in the exponential phase of the vegetative growth (around 24 h) and degraded it at the beginning of the stationary phase (de Paula et al. 2002). At this time, *gsn* expression was at its maximum. In addition, glycogen levels were highly regulated on exposure of cultures to some stress situations, such as heat shock (transfer from 30°C to 45°C) and carbon source limitation (sugar-free medium) (de Paula et al. 2002). Trehalose is another reserve carbohydrate that can be mobilized under different growth conditions. However, while *N. crassa* degraded glycogen under heat shock, trehalose was accumulated (de Pinho et al. 2001; Noventa-Jordão et al. 1996). Mycelia exposed to heat shock and carbon starvation presented changes in the expression of the gene encoding glycogen synthase (*gsn*), in glycogen accumulated and in glycogen synthase activity. There was a decrease in gene expression and a concomitant fall in the levels of glycogen and glycogen synthase activity. However, the *gsn* expression and glycogen levels were recovered when the cultures were returned to normal growth conditions (30°C and 2% sugar) (de Paula et al. 2002). These results suggested that transcriptional regulation may account for the decrease in glycogen synthase activity and subsequent glycogen mobilization observed under these conditions. On the other hand, glycogen phosphorylase was activated under heat shock showing that reversible changes in the two regulatory enzymes were observed upon temperature shifting (Noventa-Jordão et al. 1996).

The yeast *S. cerevisiae* exhibited opposite responses with respect to the environmental condition of heat shock. Compared to *N. crassa,* transcription of the genes encoding glycogen synthase *GSY1* and *GSY2* was induced (Unnikrishnan et al. 2003; Ni and LaPorte 1995) and glycogen was accumulated under a heat shock condition (Parrou et al. 1997). *GSY1* and *GSY2* mRNA induction was mediated by the STRE (STress Responsive Elements) motifs present in the promoter regions of these genes (Ni and LaPorte 1995; Unnikrishnan et al. 2003; Enjalbert et al. 2004). STRE motifs are DNA regulatory elements found in the promoter regions of genes responsive to different stressing conditions in *S. cerevisiae* (Martinez-Pastor et al. 1996). This cis-regulatory element is trans-activated by two yeast C_2H_2 zinc-finger transcription factors, which are the proteins Msn2p and Msn4p (Estruch and Carlson 1993; Schmitt and McEntee 1996). These two proteins quickly translocate to the nucleus after a stressful stimulus in a PKA-dependent way (Smith et al. 1998; Gorner et al. 1998).

The decrease in glycogen content observed in *N. crassa* cells exposed to heat stress may result from the down regulation of the *gsn* gene probably mediated by the STRE motif within the promoter region (Freitas and Bertolini 2004). The *gsn* gene has two STRE motifs, one upstream from a canonical TATA-box and another in an intron within the 5′-UTR. DNA fragments containing both motifs were specifically bound by nuclear proteins activated by heat shock (Freitas and Bertolini 2004). However, Msn2/4p orthologue proteins were not identified in the *N. crassa* database suggesting the existence of a different mechanism to regulate the heat shock response (Freitas et al. 2008). Whether STRE mediates *gsn* down regulation, it is possible to speculate that in *N. crassa* STRE motifs are involved in gene repression instead of gene activation. Although the STRE motifs in *S. cerevisiae* are frequently correlated with the activation of gene expression, the role of the STRE motif as a transcriptional repressor has also been reported in literature (de Groot et al. 2000; Vyas et al. 2005).

Another DNA cis-regulatory element that might be involved in glycogen accumulation in *N. crassa* by modulating the transcription of the *gsn* gene during heat shock is the HSE (Heat Shock Elements) motif, which is recognized and bound by the HSFs (Heat Shock Factors) transcription factors (Bienz and Pelham 1986). Yeast HSFs are constitutively bound to HSE motifs maintaining the basal levels of gene transcription even under normal temperature of growth. Heat stress only increases the transcriptional activity by promoting Hsf1p trimerization (Sorger et al. 1987; Jakobsen and Pelham 1988). In multicellular organisms, the HSFs bind to the consensus HSE only after heat stress has occurred (Kingston et al. 1987). In this context, *N. crassa* is similar to the yeast since gel shift analysis using nuclear extracts and DNA fragments containing HSE motifs showed that the proteins (including HSFs) are constitutively bound to the cis HSEs before heat shock (Meyer et al. 2000; Freitas and Bertolini 2004).

Attempts to identify the transcription factor(s) that bind to the STRE motif of the *N. crassa gsn* promoter were performed by gel shift assay experiments and mass spectrometry analysis. Five putative protein candidates acting as transcriptional regulators of the *gsn* gene expression during heat stress were identified. Analyses of their polypeptide sequences revealed the presence of protein domains usually found in proteins involved with transcription regulation. Among the proteins identified, two are noteworthy (Freitas et al. 2008). One is the protein encoded by the ORF NCU03482, annotated as a RuvB-like helicase 1 and belonging to a protein family having the AAA (ATPases Associated with diverse cellular Activities) and Tip49 (TBP-Interacting Protein 49) domains. Although the function of this protein family is not totally clear, they are supposed to play an important role in nuclear events, such as chromatin remodeling and regulation of transcription, since they interact with many nuclear proteins and show DNA helicase activity (Bellosta et al. 2005; Wood et al. 2000). The three-dimensional structure of the *N. crassa* protein was determined by molecular modeling based on the RUVBL1 structure (PDB 2c9o) from *Homo sapiens* (Matias et al. 2006). The monomer contains three domains, of which two are involved in ATP binding and one is a DNA/RNA-binding domain (unpublished results).

The other protein is the one encoded by the ORF NCU06679 annotated as histone acetyltransferase type B subunit 2, and previously annotated as a chromatin assembly factor subunit orthologue to the yeast Cac3p. This protein possesses five WD-40 domains within its polypeptide chain. The WD-40 domains are short tandem repetitions with approximately 40 amino acid residues bearing the GH dipeptide having 11–24 residues from its N-terminus and the WD dipeptide at its C-terminus (Smith et al. 1999). WD-repeat proteins are a large family found in all eukaryotes and are implicated in a variety of functions ranging from signal transduction and transcription regulation to cell cycle control, autophagy and apoptosis (Neer et al. 1994). The underlying common function of all WD-repeat containing proteins is coordinating multiprotein complex assemblies. The three-dimensional structure of the *N. crassa* protein was determined by molecular modeling based on the Nurf55 structure (PDB 2xyi) from *Drosophila melanogaster* (Nowak et al. 2011), a component of different chromatin-modifying complexes. Nurf55 forms a seven-bladed β-propeller characteristic for the WD40 family of proteins (unpublished results).

3.2 Transcription Factors Regulating Glycogen Accumulation in *N. crassa*

All living organisms need an appropriate gene regulation to express specific parts of their genomes to execute critical biological functions during development and differentiation, and to respond to different environmental signals. Gene transcription, a key step in gene expression, is a dynamic process that involves distinct steps. One is the recruitment of protein complexes to target genes, which associate with specific DNA-binding sites in response to tightly controlled physiological signaling cascades. These proteins fall into several classes, with the transcription factors being the class of sequence-specific DNA-binding proteins that mediate gene-specific transcriptional activation or repression. Thus, transcription factors constitute critical regulatory molecules that, upon a cell's signal, rapidly and transiently bind to specific cis-regulatory elements (or DNA response elements) in the chromatin, in a process associated with chromatin remodeling (Hager et al. 2009).

The release of the complete *N. crassa* genome (Galagan et al. 2003) and the establishment of a methodology for gene knockout by gene replacement (Ninomiya et al. 2004) allowed the construction of a collection of mutant strains each containing single-gene deletions to start. The availability of a mutant strain subset with each strain knocked out in genes encoding known or putative transcription factors allowed us to perform screening aimed to investigate alterations in glycogen accumulation in these mutant strains (Gonçalves et al. 2011). The glycogen accumulated by the mutant strains was quantified under normal growth temperature (30°C) and under heat shock stress (45°C) and compared to the wild-type strain. The analysis under heat stress was performed since *N. crassa* exhibits an opposite response when compared to the yeast *S. cerevisiae* concerning glycogen accumulation under this environmental condition. Many transcription factors were identified (Gonçalves et al. 2011, see Fig. 2) and most of them are annotated in the *N. crassa* database as hypothetical proteins. The *gsn* and *gpn* (gene encoding glycogen phosphorylase) expression was analyzed in all mutant strains grown under the temperatures of 30°C (vegetative growth) and 45°C (heat shock). Many of the mutant strains showed differences in gene expression when compared to the wild-type strain, in which both *gsn* and *gpn* expression decreased after heat shock (Fig. 2). In some mutant strains the gene expression could be correlated to the amount of glycogen accumulated suggesting that the transcription factors would act either directly on gene expression or indirectly, by regulating a gene whose product affects *gsn* and *gpn* expression.

Many identified transcription factors are biochemically characterized proteins, either in *N. crassa* or in other fungi. Among the transcription factors identified, it is noteworthy to cite the pH-regulator PacC (Peñalva and Arst 2002), the NIT2 nitrogen metabolism regulator (Fu and Marzluf 1990), the mycelial repressor of conidiation gene expression RCO-1 (Yamashiro et al. 1996), the light-inducible transcription factor CSP-1 (Lambreghts et al. 2009), the transcriptional activator of genes required for the use of alternative carbon sources XlnR (van Peij et al. 1998),

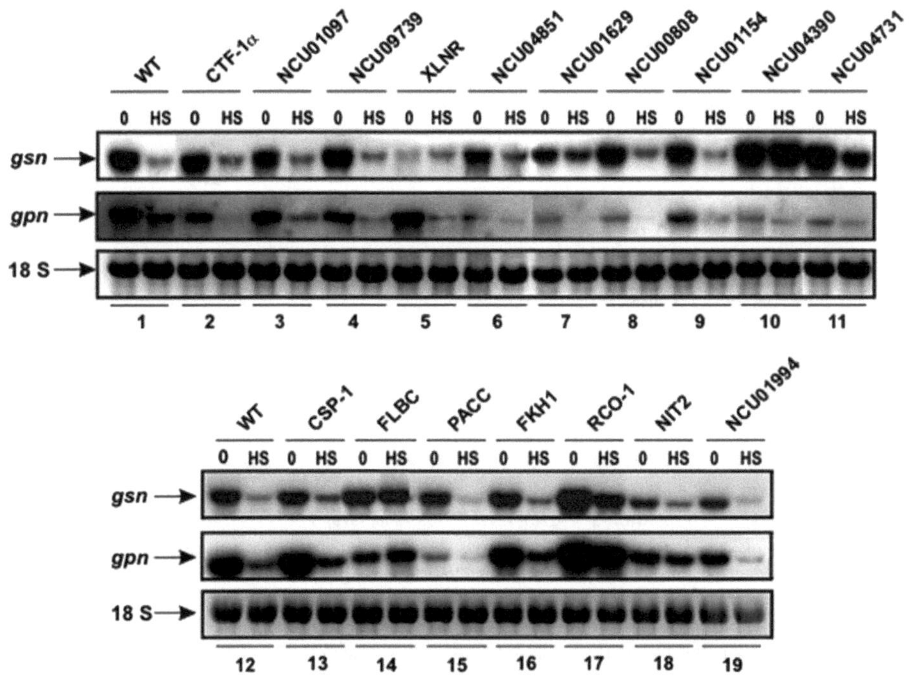

Fig. 2 Gene expression in *N. crassa* transcription-factor mutant strains. The expression of *gsn* (encoding glycogen synthase) and *gpn* (encoding glycogen phosphorylase) genes were analyzed in the transcription-factor mutant strains which exhibited a glycogen accumulation profile different from that shown by the wild-type strain. Cells from the mutant strains were cultivated for 24 h at 30°C and then subjected to a temperature shift of 45°C for 30 min. Culture samples were harvested before (0) and after the heat shock (HS) and the mycelia were used to extract total RNA. Total RNA was separated by electrophoresis in denaturing formaldehyde-agarose gels, transferred to neutral nylon membranes and the blots were probed with radio-labeled full-length *gsn* and *gpn* cDNAs. The 18 S rRNA stained with ethidium bromide was used as loading control. *WT* wild-type strain, *CTF-1α* cutinase transcription factor 1 alpha mutant strain, *XLNR* transcriptional activator xlnR mutant strain, *CSP-1* conidial separation-1 mutant strain, *FLBC* protein flbC mutant strain, *PACC* pH-response transcription factor pacC/RIM101 mutant strain, *FKH1* FKH1 protein mutant strain, *RCO-1* regulator of conidiation-1 mutant strain, *NIT2* nitrate nonutilizer-2 mutant strain. Proteins with no orthologues are represented by their respective ORFs numbers (http://www.broadinstitute.org/annotation/genome/neurospora/MultiHome.html)

among others. Search for cis DNA motifs for the transcription factors by in silico analysis of the *gsn* and *gpn* 5′-flanking regions led to the identification of putative DNA-binding sites for some proteins (Fig. 3). This information led us to start investigating the role of some transcription factors as regulators in glycogen metabolism. One is the PacC/Rim101p, which has been extensively studied in *A. nidulans* and *S. cerevisiae*, respectively, and plays an important role in the pH signaling pathway. This transcription factor mediates the cell adaptation to neutral-alkaline pH by activating genes that are preferentially expressed at alkaline pH and repressing those preferentially expressed at acidic pH (Tilburn et al. 1995).

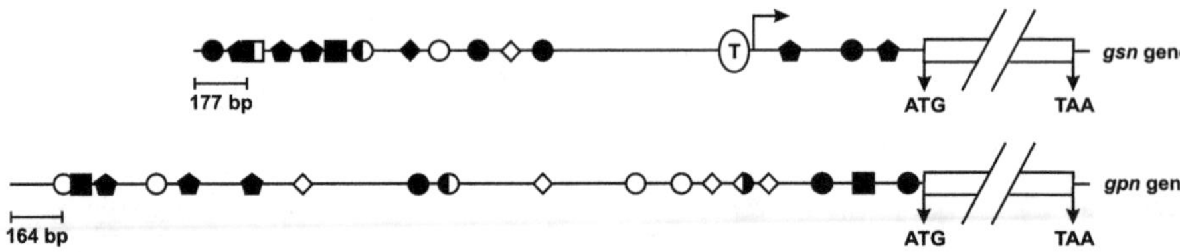

Transcription Factors	*cis* DNA element	Positions in *gsn* promoter (nt)*	Positions in *gpn* promoter (nt)*
● Mig1	5'-GGGG-3'	-2305, -2026, -1592, -1487, -317	-1610, -1564, -618, -351, -22
◆ PacC	5'-GCCAaG-	-1807	–
◇ NIT2	5'-TATCtac-3'	-1558	-2064, -1217, -660, -625, -594
■ XlnR	5'-GGCTGA-3'	-2071, -2252	-217, -2631
○ Sre1p	5'-TACA-3'	-1758, -2024	-2693, -2380, -1561, -825, -703
□ CTF1α	5'-CCGAGG-3'	-2248	–
⬟ Fkh	5'-AACA-3'	-2261, -2178, -2103, -948, -679, -214	-2548, -2327, -2115

*DNA element positions considering the ATG start codon. The Fkh DNA element positioned at -948 is coincident with the Transcription Initiation Site (TIS) at the *gsn* promoter

Fig. 3 Representation of the *gsn* and *gpn* genes 5′-flanking regions. The relative positions of the DNA motifs recognized by the transcription factors are indicated. The TATA-box sequence in the *gsn* gene is indicated by *T*. The ORFs are delimited by their ATG start codon and the TAA stop codon. The Transcription Initiation Site (*TIS*) in the *gsn* gene is represented by an *arrow*

In *A. nidulans*, PacC is activated by two successive proteolytic cleavage steps at the C-terminus, leading to the active protein $PacC^{27}$ that contains a DNA-binding domain capable of binding to the promoters of pH-regulated genes (reviewed in Arst and Peñalva 2003). The yeast protein appears to have a broader role than simply that of promoting alkaline pH-inducible responses (Lamb and Mitchell 2003). In addition, it differs from *A. nidulans* protein since it requires only a single cleavage step to be activated (Li and Mitchell 1997) and while PacC acts as a transcriptional activator under alkaline pH, Rim101p exerts its role as a repressor (Lamb and Mitchell 2003). Thus, the molecular mechanisms involved in the pH response differ among organisms.

The existence of a DNA motif for the *A. nidulans* PacC in the promoter *gsn* prompted us to investigate whether this transcription factor regulates glycogen accumulation. We demonstrated that the pH signaling pathway regulates glycogen metabolism where the *N. crassa* PACC may play a central role. Cells of the wild-type strain grown at alkaline pH (7.8) showed reduction in *gsn* transcript levels and low intracellular glycogen accumulation. In contrast, the $pacC^{KO}$ strain accumulated similar amounts of glycogen as the non-stressed wild-type strain and showed high *gsn* transcript levels at alkaline pH. Under the same conditions, *pacC* expression was up regulated. From these results, it was possible to characterize *gsn* as an acidic gene and to suggest a regulatory role for PACC in *gsn* expression. The recombinant PACC was produced in *E. coli* as a truncated protein containing the DNA-binding domain and was able to bind in vitro to a *gsn* DNA fragment containing the PacC motif. In addition, DNA-protein complexes were formed with extracts from cells grown at normal and alkaline pH, and binding was confirmed by ChIP analysis. An interesting result was that PACC present in the extracts prepared from cells grown at normal and alkaline pH had the same molecular mass, indicating that protein processing was not triggered by alkaline pH in contrast to what is described for the *A. nidulans* protein. Our results for PACC processing led us to suggest that PACC proteolysis may involve a mechanism that is different from that described for the *A. nidulans* protein (unpublished results).

Many other transcription factors are being investigated at this moment for their role as putative regulators of glycogen metabolism in *N. crassa*. The studies will provide insights into the regulatory network that is involved in the molecular mechanisms controlling glycogen metabolism in this microorganism.

4 Conclusions and/or Future Perspectives

Over the last decade, new findings in literature together with the availability of numerous genome sequences have allowed advances in understanding how glycogen metabolism is regulated in microorganisms. In addition, the establishment of strategies for gene deletion has substantially accelerated the search for genes linked to a particular phenotype. Regarding the fungus *N. crassa*, the majority of genes were individually deleted by a consortium of laboratories resulting in the creation of a deletion strain set that constitutes a powerful tool to start investigating the role of specific proteins as regulators of glycogen metabolism. We have taken the advantage of a transcription-factor deletion-strain collection to identify regulatory proteins and interesting information is now becoming available. Most of the transcription factors identified were annotated in the fungus database as hypothetical proteins. However, some of them were biochemically well characterized either in *N. crassa* or in other fungi, and some of them have DNA-binding motifs in the *gsn* and/or *gpn* 5′-flanking regions. The results suggest that glycogen metabolism could be interconnected to multiple cellular processes in this fungus.

An interesting piece of information emerging from these studies was that some transcription-factor mutant strains showing glycogen accumulation different from the wild-type strain presented impairments in cell cycle progression, contributing to the identification of connections between the core cell cycle oscillator and cell cycle transcription. The fact that some transcription factors identified in our screening are light-inducible proteins allowed us to speculate on the existence of a connection between circadian clocks and glycogen metabolism and also that circadian rhythms and energy state might be linked. Although the identification of the transcription factors did not mean that they have a direct regulatory role, they constitute a valuable group of candidate proteins acting as regulators in glycogen metabolism control. The results open up new opportunities for investigating key questions concerning glycogen metabolism regulation, such as how glycogen metabolism could be connected to cell cycle regulation, the biological clock, and other aspects of cellular metabolism. The understanding of such connections will be valuable to understand the importance of the energy balance to biological processes.

Acknowledgements The work in the author's laboratory was supported by grants and fellowships from FAPESP (Fundação de Amparo à Pesquisa do Estado de São Paulo), CNPq (Conselho Nacional de Desenvolvimento Científico e Tecnológico), and CAPES (Coorde nação de Aperfeiçoamento de Pessoal de Nível Superior).

References

Alonso MD, Lomako J, Lomako WM, Whelan WJ (1995) A new look at the biogenesis of glycogen. FASEB J 9:1126–1137

Arst HN Jr, Peñalva MA (2003) pH regulation in *Aspergillus* and parallels with higher eukaryotic regulatory systems. Trends Genet 19:224–231

Baskaran S, Chikwana VM, Contreras CJ, Davis KD, Wilson WA, DePaoli-Roach AA, Roach PJ, Hurley TD (2011) Multiple glycogen-binding sites in eukaryotic glycogen synthase are required for high catalytic efficiency toward glycogen. J Biol Chem 286:33999–34006

Bellosta P, Hulf T, Balla Diop S, Usseglio F, Pradel J, Aragnol D, Gallant P (2005) Myc interacts genetically with Tip48/Reptin and Tip49/Pontin to control growth and proliferation during *Drosophila* development. Proc Nat Acad Sci USA 102:11799–11804

Bienz M, Pelham HBR (1986) Heat shock regulatory elements function as an inducible enhancer in the *Xenopus* hsp70 gene and when linked to a heterologous promoter. Cell 45:753–760

Cameron S, Levin L, Zoller M, Wigler M (1988) cAMP-independent control of sporulation, glycogen metabolism and heat shock resistance in *S. cerevisiae*. Cell 53:555–566

Cannon JF, Pringle JR, Fiechter A, Khalil M (1994) Characterization of glycogen-deficient glc mutants of *Saccharomyces cerevisiae*. Genetics 136:485–503

Cao Y, Mahrenholz AM, DePaoli-Roach AA, Roach PJ (1993a) Characterization of rabbit skeletal muscle glycogenin. Tyrosine 194 is essential for function. J Biol Chem 268:14687–14693

Cao Y, Skurat AV, DePaoli-Roach AA, Roach PJ (1993b) Initiation of glycogen synthesis. Control of glycogenin by glycogen phosphorylase. J Biol Chem 268:21717–21721

Cheng C, Mu J, Farkas I, Huang D, Goebl MG, Roach PJ (1995) Requirement of the self-glucosylating initiator proteins Glg1p and Glg2p for glycogen accumulation in *Saccharomyces cerevisiae*. Mol Cell Biol 15:6632–6640

Cheng C, Huang D, Roach PJ (1997) Yeast PIG genes: PIG1 encodes a putative type 1 phosphatase subunit that interacts with the yeast glycogen synthase Gsy2p. Yeast 13:1–8

de Groot E, Bebelman JP, Mager WH, Planta RJ (2000) Very low amounts of glucose cause repression of the stress responsive gene HSP12 in *Saccharomyces cerevisiae*. Microbiology 146:367–375

de Paula R, de Pinho CA, Terenzi HF, Bertolini MC (2002) Molecular and biochemical characterization of the *Neurospora crassa* glycogen synthase encoded by the gsn cDNA. Mol Genet Genomics 267:241–253

de Paula RM, Wilson WA, Terenzi HF, Roach PJ, Bertolini MC (2005a) GNN is a self-glucosylating protein involved in the initiation step of glycogen biosynthesis in *Neurospora crassa*. Arch Biochem Biophys 435:112–124

de Paula RM, Wilson WA, Roach PJ, Terenzi HF, Bertolini MC (2005b) Biochemical characterization of *Neurospora crassa* glycogenin (GNN), the self-glucosylating initiator of glycogen synthesis. FEBS Lett 10:2208–2214 de Pinho CA, Polizeli MLTM, Jorge JA, Terenzi HF (2001) Mobilization of trehalose in mutants of the cyclic AMP signaling pathway, cr-1 (CRISP-1) and mcb (microcycle conidiation), of *Neurospora crassa*. FEMS Microbiol Lett 199:85–89

Enjalbert B, Parrou JL, Teste MA, François J (2004) Combinatorial control by the protein kinases PKA, PHO85 and SNF1 of the transcriptional induction of the *Saccharomyces cerevisiae GSY2* gene at the diauxic shift. Mol Genet Genomics 271:697–708

Estruch F, Carlson M (1993) Two homologous zinc finger genes identified by multicopy suppression in a SNF1 protein kinase mutant of *Saccharomyces cerevisiae*. Mol Cell Biol 13:3872–3881

Farkas I, Hardy TA, DePaoli-Roach AA, Roach PJ (1990) Isolation of the GSY1 gene encoding glycogen synthase and evidence for the existence of a second gene. J Biol Chem 265:20879–20886

Farkas I, Hardy TA, Goebl MG, Roach PJ (1991) Two glycogen synthase isoforms in *Saccharomyces cerevisiae* are coded by distinct genes that are differentially controlled. J Biol Chem 266:15602–15607

Fletterick RJ, Madsen NB (1980) The structures and related functions of phosphorylase a. Annu Rev Biochem 49:31–61

François J, Parrou JL (2001) Reserve carbohydrates metabolism in the yeast *Saccharomyces cerevisiae*. FEMS Microbiol Rev 25:125–145

Freitas FZ, Bertolini MC (2004) Genomic organization of the *Neurospora crassa* gsn gene: possible involvement of the STRE and HSE elements in the modulation of transcription during heat shock. Mol Genet Genomics 272:550–561

Freitas FZ, Chapeaurouge A, Perales J, Bertolini MC (2008) A systematic approach to identify STRE-binding proteins of the *gsn* glycogen synthase gene promoter in *Neurospora crassa*. Proteomics 8:2052–2061

Freitas FZ, de Paula RM, Barbosa LC, Terenzi HF, Bertolini MC (2010) cAMP signaling pathway controls glycogen metabolism in *Neurospora crassa* by regulating the glycogen synthase gene expression and phosphorylation. Fungal Genet Biol 47:43–52

Fu YH, Marzluf GA (1990) *nit-2*, the major nitrogen regulatory gene of *Neurospora crassa*, encodes a protein with a putative zinc finger DNA-binding domain. Mol Cell Biol 10:1056–1065

Galagan JE, Calvo SE, Borkovich KA, Selker EU et al (2003) The genome sequence of the filamentous fungus *Neurospora crassa*. Nature 422:859–868

Galagan JE, Henn MR, Ma LJ, Cuomo CA, Birren B (2005) Genomics of the fungal kingdom: insights into eukaryotic biology. Genome Res 15:1620–1631

Gibbons BJ, Roach PJ, Hurley TD (2002) Crystal structure of the autocatalytic initiator of glycogen biosynthesis, glycogenin. J Mol Biol 319:463–477

Gonçalves RD, Cupertino FB, Freitas FZ, Luchessi AD, Bertolini MC (2011) A genome-wide screen for *Neurospora crassa* transcription factors regulating glycogen metabolism. Mol Cell Proteomics 10(11). doi:10.1074/mcp.M111.007963

Görner W, Durchschlag E, Martinez-Pastor MT, Estruch F, Ammerer G, Hamilton B, Ruis H, Schüller C (1998) Nuclear localization of the C_2H_2 zinc finger protein Msn2p is regulated by stress and protein kinase A activity. Genes Dev 12:586–597

Hager GL, McNally JG, Misteli T (2009) Transcription dynamics. Mol Cell 35:741–753

Hardy TA, Roach PJ (1993) Control of yeast glycogen synthase-2 by COOH-terminal phosphorylation. J Biol Chem 268:23799–23805

Huang D, Moffat J, Wilson WA, Moore L, Cheng C, Roach PJ, Andrews B (1998) Cyclin partners determine Pho85 protein kinase substrate specificity in vitro and in vivo: control of glycogen biosynthesis by Pcl8 and Pcl10. Mol Cell Biol 18:3289–3299

Jakobsen BK, Pelham HRB (1988) Constitutive binding of yeast heat shock factor to DNA in vivo. Mol Cell Biol 8:5040–5042

Johnston M, Carlson M (1992) Regulation of carbon and phosphate utilization. In: Jones EW, Pringle JR, Broach JR (eds) The molecular and cellular biology of the yeast *Saccharomyces*. Cold Spring Harbor Laboratory Press, Cold Spring Harbor, pp 193–281

Kingston RE, Schuetz TJ, Larin Z (1987) Heat-inducible human factor that binds to a human hsp70 promoter. Mol Cell Biol 7:1530–1534

Krisman CR, Barengo R (1975) A precursor of glycogen biosynthesis: alpha-1,4-glucan-protein. Eur J Biochem 52:117–123

Lamb TM, Mitchell AP (2003) The transcription factor Rim101p governs ion tolerance and cell differentiation by direct repression of the regulatory genes NRG1 and SMP1 in *Saccharomyces cerevisiae*. Mol Cell Biol 23:677–686

Lambreghts R, Shi M, Belden WJ, Decaprio D, Park D, Henn MR, Galagan JE, Bastürkmen M, Birren BW, Sachs MS, Dunlap JC, Loros JJ (2009) A high-density single nucleotide polymorphism map for *Neurospora crassa*. Genetics 181:767–781

Li W, Mitchell AP (1997) Proteolytic activation of Rim1p, a positive regulator of yeast sporulation and invasive growth. Genetics 145:63–73

Lomako J, Lomako WM, Whelan WJ (1988) A self-glucosylating protein is the primer for rabbit muscle glycogen biosynthesis. FASEB J 2:3097–3103

Martinez-Pastor M, Marchler G, Schuller C, Marchler BA, Ruis H, Estruch F (1996) The *Saccharomyces cerevisiae* zinc finger proteins Msn2p and Msn4p are required for transcriptional induction through the stress response element (STRE). EMBO J 15:2227–2235

Matias PM, Gorynia S, Donner P, Carrondo MA (2006) Crystal structure of the human AAA+ protein RuvBL1. J Biol Chem 281:38918–38929

Meyer U, Monnerjahn C, Techel D, Rensing L (2000) Interaction of the *Neurospora crassa* heat shock factor with the heat shock element during heat shock and different developmental stages. FEMS Microbiol Lett 185:255–261

Mu J, Cheng C, Roach PJ (1996) Initiation of glycogen synthesis in yeast. Requirement of multiple tyrosine residues for function of the self-glucosylating Glg proteins in vivo. J Biol Chem 271:26554–26560

Neer EJ, Schmidt CJ, Nambudripad R, Smith TF (1994) The ancient regulatory-protein family of WD-repeat proteins. Nature 371:297–300

Ni HT, LaPorte DC (1995) Response of a yeast glycogen synthase gene to stress. Mol Microbiol 16:1197–1205

Ninomiya Y, Suzuki K, Ishii C, Inoue H (2004) Highly efficient gene replacements in *Neurospora* strains deficient for nonhomologous end-joining. Proc Natl Acad Sci USA 101:12248–12253

Noventa-Jordão MA, Polizeli MLTM, Bonini BM, Jorge JA, Terenzi HF (1996) Effects of temperature shifts on the activities of *Neurospora crassa* glycogen synthase, glycogen phosphorylase and trehalose-6-phosphate synthase. FEBS Lett 378:32–36

Nowak AJ, Alfieri C, Stirnimann CU, Rybin V, Baudin F, Ly-Hartig N, Lindner D, Müller CW (2011) Chromatin-modifying complex component Nurf55/p55 associates with histones H3 and H4 and polycomb repressive complex 2 subunit Su(z)12 through partially overlapping binding sites. J Biol Chem 286:23388–23396

Parrou JL, Teste MA, François J (1997) Effects of various types of stress on the metabolism of reserve carbohydrates in *Saccharomyces cerevisiae*: genetic evidence for a stress-induced recycling of glycogen and trehalose. Microbiology 143:1891–1900

Peñalva MA, Arst HN Jr (2002) Regulation of gene expression by ambient pH in filamentous fungi and yeasts. Microbiol Mol Biol Rev 66:426–446

Pitcher J, Smythe C, Campbel DG, Cohen P (1987) Identification of the 38-kDa subunit of rabbit skeletal muscle glycogen synthase as glycogenin. Eur J Biochem 169:497–502

Preiss J, Romeo T (1994) Molecular biology and regulatory aspects of glycogen biosynthesis in bacteria. Prog Nucleic Acid Res Mol Biol 47:299–329

Roach PJ, Skurat AV, Harris RA (2001) Regulation of glycogen metabolism. In: Jefferson LS, Cherrington AD (eds) Handbook of physiology. The endocrine pancreas and regulation of metabolism, vol II. Oxford University Press, New York, pp 609–647

Schmitt AP, McEntee K (1996) Msn2p, a zinc finger DNA-binding protein, is the transcriptional activator of the multistress response in *Saccharomyces cerevisiae*. Proc Natl Acad Sci USA 93:5777–5782

Skurat AV, Dietrich AD, Zhai L, Roach PJ (2002) GNIP, a novel protein that binds and activates glycogenin, the self-glucosylating initiator of glycogen biosynthesis. J Biol Chem 277:19331–19338

Skurat AV, Dietrich AD, Roach PJ (2006) Interaction between glycogenin and glycogen synthase. Arch Biochem Biophys 456:93–97
Smith A, Ward MP, Garrett S (1998) Yeast PKA represses Msn2p/Msn4p-dependent gene expression to regulate growth, stress response and glycogen accumulation. EMBO J 17:3556–3564
Smith TF, Gaitatzes C, Saxena K, Neer EJ (1999) The WD repeat: a common architecture for diverse functions. Trends Biochem Sci 24:181–185
Smythe C, Caudwell FB, Ferguson M, Cohen P (1988) Isolation and structural analysis of a peptide containing the novel tyrosyl-glucose linkage in glycogenin. EMBO J 7:2681–2686
Smythe C, Watt P, Cohen P (1990) Further studies on the role of glycogenin in glycogen biosynthesis. Eur J Biochem 189:199–204
Sorger PK, Lewis MJ, Pelham HR (1987) Heat shock factor is regulated differently in yeast and HeLa cells. Nature 329:81–84
Téllez-Iñón MT, Torres HN (1970) Interconvertible forms of glycogen phosphorylase in *Neurospora crassa*. Proc Natl Acad Sci USA 66:459–463
Téllez-Iñón MT, Terenzi H, Torres HN (1969) Interconvertible forms of glycogen synthetase in *Neurospora crassa*. Biochim Biophys Acta 191:765–768
Tilburn J, Sarkar S, Widdick DA, Espeso EA, Orejas M, Mungroo J, Peñalva MA, Arst HN Jr (1995) The *Aspergillus* PacC zinc finger transcription factor mediates regulation of both acid- and alkaline-expressed genes by ambient pH. EMBO J 14:779–790
Torija MJ, Novo M, Lemassu A, Wilson WA, Roach PJ, François J, Parrou JL (2005) Glycogen synthesis in the absence of glycogenin in the yeast *Saccharomyces cerevisiae*. FEBS Lett 18:3999–4004
Unnikrishnan I, Miller ST, Meinke M, LaPorte DC (2003) Multiple positive and negative elements involved in the regulation of expression of GSY1 in *Saccharomyces cerevisiae*. J Biol Chem 278:26450–26457
van Peij NN, Gielkens MM, de Vries RP, Visser J, de Graaff LH (1998) The transcriptional activator XlnR regulates both xylanolytic and endoglucanase gene expression in *Aspergillus niger*. Appl Environ Microbiol 64:3615–3619
Viskupic E, Cao Y, Zhang W, Cheng C, DePaoli-Roach AA, Roach PJ (1992) Rabbit skeletal muscle glycogenin. Molecular cloning and production of fully functional protein in *Escherichia coli*. J Biol Chem 267:25759–25763
Vyas VK, Berkey CD, Miyao T, Carlson M (2005) Repressors Nrg1 and Nrg2 regulate a set of stress-responsive genes in *Saccharomyces cerevisiae*. Eukaryot Cell 4:1882–1891
Wang Z, Wilson WA, Fujino MA, Roach PJ (2001) Antagonistic control of autophagy and glycogen accumulation by Snf1p, the yeast homolog of AMP-activated protein kinase and the cyclin-dependent kinase Pho85p. Mol Cell Biol 21:5742–5752
Wang TY, He F, Hu QW, Zhang Z (2011) A predicted protein-protein interaction network of the filamentous fungus *Neurospora crassa*. Mol Biosyst 7:2278–2285
Wilson WA, Roach PJ, Montero M, Baroja-Fernández E, Muñoz FJ, Eydallin G, Viale AM, Pozueta-Romero J (2010) Regulation of glycogen metabolism in yeast and bacteria. FEMS Microbiol Rev 34:952–985
Wood MA, McMahon SB, Cole MD (2000) An ATPase/helicase complex is an essential cofactor for oncogenic transformation by c-Myc. Mol Cell 5:321–330
Yamashiro CT, Ebbole DJ, Lee BU, Brown RE, Bourland C, Madi L, Yanofsky C (1996) Characterization of *rco-1* of *Neurospora crassa*, a pleiotropic gene affecting growth and development that encodes a homolog of Tup1 of *Saccharomyces cerevisiae*. Mol Cell Biol 16:6218–6228
Zhai L, Dietrich A, Skurat AV, Roach PJ (2004) Structure-function analysis of GNIP, the glycogenin-interacting protein. Arch Biochem Biophys 421:236–242

Epigenetic Regulation of Secondary Metabolite Biosynthetic Genes in Fungi

Robert Cichewicz

Abstract Many fungi are prolific producers of secondary metabolites. These compounds are thought to fill a variety of ecologically relevant functions including participating in chemical sensing systems and communicating with other organisms in their surroundings. In order for natural products to operate effectively, the host organisms must maintain control over their expression. Epigenetic processes have emerged as important contributors to the regulation of secondary metabolite expression in fungi. Several examples highlighting the range of secondary metabolites whose production is controlled wholly or in part by epigenetic processes are presented. Recently reported cases demonstrating how insights into epigenetic control mechanisms can improve the production and diversification of secondary metabolite production are also examined.

1 Introduction

Although the central dogma (DNA → RNA → proteins) provides a useful framework for understanding the functional organization of cells, its applicability to natural products chemistry requires that a fourth step (DNA → RNA → proteins → secondary metabolites) be added, which recognizes the genomic origins of these unique biomolecules. It is widely acknowledged that natural products are not inert byproducts of a cell's biosynthetic machinery, but rather play dynamic roles regulating intracellular and intercellular gene expression and other physiological

R. Cichewicz (✉)
Natural Products Discovery Group, Department of Chemistry and Biochemistry, Stephenson Life Sciences Research Center, University of Oklahoma, 101 Stephenson Parkway, Room 1000, Norman, OK 73019-5251, USA

Ecology and Evolutionary Biology Program, University of Oklahoma, Norman, OK 73019-5251, USA
e-mail: rhcichewicz@ou.edu

G. Witzany (ed.), *Biocommunication of Fungi*, DOI 10.1007/978-94-007-4264-2_4,

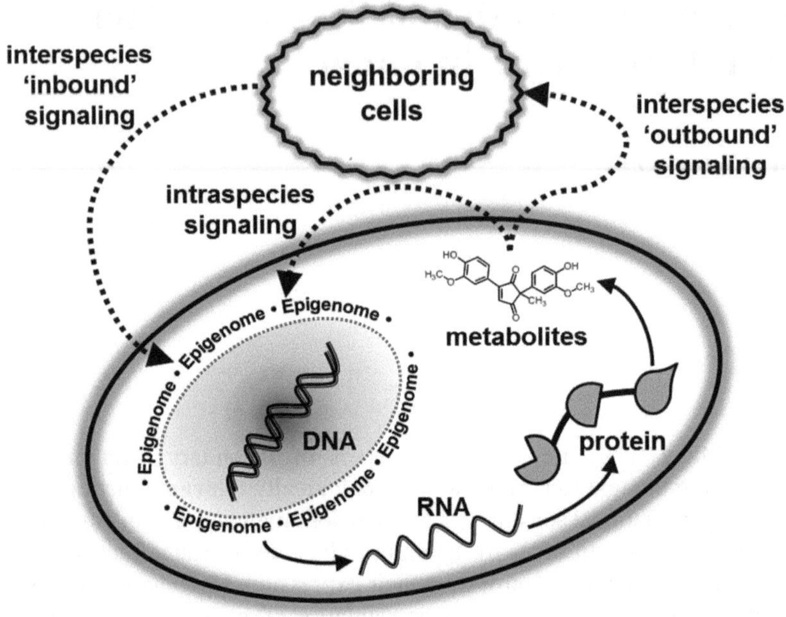

Fig. 1 Secondary metabolites are believed to be important components of the diverse lexicon of fungi. The epigenome functions as a molecular filter that serves to amplify and/or suppress a variety of 'inbound' and 'outbound' intraspecies and interspecies signals

functions (Davies and Ryan 2011). Moreover, many of the biosynthetic pathways leading to the production of secondary metabolites can be recalcitrant to activation suggesting that the products of these pathways may be involved in carefully regulated intraspecies and interspecies interactions (Brakhage and Schroeckh 2011; Pettit 2011; Scherlach and Hertweck 2009). More recently, the complexity of the chemical interplay between natural products and cells has been shown to extend into a new realm involving the epigenome (Fig. 1) (Cichewicz 2010).

The epigenome is defined as the assortment of biochemical features that serve to alter the transcription of a gene or genes, but do not directly alter the composition of DNA. In many ways, the epigenome functions as a biological filter (Fig. 1) that is responsible for modulating the receptivity of cells to 'inbound' interspecies and intraspecies signaling events. Likewise, the epigenome also has the capacity to act as an 'outbound' filter that is capable of blocking DNA transcription and thereby suppressing signal generation. Consequently, the epigenome plays an important set of regulatory roles determining how cells both sense and respond to a range of chemical stimuli in their surroundings. Determining the specific molecular events associated with how a cell's epigenome filters these signals is crucial for enabling researchers to navigate the milieu of complex interactions that characterize an organism's native environment.

Small molecules have demonstrated tremendous promise for their abilities to directly interfere with epigenetic processes (Cichewicz 2010). Although many

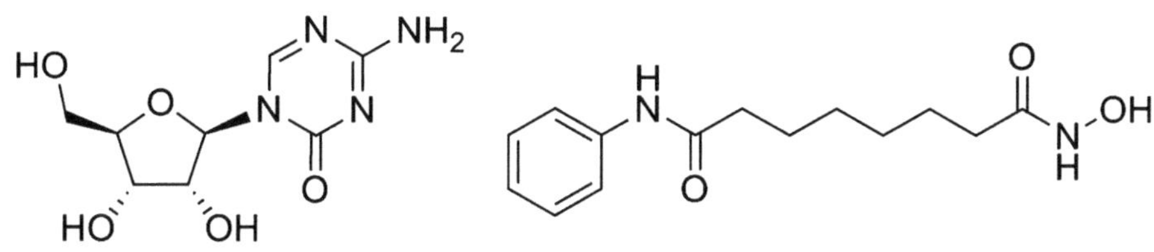

Fig. 2 Two important and widely used chemical epigenetic modifiers: 5-azacytidine inhibits DNA methyltransferase activity and suberoylanilide hydroxamic acid inhibits histone deacetylases

epigenome modulators have been reported that can interfere with a wide range of molecular targets, two groups of these have become established as the major focus of small-molecule interference strategies. The first group consists of histone modifying agents, which are responsible for altering the DNA-binding interactions of histones (e.g., inhibit histone acetylation). A well-established example of a small-molecule inhibitor that targets histone deacetylases is suberoylanilide hydroxamic acid (Fig. 2). The second group targets the addition of covalent markers to DNA (e.g., DNA methylation). The nucleotide analog 5-azacytidine represents a widely used inhibitor of DNA methyltransferase activity (Fig. 2). An extraordinary body of knowledge has developed in recent years concerning the epigenomes of many organisms including fungi. The interested reader is urged to consult the variety of previously published papers and reviews concerning the epigenomic mechanisms of altered gene expression (Brosch et al. 2008; Emre and Berger 2006; Martienssen et al. 2008; Palmer and Keller 2010; Reyes-Dominguez et al. 2010; Stimpson and Sullivan 2010; Strauss and Reyes-Dominguez 2011; Suzuki and Bird 2008). Within the last several years, we (Cichewicz 2010; Wang et al. 2010; Williams et al. 2008) and others (Bok et al. 2009; Chiang et al. 2009; Hertweck 2009; Strauss and Reyes-Dominguez 2011) have noted an intriguing link between the alteration of a cell's epigenome and the production of secondary metabolites. Although these studies have focused exclusively on fungi, it is reasonable to speculate that the same principals are applicable to other eukaryotic organisms including plants and protists. This chapter will focus on some of the recent developments concerning the application of epigenetic approaches to modulating secondary metabolite production in fungi.

2 Chemical Epigenetics Elicits the Production of New Molecules from Fungi

The general disparity between the relatively large numbers of secondary metabolite biosynthetic pathways encoded within the genomes of many fungi versus the relatively small numbers of compounds that they typically generate under laboratory

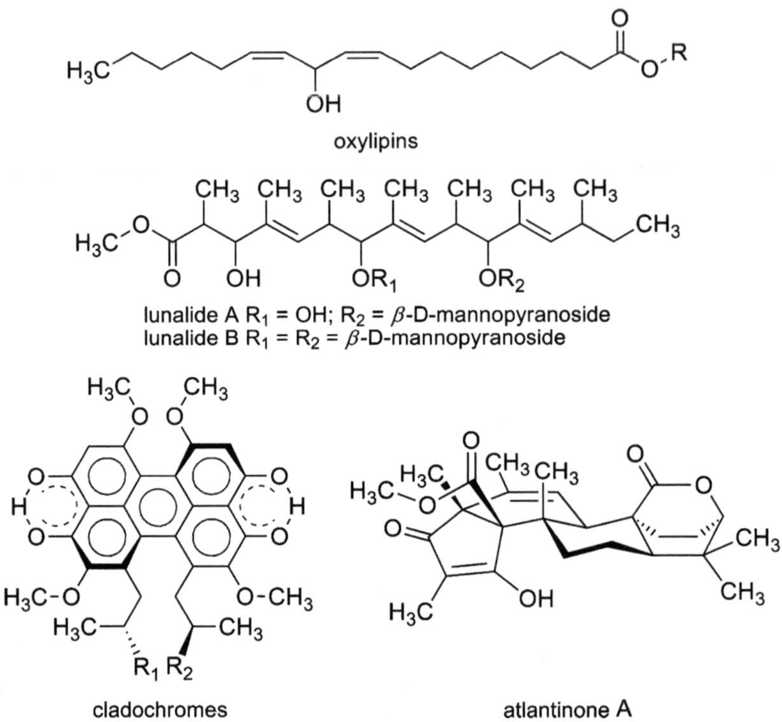

Fig. 3 Examples of fungal metabolites obtained following administration of epigenetic modifiers. Cladochromes were obtained following treatment of *Cladosporium cladosporioides* with suberoylanilide hydroxamic acid (R groups consist of various hydroxybenzoate and hydroxycarbonate groups). Oxylipins were produced from the same fungus upon treatment with 5-azacytidine (the R group consists of the free acid and esters). Lunalides and atlantinones were made by a *Diatrype* sp. and *P. citrenum*, respectively, administered 5-azacytidine

culture conditions has been previously noted (Cichewicz et al. 2010). In view of this discrepancy, several methods have been proposed as tools for accessing the products of these "silent" or "cryptic" pathways (Cichewicz et al. 2010). Chemical epigenetics has emerged as a useful tool for simply and affordably probing the secondary metabolite diversity potential of many fungal species. This led to the procurement of several new and previously known compounds by my research group (e.g., cladochromes, oxylipins, lunalides, and atlantinones; refer to Fig. 3) (Wang et al. 2010; Williams et al. 2008). The atlantinones (Wang et al. 2010) are particularly intriguing given their occurrence in the guttate of an epigenetically modified (5-azacytidine, a DNA methyltransferase inhibitor) *Penicillium citrenum* culture.

Recently, the Crews lab has reported the purification and structure determination of EGM-556 (Fig. 4) from a marine-derived *Microascus* isolate (Vervoort et al. 2010). This cyclodepsipeptide was obtained only after treatment of the culture with

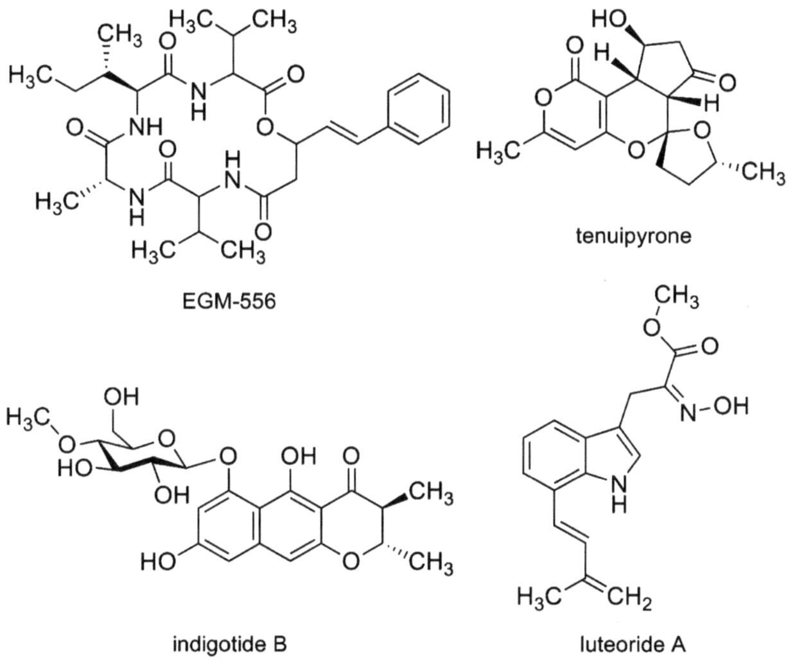

Fig. 4 EGM-556, tenuipyrone, indigotide B, luteorides serve as additional important examples of the capacity for DNA methyltransferase and histone deacetylase inhibitors to generated new and structurally diverse secondary metabolites from fungi

the histone deacetylase inhibitor suberoylanilide hydroxamic acid. In another case, the combination of histone deacetylase and DNA methyltransferase inhibitors was applied to an entomopathogenic *Isaria tenuipes* isolate, which led to the generation of the novel tetracyclic polyketide tenuipyrone (Fig. 4) (Asai et al. 2011a). Chemical epigenetic induction of additional compounds including the polyketide indigotide B (Fig. 4) from *Cordyceps indigotica* (Asai et al. 2012) and prenylated tryptophan metabolites luteorides A-C (Fig. 4) from *Torrubiella luteorostrata* (Asai et al. 2011b) have been reported following treatment of fungal cultures with 5-azacytidine and suberoyl *bis*-hydroxamic acid, respectively. The use of chemical epigenetic modifiers has recently been extended to modulating the production of volatile natural products as reported for the endophytic fungus *Hypoxylon* sp. (Ul-Hassan et al. 2012). The role of chemical epigenetic modulators in facilitating the production of natural products from fungi is further supported by a study in which the expression of secondary-metabolite-encoding biosynthetic genes in *Aspergillus niger* was tracked following treatment of the fungus with 5-azacytidine or suberoylanilide hydroxamic acid (Fisch et al. 2009). The fungus exhibited substantial up-regulation in the expression of dozens of polyketide, nonribosomal peptide, and hybrid polyketide-nonribosomal peptide biosynthetic pathways. These data provide compelling evidence that chemical epigenetic elicitation of silent biosynthetic pathways is an effective approach for securing new fungal natural products.

3 Epigenetic Mechanisms and Marks Involved in Secondary Metabolite Biosynthesis

Epigenomic control of secondary metabolite production in fungi is highly complex with a multitude of proteins suggested to play a role in this process. But beyond the level of proteins, even the location of secondary metabolite gene clusters in the fungal genome appears to have a considerable influence on transcriptional regulation. A recent review of this topic eloquently detailed several established and emerging details linking secondary metabolite gene cluster position (both chromosomal location and positional effects within gene clusters) with epigenetic features (Palmer and Keller 2010). For example, a substantial body of convincing data draw a link between the function of epigenetic proteins like LaeA and their abilities to impact the production of natural products whose biosynthetic genes are located in subtelomeric regions (McDonagh et al. 2008; Perrin et al. 2007).

In the fungi *Aspergillus nidulans, Aspergillus fumigatus,* and *Fusarium graminearum*, several chromatin modifying proteins have come to light whose functions are directly involved in regulating secondary metabolite production (Lee et al. 2009; Palmer and Keller 2010; Reyes-Dominguez et al. 2010, 2011). Chief among these are proteins that exhibit histone deacetylase and methyltransferase activities. Histone 3 lysine 9 trimethylation appears to be an important repressive marker for limiting secondary metabolite biosynthetic gene cluster function in multiple fungal species (Reyes-Dominguez et al. 2010, 2011). In some cases, activating marks such as histone 3 acetylation have also been reported (Reyes-Dominguez et al. 2010). Additional epigenetic marks that play varying roles in secondary metabolite biosynthetic pathway expression have been previously reviewed (Cichewicz 2010).

4 Bacterial Contact Impacts the Fungal Epigenome and Alters Secondary Metabolite Production

Hertweck and Brakhage have recently reported on a fascinating new dimension concerning interspecies interactions between fungi and bacteria (Schroeckh et al. 2009). It was observed that cell-cell contact between the model fungus *A. nidulans* and *Streptomyces rapamycinicus* led to several dramatic changes in the secondary metabolite profile of the fungal species. Prominent among these changes was the generation of several previously undetectable polyketides including orsellinic acid, lecanoric acid, F-9775A, and F-9775B (Fig. 5). One of the most striking findings from this study was that no diffusible signaling elements were detected by the authors that could be linked to the induced polyketide production. Instead, direct physical contact between the fungus and bacteria was required for upregulating the expression of the biosynthetic genes.

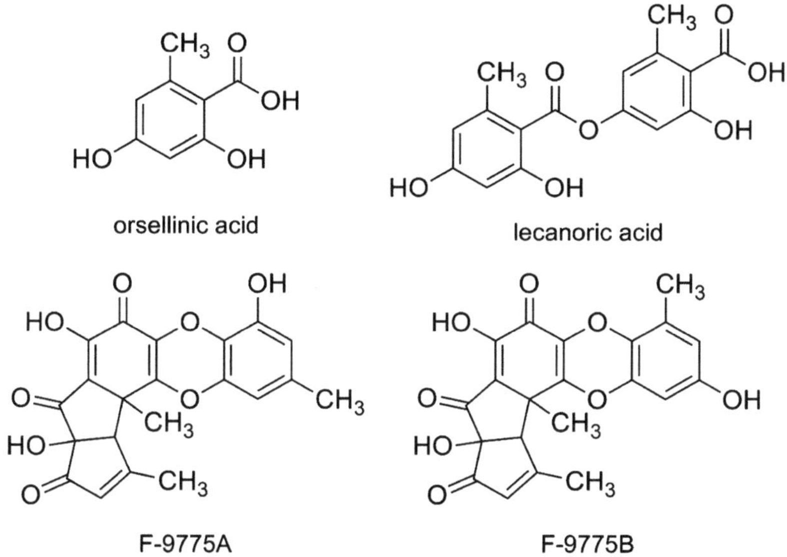

Fig. 5 The altered biosynthesis of orsellinic acid, lecanoric acid, F-9775A, and F-9775B by *A. nidulans* during physical contact with the bacteria *S. rapamycinicus* offers several fascinating lessons concerning the epigenetic regulation of secondary metabolite production in fungi

While investigating a potential role for histone posttranslational modification in regulating polyketide production in *A. nidulans*, it was observed that anacardic acid (a histone acetyltransferase inhibitor) suppressed transcription of the orsellinic acid biosynthetic gene, *orsA* (Nützmann et al. 2011). Conversely, treatment with suberoylanilide hydroxamic acid resulted in upregulated transcription of *orsA* and accumulation of orsellinic acid in the culture medium. These data suggested that epigenome modifications were required to facilitate the production of *A. nidulans* polyketides upon contact with *S. rapamycinicus*. Using a genomic-based search strategy, a total of 40 acetyltransferase genes were detected in the fungus and individual knock out mutants tested for their transcriptional responses to *S. rapamycinicus* contact. From this study, the *gcnE* (a histone acetyltransferase ortholog similar to GCN5 in *Saccharomyces cerevisiae*) mutant was identified as having suppressed *orsA* transcription even while in intimate contact with *S. rapamycinicus*. Orthologs of GCNE are reported as known members of Saga, Ada, and NuA4 complexes, which are involved in histone acetylation and overall restructuring of chromatin in fungi. As an extension of these findings, an *adaB* mutant was prepared to test the role of the Saga/Ada complex in controlling orsellinic acid biosynthetic. The *adaB* mutant exhibited no detectable orsellinic acid production following incubation with *S. rapamycinicus*, which further supported the idea that epigenome-level regulation is involved in controlling secondary metabolite production under bacterial-fungal coculture conditions.

Chromatin immunoprecipitation studies were also used to confirm that altered histone acetylation patterns were incurred by the fungus upon contact with

S. rapamycinicus. Antibodies specific for histone 3 acetylation of lysine 9 and lysine 14 showed that wild-type *A. nidulans* experienced a significant increase in histone acetylation at these key position, which was in synchrony with secondary metabolite production. In comparison, the *gcnE* and *adaB* mutants exhibited substantially reduced levels of histone 3 lysine 9 and lysine 14 acetylation. It was further noted that increased levels of epigenome-wide histone 3 lysine 14 acetylation was associated with secondary metabolite production, whereas histone 3 lysine 9 acetylation was required for gene-specific transcriptional promotion.

Examination of additional *A. nidulans* secondary metabolite pathways showed that both the *gcnE* and *adaB* mutants exhibited greatly impaired production of sterigmatocystin, terrequinone, and penicillin. In contrast, a *laeA* mutant, whose gene product has been proposed as a key global regulator or secondary metabolite production in *Aspergillus* spp., showed no change in orsellinic acid or lecanoric acid production in response to *S. rapamycinicus* cocultivation. Considering the unaltered expression of *laeA* during bacterial-fungal interactions, despite substantial changes in secondary metabolite production by *A. nidulans*, the proposed role of this putative global secondary metabolite regulator may need to be reevaluated.

5 Induction of Fungal Polyketides by Chemical Epigenetic Modulation and Biotic Challenge

In 2008, our group reported the discovery of lunalides A and B from a *Diatrype*-like fungal species (Williams et al. 2008). The fungus had been obtained from a fifth instar luna moth (*Actias luna*; Saturniidae) larva. Later, our group also isolated the fungus from adult moths, as well as surface sterilized eggs. Whereas the initial shake flask cultures of the fungus were relatively devoid of natural products, the fungus treated with 5-azacytidine showed a substantial change in its secondary metabolite profile. These changes were highlighted by the appearance of two prominent peaks. Purification and subsequent structure determination of the compounds afforded the lunalides in good yield. Another notable change in the appearance of the 5-azacytidine treated fungus was the induction of an intense black pigmentation throughout the fungal mycelia (Fig. 6).

While screening for other growth conditions that promoted the production of lunalides, we found that the addition of *Escherichia coli* to the fungal culture resulted in the generation of these polyketides, but in greatly reduced yields. Moreover, the phenotype of the *E. coli* treated culture markedly differed from the 5-azacytidine dosed culture suggesting that *E. coli* was less effective at stimulating lunalide and pigment production. This led our group to screen the stimulatory potential of several additional biotic sources including bacteria, fungi, protists, and nematodes. Representative examples from this test are illustrated in Fig. 6. Amazingly, the addition of nematodes (*Caenorhabditis elegans*) to the culture medium resulted in the accumulation of lunalides and generation of black pigmentation similar to the 5-azacytidine treatment.

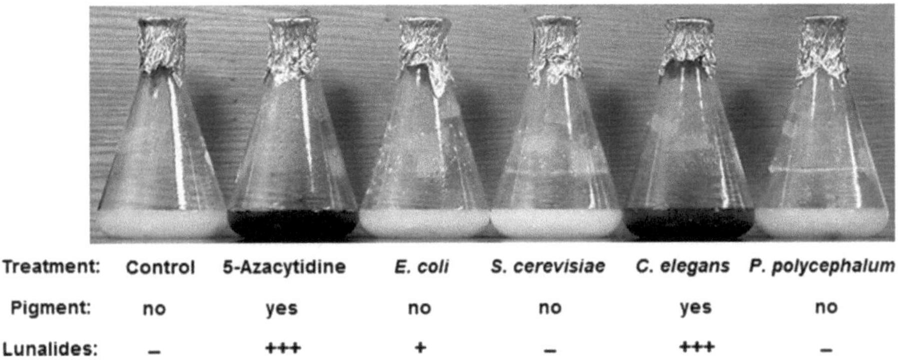

Fig. 6 The *Diatrype* sp. isolate was highly responsive to treatment with 5-azacytidine resulting in the production of a black pigment and new polyketides metabolites, lunalides. Elicitation of 250 mL *Diatrype* sp. cultures was performed by adding 1 mL of autoclaved *Escherichia coli* (bacterium) broth culture, *Saccharomyces cerevisiae* (yeast) broth culture, *Caenorhabditis elegans* (nematode) culture plate wash, or *Physarum polycephalum* (mycetozoan – slime mold) culture plate wash. Treatment with bacteria yielded modest amounts of lunalides, but no pigment. Addition of *C. elegans* resulted in significant changes in the secondary metabolite profile of the *Diatrype* sp. isolate, which were remarkably similar to the changes in the culture treated with 5-azacytidine. None of the other treatments showed any changes in secondary metabolites relative to the control culture

Even more striking was the finding that lunalide and lunalide-like polyketide metabolites possessed substantial nematicidal activities. In a previous screen for nematicidal agents, the mannose conjugated polyketide MK7924 had been detected as a potent inhibitor of *C. elegans* (Kumazawa et al. 2003). This natural product, which is structurally similar to lunalides, displayed activity against a range of different organisms. More recently, the roselipin analogs roselipins 3A-3E were found as inhibitors of the parasitic worm *Haemonchus contortus* (Ayers et al. 2010). These data support the intriguing possibility that the *Diatrype*-like fungal specimen requires a nematode-specific cue for inducing the production of lunalides and that chemical epigenetic stimulation was capable of overriding this highly controlled expression system. Further experiments will be needed to address these questions in detail.

6 Chemical Epigenetic Revamping of Secondary Metabolite Biosynthesis

In a new and unique application of the chemical epigenetic methodology, the Cox lab and its collaborators used both histone deacetylase (suberoyl *bis*-hydroxamic acid) and DNA methyltransferase (5-azacytidine) inhibitors to probe the synthetic plasticity of the tenellin biosynthetic gene cluster (Yakasai et al. 2011). The biosynthesis of tenellin and many of its derivatives has been investigated in

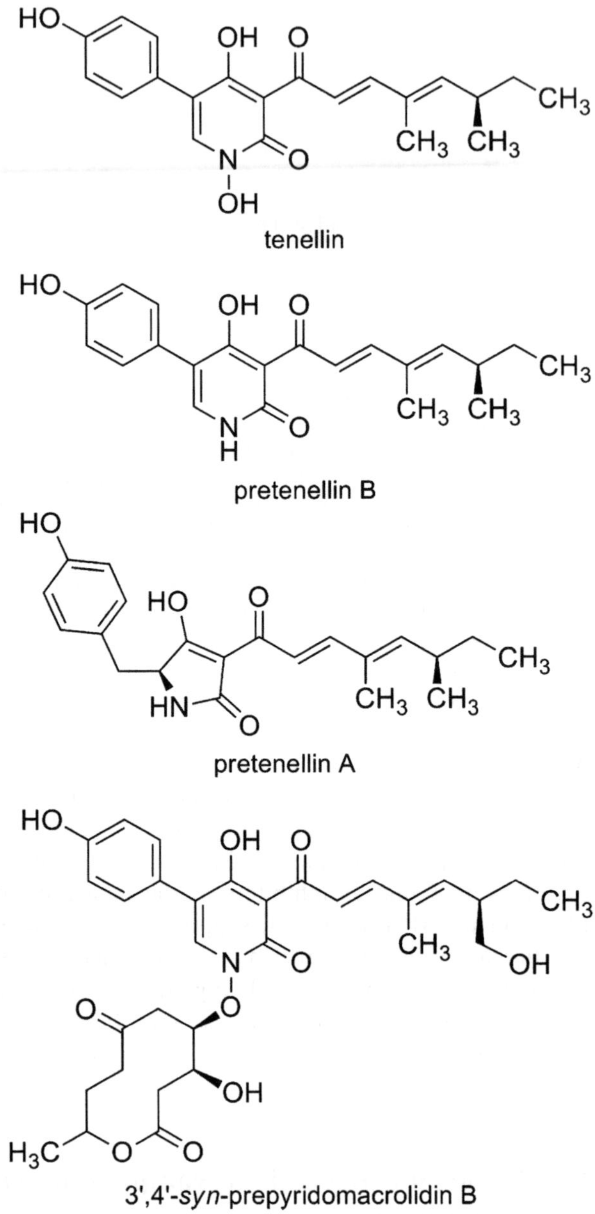

Fig. 7 Using a combination of RNA interference and chemical epigenetic modifiers, the biosynthesis of tenellin was substantially revamped in *B. bassiana* leading to the production of numerous new and known tenellin derivatives

the fungus *Beauveria bassiana* CBS110.25. Not surprisingly, the addition of epigenetic modifiers to fungal cultures resulted in the production of several new tenellin derivatives including new conjugated products of tenellin with cephalosporolide B (Fig. 7). Of more considerable interest, was the series of new tenellin analogs, which were produced using a combination of epigenetic modifiers with RNA interference of the tenellin biosynthetic genes *tenA* and *tenB* (responsible for the conversion of pretenellinA and pretenellin B, respectively, to tenellin; refer to Fig. 7). In total, 22 tenellin metabolites were characterized with over half of these compounds noted as new. These results confirm several important lessons concerning the range of valuable outcomes that are realized with the application of chemical epigenetic modifiers to fungal secondary metabolite production. First, enhanced titers of known secondary metabolites were often observed (Williams et al. 2008) and this can be very important for providing immediate access to compounds that are otherwise difficult to obtain. Second, the diversity of secondary metabolite families can be potentially diversified to create new or uncommon analogs for augmenting bioactivity based screening programs. Third, chemical epigenetic modifiers have the potential to impact the expression of a wide range of genes beyond those directly associated with a secondary metabolite's core biosynthetic genes. Consequently, chemical epigenetic modifiers have the potential to significantly revamp constitutive biosynthetic processes in new and unexpected ways. This approach promises to further enhance the underlying potential for fungi to generate chemically diverse assemblages of compounds from each core biosynthetic pathways.

7 Conclusions

Secondary metabolites are assumed to play important roles for enabling fungi to carryout intraspecies and interspecies interactions. The epigenome is proposed to function as a vital filter that helps to attenuate, as well as amplify the variety of chemical signals that fungi both receive and transmit. Understanding the cellular basis of epigenetic gene regulation is important for fully appreciating the extensive 'linguistic' capacity of fungi. Moreover, overcoming powerful epigenetic silencing factors is necessary for accessing the rich chemical lexicon of fungi. It is proposed that new and intriguing regulatory roles for epigenetic processes will continue to emerge as greater insights into fungal signaling networks evolve.

References

Asai T, Chung Y-M, Sakurai H, Ozeki T, Chang F-R, Yamashita K, Oshima Y (2011a) Tenuipyrone, a novel skeletal polyketide from the entomopathogenic fungus, *Isaria tenuipes*, cultivated in the presence of epigenetic modifiers. Org Lett 14(2):513–515. doi:10.1021/ol203097b

Asai T, Yamamoto T, Oshima Y (2011b) Histone deacetylase inhibitor induced the production of three novel prenylated tryptophan analogs in the entomopathogenic fungus, *Torrubiella luteorostrata*. Tetrahedron Lett 52:7042–7045

Asai T, Yamamoto T, Chung Y-M, Chang F-R, Wu Y-C, Yamashita K, Oshima Y (2012) Aromatic polyketide glycosides from an entomopathogenic fungus, *Cordyceps indigotica*. Tetrahedron Lett 53:277–280

Ayers S, Zink DL, Mohn K, Powell JS, Brown CM, Bills G, Grund A, Thompson D, Singh SB (2010) Anthelmintic constituents of *Clonostachys candelabrum*. J Antibiot 63:119–122
Bok JW, Chiang Y-M, Szewczyk E, Reyes-Dominguez Y, Davidson AD, Sanchez JF, Lo H-C, Watanabe K, Strauss J, Oakley BR, Wang CCC, Keller NP (2009) Chromatin-level regulation of biosynthetic gene clusters. Nat Chem Biol 5:462–464
Brakhage AA, Schroeckh V (2011) Fungal secondary metabolites – strategies to activate silent gene clusters. Fungal Genet Biol 48:15–22
Brosch G, Loidl P, Graessle S (2008) Histone modifications and chromatin dynamics: a focus on filamentous fungi. FEMS Microbiol Rev 32:409–439
Chiang YM, Lee KH, Sanchez JF, Keller NP, Wang CC (2009) Unlocking fungal cryptic natural products. Nat Prod Commun 4:1505–1510
Cichewicz RH (2010) Epigenome manipulation as a pathway to new natural product scaffolds and their congeners. Nat Prod Rep 27:11–22
Cichewicz RH, Henrikson JC, Wang X, Branscum KM (2010) Strategies for accessing microbial secondary metabolites from silent biosynthetic pathways. In: Baltz RH, Demain AL, Davies JE (eds) Manual of industrial microbiology and biotechnology, 3rd edn. ASM Press, Washington, DC, pp 78–95
Davies J, Ryan KS (2011) Introducing the parvome: bioactive compounds in the microbial world. ACS Chem Biol 7:252–259. doi:10.1021/cb200337h
Emre NC, Berger SL (2006) Histone post-translational modifications regulate transcription and silent chromatin in *Saccharomyces cerevisiae*. Ernst Schering Res Found Workshop 57:127–153
Fisch K, Gillaspy A, Gipson M, Henrikson J, Hoover A, Jackson L, Najar F, Wägele H, Cichewicz R (2009) Chemical induction of silent biosynthetic pathway transcription in *Aspergillus niger*. J Ind Microbiol Biotechnol 36:1199–1213
Hertweck C (2009) Hidden biosynthetic treasures brought to light. Nat Chem Biol 5:450–452
Kumazawa S, Kanda M, Utagawa M, Chiba N, Ohtani H, Mikawa T (2003) MK7924, a novel metabolite with nematocidal activity from *Coronophora gregaria*. J Antibiot 56:652–654
Lee I, Oh J-H, Keats Shwab E, Dagenais TRT, Andes D, Keller NP (2009) HdaA, a class 2 histone deacetylase of *Aspergillus fumigatus*, affects germination and secondary metabolite production. Fungal Genet Biol 46:782–790
Martienssen RA, Kloc A, Slotkin RK, Tanurdžić M (2008) Epigenetic inheritance and reprogramming in plants and fission yeast. Cold Spring Harb Symp Quant Biol 73:265–271
McDonagh A, Fedorova ND, Crabtree J, Yu Y, Kim S, Chen D, Loss O, Cairns T, Goldman G, Armstrong-James D, Haynes K, Haas H, Schrettl M, May G, Nierman WC, Bignell E (2008) Sub-telomere directed gene expression during initiation of invasive aspergillosis. PLoS Pathog 4:e1000154
Nützmann H-W, Reyes-Dominguez Y, Scherlach K, Schroeckh V, Horn F, Gacek A, Schümann J, Hertweck C, Strauss J, Brakhage AA (2011) Bacteria-induced natural product formation in the fungus *Aspergillus nidulans* requires Saga/Ada-mediated histone acetylation. Proc Natl Acad Sci USA 108:14282–14287
Palmer JM, Keller NP (2010) Secondary metabolism in fungi: does chromosomal location matter? Curr Opin Microbiol 13:431–436
Perrin RM, Fedorova ND, Bok JW, Cramer RA Jr, Wortman JR, Kim HS, Nierman WC, Keller NP (2007) Transcriptional regulation of chemical diversity in *Aspergillus fumigatus* by LaeA. PLoS Pathog 3:e50
Pettit RK (2011) Small-molecule elicitation of microbial secondary metabolites. Microb Biotechnol 4:471–478
Reyes-Dominguez Y, Bok JW, Berger H, Shwab EK, Basheer A, Gallmetzer A, Scazzocchio C, Keller N, Strauss J (2010) Heterochromatic marks are associated with the repression of secondary metabolism clusters in *Aspergillus nidulans*. Mol Microbiol 76:1376–1386
Reyes-Dominguez Y, Boedi S, Sulyok M, Wiesenberger G, Stoppacher N, Krska R, Strauss J (2011) Heterochromatin influences the secondary metabolite profile in the plant pathogen *Fusarium graminearum*. Fungal Genet Biol 49:39–47. doi:10.1016/j.fgb.2011.11.002

Scherlach K, Hertweck C (2009) Triggering cryptic natural product biosynthesis in microorganisms. Org Biomol Chem 7:1753–1760

Schroeckh V, Scherlach K, Nützmann H-W, Shelest E, Schmidt-Heck W, Schuemann J, Martin K, Hertweck C, Brakhage AA (2009) Intimate bacterial–fungal interaction triggers biosynthesis of archetypal polyketides in *Aspergillus nidulans*. Proc Natl Acad Sci USA 106:14558–14563

Stimpson KM, Sullivan BA (2010) Epigenomics of centromere assembly and function. Curr Opin Cell Biol 22:772–780

Strauss J, Reyes-Dominguez Y (2011) Regulation of secondary metabolism by chromatin structure and epigenetic codes. Fungal Genet Biol 48:62–69

Suzuki MM, Bird A (2008) DNA methylation landscapes: provocative insights from epigenomics. Nat Rev Genet 9:465–476

Ul-Hassan SR, Strobel GA, Booth E, Knighton B, Floerchinger C, Sears J (2012) Modulation of volatile organic compound formation in the mycodiesel-producing endophyte – *Hypoxylon* sp. CI-4. Microbiology 158:465–473. doi:10.1099/mic.0.054643-0

Vervoort HC, Drašković M, Crews P (2010) Histone deacetylase inhibitors as a tool to up-regulate new fungal biosynthetic products: isolation of EGM-556, a cyclodepsipeptide, from *Microascus* sp. Org Lett 13:410–413

Wang X, Sena Filho JG, Hoover AR, King JB, Ellis TK, Powell DR, Cichewicz RH (2010) Chemical epigenetics alters the secondary metabolite composition of guttate excreted by an Atlantic-Forest-soil-derived *Penicillium citreonigrum*. J Nat Prod 73:942–948

Williams RB, Henrikson JC, Hoover AR, Lee AE, Cichewicz RH (2008) Epigenetic remodeling of the fungal secondary metabolome. Org Biomol Chem 6:1895–1897

Yakasai AA, Davison J, Wasil Z, Halo LM, Butts CP, Lazarus CM, Bailey AM, Simpson TJ, Cox RJ (2011) Nongenetic reprogramming of a fungal highly reducing polyketide synthase. J Am Chem Soc 133:10990–10998

Genes from Double-Stranded RNA Viruses in the Nuclear Genomes of Fungi

Jeremy Bruenn

Abstract The sequencing of large numbers of eukaryotic genomes has demonstrated the widespread occurrence of viral genes in nuclear (and mitochondrial) genomes. Essentially all the families of RNA viruses are represented, and we have coined the term NIRV (non-retroviral integrated RNA virus) for these sequences. Some 3% of sequenced eukaryotic genomes have NIRVs and the fungi are well represented with both complete and partial copies of RNA viral genes. The fungal NIRVs are the best characterized and demonstrate that, with the most widespread fungal viruses, the dsRNA totiviruses and partitiviruses, which generally exist as stable persistent infections in their hosts, gene transfers between host and virus have taken place in both directions. Selection has preserved those events with adaptive value: hosts with NIRVs have become immune to infection with the cognate virus and viruses with cellular toxin genes provide an advantage to their hosts by killing cells without the virus.

1 Introduction

While genomic transfers from retrovirus to eukaryotic host are well known and expected, horizontal transfer of non-retroviral RNA viruses is unexpected, since non-retroviral RNA viruses lack reverse transcriptase and the integration machinery for successful transfer to DNA genomes. However, in 2004, Crochu et al. (2004) convincingly demonstrated transfer of flavivirus-like (+ssRNA) elements to mosquito genomes by sequencing across integration boundaries. Holmes (2009) identified this discovery as "one of the most remarkable observations in viral evolution of recent years." But this finding marked only the beginning of discovery

J. Bruenn (✉)
Department of Biological Sciences, The State University of New York at Buffalo, Buffalo, NY 14260, USA
e-mail: cambruen@buffalo.edu

G. Witzany (ed.), *Biocommunication of Fungi*, DOI 10.1007/978-94-007-4264-2_5,

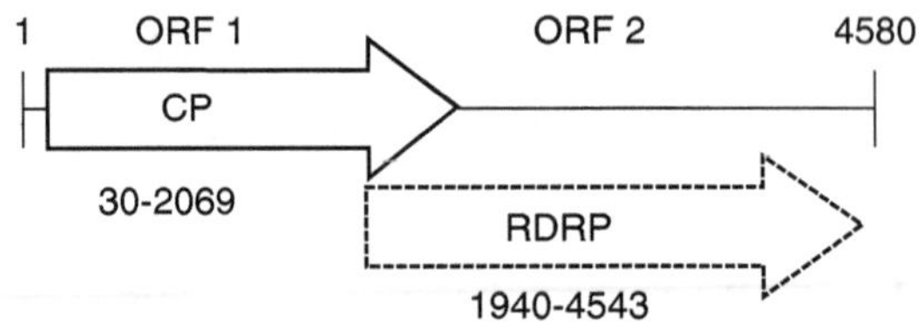

Fig. 1 Genomic structure of the ScVL1 (ScVLA) totivirus

for a new type of genomic interaction. In 2009 we showed that the process was in fact widespread by sequencing across putative viral (Totiviruses: dsRNA) integration sites and carrying out detailed evolutionary and functional analyses among such integrated elements in yeast genomes. We coined the term NIRVs (non-retroviral integrated RNA viruses) for this genomic interaction and this term has been accepted (Horie et al. 2010; Koonin 2010). Independently, Frank et al. reported some of the NIRVs of Totiviruses reported by us using BLAST evidence (Frank and Wolfe 2009). Horie et al. (2010) reported that NIRVs of Bornaviruses (−ssRNA) are common in mammals including humans. We also found that Ebola and Marburg-like elements (−ssRNA) are integrated in several mammalian genomes (Taylor et al. 2010). Belyi et al. (2010) found additional elements and genomes in the Bornavirus-mammal and Filovirus-mammal NIRV systems. Recently, Liu et al. (2010) reported that NIRVs of Partitiviruses and Totiviruses are widespread in eukaryotes (animals, plants and fungi). NIRVs appear to include viruses from most families of RNA viruses, to be widespread in animal genomes and plant genomes (Chiba et al. 2011), and to have originated many millions of years ago (Katzourakis and Gifford 2010). Geuking et al. (2009) demonstrated a mechanism in mice whereby host integration enzymes can act to form a NIRV from a non-retroviral RNA virus. Several studies have now implicated and identified the signatures of retrotransposon activity in association with NIRV formation (Crochu et al. 2004; Horie et al. 2010; Liu et al. 2010; Taylor and Bruenn 2009; Taylor et al. 2010).

The two major families of fungal viruses for which NIRVs are known are the totiviruses and the partitiviruses. Totiviruses have a single viral double-stranded RNA (dsRNA) which encodes a viral capsid protein and a viral RNA-dependent RNA polymerase (RdRp). The partitiviruses have at least two viral dsRNAs, separately encapsidated, which encode a viral capsid polypeptide and a RdRp. The genome of a typical totivirus is shown in Fig. 1. All but one of the known NIRVs in fungi are totivirus NIRVs, so we shall consider these first.

2 Origin of NIRVs: Horizontal Transfer from Virus to Host

Totivirus NIRVs occur in fungi, plants, insects, and nematodes. In each case, the NIRV is most closely related to a totivirus infecting a closely related organism: plant totivirus NIRVs are most closely related to plant totiviruses; fungal totivirus NIRVs to fungal totiviruses, animal totivirus NIRVs to animal totiviruses etc.

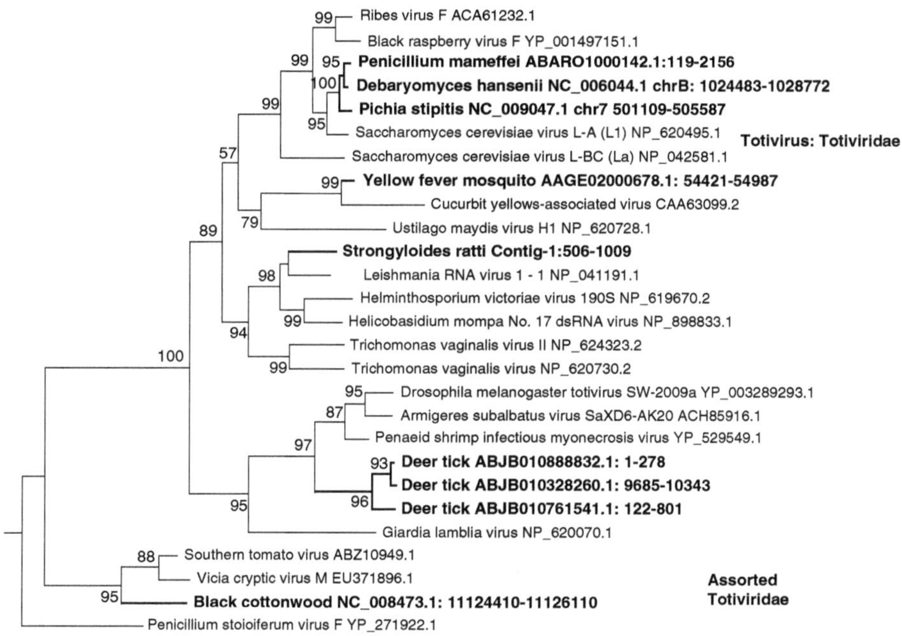

Fig. 2 Phylogeny of totiviruses (based on RdRp sequences) and totivirus-like NIRVs (shown in *bold* and modified from Liu et al. 2010). The genus Totivirus extends from Ribes virus to the *Ustilago maydis* virus

(Liu et al. 2010). This is illustrated in Fig. 2 and is consistent with NIRV origination by reverse transcription of endogenous viruses. There is now direct evidence for this mechanism of formation of NIRVs in a NIRV of recent origin in Drosophila yakuba (Ballinger et al. 2011). In addition, the viruses most prominently represented by NIRVs and those organisms in which the NIRVs occur are exactly those expected by such a model: those virus–host systems for which persistent infection is common. The majority of NIRVs discovered so far are of viruses that persist in their hosts: bornaviruses in mammals (Horie et al. 2010); flaviviruses, mononegavirales, orthomyxoviruses, picornaviruses, and rhabdoviruses in insects (Ballinger et al. 2011; Katzourakis and Gifford 2010); totiviruses in fungi (Taylor and Bruenn 2009). Presumably, the longer viral RNAs exist in viable cells, the more likely they are to be erroneously reverse transcribed by endogenous retrotransposons and adventitiously integrated into host genomes (and into the germ line in metazoans).

The other major group of fungal viruses for which NIRVs is known is the partitiviruses. Partitivirus NIRVs most closely related to fungal viruses occur in arthropods, including *Drosophila grimshawi*, *Rhodnius prolixus*, *Ixodes scapularis*, and the protozoan *Entamoeba histolytica* (Liu et al. 2010). This seems most likely to have occurred by horizontal transfer from virus-infected fungal parasites to their hosts, as with other host–parasite systems (Gilbert et al. 2010; Moran and Jarvik 2010; Yoshida et al. 2010). For example, the partitivirus NIRV closest to

Sclerotinia sclerotiorum partitivirus S occurs in *Arabidopsis thaliana* (Liu et al. 2010). *Sclerotinia sclerotiorum* is a plant parasitic fungus. No examples of partitivirus NIRVs in fungal genomes were reported. However, *Wickerhamomyces anomalus*, a wine yeast, does have at least two partial copies of a partitivirus RdRp (AEGI01002047,3803-4354; AEGI01001398,4000-3290; AEGI01000208,3740-4336; and AEGI01000553, 4025-3261). Since the contigs for this assembly are rather short, it is not clear that four copies of the NIRV occur. AEGI01002047 and AEGI01000208 are nearly identical, and AEGI0100138 and AEGI01000553 are nearly identical. The longest of these ORFs has 214 amino acids related to the partitivirus RdRp (closest relative Aspergillus ochraceous virus, E value of 2×10^{-13}). This NIRV preserves the F-A-B region of the RdRp, or about half the most highly conserved region of the protein.

3 Function of NIRVs: The Cell

The preservation of NIRVs over long periods of evolution clearly implies a function. This is obvious for totivirus NIRVs in the fungi. Some of these NIRVs preserve open reading frames of nearly 1,500 codons, implying that they continue to encode proteins and that these proteins have an advantageous function. In fact, our recent results show for the first time that NIRVs are expressed as protein, at least in the case of one fungal totivirus NIRV (Bruenn and Taylor 2012).

Analysis of totivirus NIRVs in fungi shows that they have novel (non-viral) protein mediated functions. First, the complete open reading frames of the cap and RdRp genes are preserved (with open reading frames as long as 1,441 amino acids in the cap-pol fusion in the *P. stipitis* NIRV), and only the coding sequences are preserved. The signals for frameshifting and protein processing have been lost, and nothing outside the coding sequences is preserved. For instance, the complete totivirus genomes encoded by *Shefferomyces stipitis* and *Debaryomyces hansenii* (the RNA sequences) are compared in Fig. 3. The region between the conserved regions of the cap and pol genes is not conserved and the region outside the coding regions also shows no conservation. The major conserved motifs characteristic of RdRps (and those peculiar to the totiviruses) are the regions showing the most conservation, again consistent with protein expression. Second, there are cellular RNAPII transcripts of each gene, and the transcripts of the RdRps are initiated at new positions just upstream of the RdRp coding sequences, rather than at the 5′ end of the cap gene as in the initiation site for the viral RdRp. Third, the evolution of NIRV sequences is consistent with purifying selection, rather than neutral evolution. Finally, six or more of the eight biochemically conserved residues important to the mRNA decapping function of cap are altered in three of the cap NIRVs of *P. stipitis*, suggesting that these cap NIRVs have a function different from that in the virus (Taylor and Bruenn 2009). There are no totivirus viral

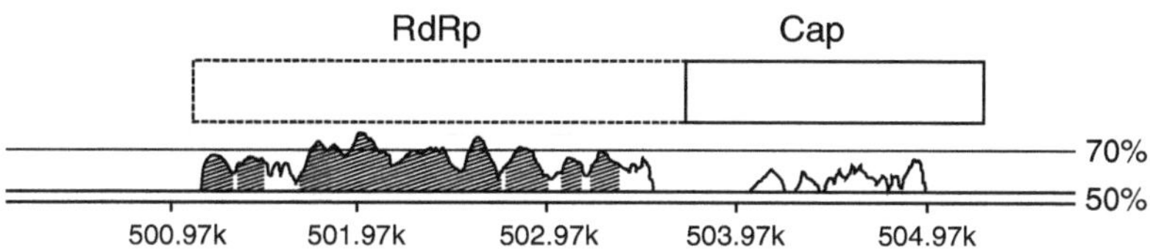

Fig. 3 Comparison of *Shefferomyces stipitis* chromosome 7 and *Debaryomyces hansenii* chromosome B in the region encoding the complete totivirus genome. Numbers are from *S. stipitis* chromosome 7. Visualization of similarity by ViSTA (Mayor et al. 2000). *Horizontal lines* show 50% and 70% nucleotide sequence identity

particles present in any of the yeasts with totivirus NIRVs (Taylor and Bruenn 2009; Bruenn and Taylor 2012).

The most obvious function for NIRVs would be conferring host resistance to their cognate viruses. It has been known for decades that overexpression of RNA viral cap proteins (or in some cases their RNAs) confers resistance to cognate virus infection in plants (Abel et al. 1986; Reimann-Philipp 1998; Smith et al. 1994; Yie et al. 1992). Similarly, overexpression of the cap or RdRp proteins of a fungal totivirus will cure cells of the virus (Valle and Wickner 1993), as will overproduction of portions of the cap protein (Yao and Bruenn 1995). It is not clear how similar an overexpressed cap or RdRp protein must be to its viral equivalent to prevent viral infection, but ScVL1 and ScVLa sequences (22–31% identical) have no cross-reaction (Yao and Bruenn 1995). It is worth noting that even at 22% identity (in the majority of the cap sequence), alignment is unequivocal (E value of 10^{-13}). If the function of NIRVs is to prevent viral infection, we would expect that fungal species blessed with totivirus NIRVs would not be infected with closely related totiviruses. As far as we can tell, this appears to be the case (Table 1). We have listed all fungi known to be infected with totiviruses or partitiviruses whose cellular genomes have been sequenced; none have NIRVs. We have listed all fungi known to have NIRVs; none has totivirus or partitivirus infections.

The known anti-viral defenses of fungi are not as effective as NIRVs appear to be. Eukaryotes are known to control viral reproduction by the synthesis of small RNAs complementary to viral mRNAs whose hybridization to their target RNAs results in mRNA degradation (Bivalkar-Mehla et al. 2011). This process requires a complex of proteins anchored by a member of the argonaute family (Czech and Hannon 2011). Many of the fungi lack this key protein and hence have no siRNA system. However, there appears to be no direct correlation between the presence of a siRNA synthesis system and the presence or absence of totivirus or partitivirus infection (Table 1). Introduction of argonaute to Saccharomyces cerevisie does cure it of one of its totiviruses, but not the other (Drinnenberg et al. 2011), and there are cases in which the siRNA system modulates, rather than eliminates totivirus infection (Hammond et al. 2008). Remarkably, though, so far there is only one species in which there are both NIRVs and argonaute, *P. marnefeii* (Table 1).

Table 1 NIRVs are incompatible with totivirus or partitivirus infection

Host	Totivirus or partitivirus	Totivirus or partitivirus NIRV	Argonaute
Candida parapsilosis	–	+	–
Debaryomyces hansenii	–	+	–
Shefferomyces stipitis	–	+	–
Candida coipomoensis	–	+	–
Wickerhamomyces anomalus	?	+	–
Candida albicans	–	–	+
Candida parapsilosis	–	–	+
Candida tropicalis	–	–	+
Lodderomyces elongisporus	–	–	+
Pichia spartinae	–	–	–
Botryotinia fuckeliana	+	–	+
Drosophila melanogaster	+	–	+
Epichloe festucae	+	–	+
Giardia lamblia	+	–	+
Leishmania major	+	–	+
Pichia segobiensis	+	?	?
Saccharomyces cerevisiae	+	–	–
Trichomonas vaginalis	+	–	+
Ustilago maydis	+	–	–
Aspergillus fumigatus	+	–	+
Rhizoctonia solani	+	–	–
Rosellinia necatrix	+	–	–
Sclerotinia sclerotiorum	+	–	+
Penicillium marneffei	?	+	+

– is not present; + is present; ? is not known

4 Origin of NIRVs: Horizontal Transfer from Host to Virus

There is another collection of totivirus genes with widespread occurrence in fungal genomes: genes with significant similarity to the totivirus killer toxins. Totiviruses often have satellite viruses associated with them. These consist of smaller segments of dsRNA separately packaged in the helper virus capsid polypeptides. Quite often, these encode secreted cellular toxins. Often, the preprotoxin itself confers immunity to the toxin on the host cell, there by providing a selective advantage to cells hosting the totivirus. All that appears to be necessary for packaging of satellite RNAs is the inclusion of a packaging signal, which can be as small as 20 bases (Boone et al. 1986; Bostian et al. 1984; Breinig et al. 2006; Bruenn 1986, 1999, 2000; Li et al. 1999; Lolle et al. 1984; Park et al. 1996; Peery et al. 1987; Ribas et al. 1994; Schmitt and Breinig 2002, 2006; Schmitt and Neuhausen 1994; Schmitt et al. 1997; Schmitt and Tipper 1990, 1995; Sommer and Wickner 1982; Steinlauf et al. 1988; Sturley et al. 1986; Tao et al. 1990; Tipper and Schmitt 1991; Weiler et al. 2002; Weiler and Schmitt 2003; Wickner 1979, 1983, 1996; Wickner et al. 1986; Yao et al. 1995, 1997; Zhu and Bussey 1991).

A summary of presently known occurrences of the eight well-characterized totivirus killer toxin genes (ScV k1, k2, k28 and Klus; UmV KP1, KP4, and KP6; ZmV zygocin) in the nuclear genomes of eucaryotes is given in Table 2. Some of the k2 (Frank and Wolfe 2009), KP4 (Bruenn 2008) and Klus (Rodriguez-Cousino et al. 2011) NIRVs have been described previously. Contrary to a previous report (Frank and Wolfe 2009), there is no evidence for a relationship between the *S. cerevisiae* KHS1 gene (P39690) and ScV k2 or for k2 NIRVs in *Vanderwaltozyuma polyspora*.

Although none of these genes has a known cellular function, there are reasons to suppose that these genes represent the ancestors of the totivirus satellite killer viruses. Of the eight well characterized killer toxin satellite dsRNAs, five have internal polyA-rich regions immediately following the preprotoxin coding sequence on the plus strand. Four of these have been sequenced: the S. cerevisiae k1 dsRNA polyA region is 92 bp long (Russell et al. 1997); the *S. cerevisiae* k28 dsRNA polyA region is 63 bp long (Schmitt and Tipper 1995); the *Zygosaccharomyces bailii* zygocin dsRNA polyA region is 262 bp long (Weiler et al. 2002); the *S. cerevisiae* k2 dsRNA poly A region is of undetermined length (Dignard et al. 1991); and the Klus dsRNA has two internal A-rich regions of 106 and 100 bp (Rodriguez-Cousino et al. 2011). The packaging sites are all 3′ to the polyA regions on the plus strand, just as they are downstream of the coding region in the totivirus genome (Fujimura et al. 1990; Shen and Bruenn 1993; Yao et al. 1995, 1997). Since none of the totiviruses encode or synthesize polyA 3′ ends on their mRNAs, the origin of these polyA regions is probably cellular. The most obvious possible origin of these satellite dsRNAs is by way of an erroneous packaging of a normal, polyadenylated cellular mRNA followed by a replication event in which the 5′ end of the minus strand is copied from an authentic viral plus strand and template switching to an incorrectly packaged cellular mRNA results in a hybrid minus strand with the complete cellular mRNA at its 3′ end. Transcription of this within the viral particle would result in an mRNA encoding the toxin, the polyA, and the packaging signal, which would then become a satellite dsRNA upon subsequent packaging.

Where there are adequate numbers of nuclear genome toxin genes (e.g. Klus), the phylogeny supports the viral gene having diverged from a cellular gene much more recently than the origin of the cellular gene (Fig. 4). Cellular genes closely related to Klus are widespread, but the Klus preprotoxin gene is most closely related to genes within the Saccharomyces genus, as is the virus (ScVL1) which has captured it (Fig. 4). As yet, there are no known NIRVs for the UmV KP1 and KP6 toxins or for the ScV k1 toxin, but given the number of unsequenced fungal genomes, we should be patient. Note that the KP4 toxin gene appears to have been horizontally transferred from a fungal pathogen to a plant (*Physocomitrella patens*), as well as from host to virus, and the Klus gene seems to have been horizontally transferred to a protozoan (*Plasmodium vivax*) either from a totivirus (several of which are known in protozoans) or from a fungal parasite (Table 2; Fig. 4).

Table 2 Killer toxin NIRVs

Toxin	Reference	Genome	Reference[a]	*E* value
ScV k2	CAA39941	*Kluyveromyces lactis*	XP_451522	4×10^{-42}
	CAA39941	*Debaryomyces hansenii*	XP_462365	1×10^{-13}
	CAA39941	*Sclerotinia sclerotiorum*	XP_001590644	1×10^{-10}
	CAA39941	*Kluyveromyces aestuarii*	AEAS01000241	3×10^{-9}
	CAA39941	*Candida tropicalis*	XP_002545278	5×10^{-6}
ScV k28	2205370A	*Saccharomyces paradoxus*	AABY01000124	4×10^{-80}
ScVKlus	GU723494	*Kluyveromyces aestuarii*	AEAS01000139.1	3×10^{-46}
	GU723494	*Vanderwaltozyma polyspora*	AAZN01000326.1	2×10^{-38}
	GU723494	*Saccharomyces cerevisiae*	EDZ72392.1	5×10^{-32}
	GU723494	*Saccharomyces pastorianus*	ABPO01000468.1	6×10^{-28}
	GU723494	*Saccharomyces paradoxus*	AABY01000021.1	7×10^{-28}
	GU723494	*Wickerhamomyces anomalus*	AEGI02001228.1	3×10^{-26}
	GU723494	*Naumovozyma dairenensis*	CCD27497.1	3×10^{-26}
	GU723494	*Saccharomyces mikatae*	AACH01000177.1	7×10^{-22}
	GU723494	*Clavispora lusitaniae*	XP_002614128.1	4×10^{-18}
	GU723494	*Saccharomyces kudriavzevii*	AACI02000605.1	4×10^{-15}
	GU723494	*Candida tropicalis*	XP_002550812.1l	1×10^{-11}
	GU723494	*Meyerozyma guilliermondii*	AAFM01000062.1	7×10^{-11}
	GU723494	*Kluyveromyces lactis*	XP_002999357.1	9×10^{-8}
	GU723494	*Fusarium oxysporum*	EGU88280.1	1×10^{-6}
	GU723494	*Neosartorya fischeri*	XP_001259100.1	2×10^{-5}
	GU723494	*Sclerotinia sclerotiorum*	XP_001588904.1	4×10^{-5}
	GU723494	*Plasmodium vivax*	AAKM01001386.1	5×10^{-5}
	GU723494	*Aspergillus fumigatus*	XP_748667.2	5×10^{-5}
ZbVzygocin	AF515592	*Kluyveromyces wickerhamii*	AEAV01000016	2×10^{-26}
UmV KP4	AAA89185	*Gibberella zeae*	XP_380238	2×10^{-13}
	AAA89185	*Metarhizium anisopliae*	EFY95639	7×10^{-12}
	AAA89185	*Metarhizium acridum*	EFY89895	2×10^{-11}
	AAA89185	*Gibberrella zeae*	XP_380237	4×10^{-11}
	AAA89185	*Grosmannia clavigera*	EFX06550	2×10^{-11}
	AAA89185	*Moniliophthora perniciosa*	XP_002388124	2×10^{-10}
	AAA89185	*Gibberella zeae*	XP_390727	2×10^{-10}
	AAA89185	*Metarhizium anisopliae*	EFZ00876	2×10^{-10}
	AAA89185	*Metarhizium acridum*	EFY91383	3×10^{-10}
	AAA89185	*Trichoderma reesei*	EGR50067	3×10^{-10}
	AAA89185	*Nectria haemotococca*	XP_003049293	6×10^{-10}
	AAA89185	*Gibberella moniliformis*	AAIM02000051	2×10^{-9}
	AAA89185	*Grosmannia clavigera*	EFX06369	9×10^{-9}
	AAA89185	*Trichoderma atroviride*	ABDG01000150	2×10^{-8}
	AAA89185	*Gibberella zeae*	XP_380236	4×10^{-8}
	AAA89185	*Fusarium oxysporum*	AAXH01001272	5×10^{-8}
	AAA89185	*Trichoderma virens*	ABDF01000487	6×10^{-8}
	AAA89185	*Aspergillus clavatus*	XP_001272264	9×10^{-8}
	AAA89185	*Grosmannia clavigera*	ACXQ02000012	1×10^{-7}
	AAA89185	*Neotyphodium gansuense*	AFRE01000069	3×10^{-7}
	AAA89185	*Epichloe brachyelytri*	AFRB01000168	7×10^{-7}
	AAA89185	*Epichloe festucae*	ADFL01000344	7×10^{-7}

(continued)

Table 2 (continued)

Toxin	Reference	Genome	Reference[a]	*E* value
	AAA89185	*Metarhizium anisopliae*	EFY94888	9×10^{-7}
	AAA89185	*Epichloe amarillans*	AFRF01000007	1×10^{-6}
	AAA89185	*Glomerella graminocola*	EFQ34120	1×10^{-6}
	AAA89185	*Epichloe typhina*	AFSE01000712	3×10^{-6}
	AAA89185	*Metarhizium anisopliae*	EFY95638	3×10^{-6}
	AAA89185	*Metarhizium acridum*	EFY89896	3×10^{-6}
	AAA89185	*Periglandula ipomoeae*	AFRD01000329	4×10^{-6}
	AAA89185	*Trichoderma reesei*	EGR47378	5×10^{-6}
	AAA89185	*Aspergillus flavus*	XP_002374615	5×10^{-6}
	AAA89185	*Ajellomyces dermatitidis*	EGE79772	3×10^{-5}
	AAA89185	*Aspergillus oryzae*	XP_001819862	3×10^{-5}
	AAA89185	*Ajellomyces dermatitidis*	EEQ86982	3×10^{-5}
	AAA89185	*Physcomitrella patens*	XP_001777519	5×10^{-5}
	AAA89185	*Physcomitrella patens*	XP_001777496	6×10^{-5}
	AAA89185	*Aspergillus clavatus*	XP_001272265	7×10^{-5}
	AAA89185	*Ajellomyces dermatidis*	XP_002625732	2×10^{-4}

[a]Cellular homologs of killer toxin genes encoded by totivirus satellite dsRNAs. Homologs are limited to one per organism except where there is clearly more than a single copy of a gene. Searches encompassed both the non-redundant protein database and the whole genome shotgun sequence database from Genbank

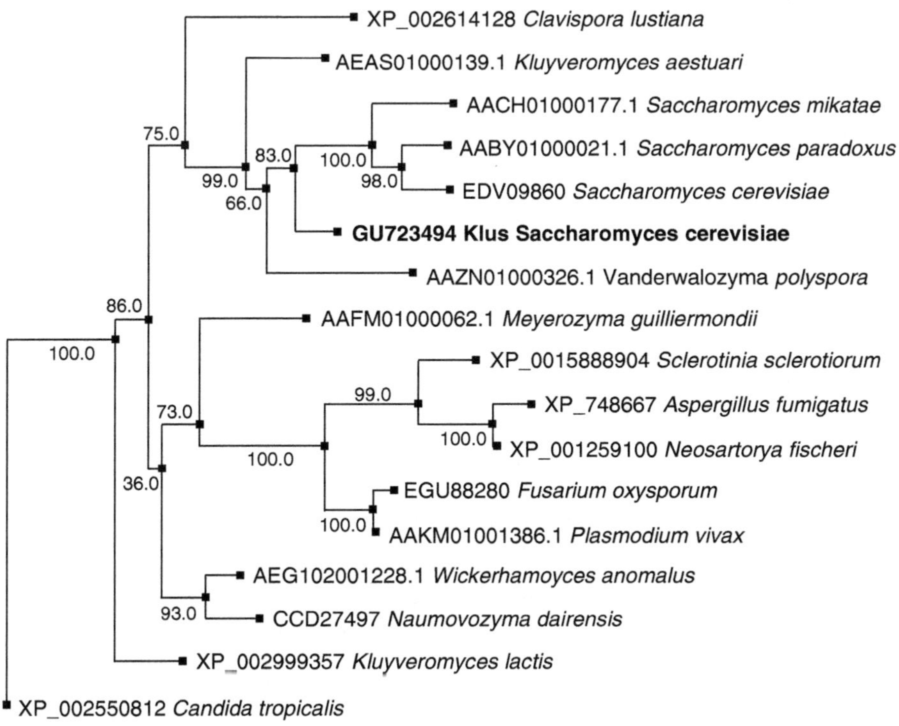

Fig. 4 Cellular genes related to the Klus preprotoxin, encoded by a totivirus satellite dsRNA. The preprotoxin is in *bold*

5 Function of NIRVs: The Virus

Just as their hosts may derive a selective advantage from incorporation of viral genes (immunity to the virus), the virus may derive a selective advantage from incorporation of cellular genes. It is not simply chance that the only known cases of horizontal transfer of host genes to virus in the totiviruses are of secreted cellular toxin genes. Secretion of protein toxins is a common strategy for eukaryotes (Breinig et al. 2006; Diamond et al. 2009; Hoffmann 2003; Palffy et al. 2009; Schmitt and Breinig 2002, 2006) and prokaryotes (Blower et al. 2011; Russell et al. 2011). Some secreted protein toxins target a wide variety of organisms while others have a narrower target specificity. The yeast killer toxins tend to be specific for yeast species closely related to the secreting cells and in most cases provide immunity to the toxin in the same precursor that is processed into a mature toxin (Breinig et al. 2006). Hence capture of killer toxin genes by totiviruses provides infected hosts with a selective advantage over non-infected hosts, conferring, by proxy, a selective advantage on the virus (Wloch-Salamon et al. 2008).

In summary, the most widespread fungal viruses, the dsRNA totiviruses and partitiviruses, which generally exist as stable persistent infections in their hosts, commonly swap genes with their hosts, and this transfer has taken place in both directions. Selection has preserved those events with adaptive value: hosts with NIRVs have become immune to infection with the cognate virus and viruses with cellular toxin genes provide an advantage to their hosts by killing cells without the virus.

References

Abel PP, Nelson RS, De B, Hoffmann N, Rogers SG, Fraley RT, Beachy RN (1986) Delay of disease development in transgenic plants that express the tobacco mosaic virus coat protein gene. Science 232:738–743

Ballinger MJ, Bruenn JA, Taylor DJ (2012) Evolution and expression of sigma virus-like genes in Drosophila after integration by an LTR retroelement (submitted)

Belyi VA, Levine AJ, Skalka AM (2010) Unexpected inheritance: multiple integrations of ancient bornavirus and ebolavirus/marburgvirus sequences in vertebrate genomes. PLoS Pathog 6: e1001030

Bivalkar-Mehla S, Vakharia J, Mehla R, Abreha M, Kanwar JR, Tikoo A, Chauhan A (2011) Viral RNA silencing suppressors (RSS): novel strategy of viruses to ablate the host RNA interference (RNAi) defense system. Virus Res 155:1–9

Blower TR, Salmond GP, Luisi BF (2011) Balancing at survival's edge: the structure and adaptive benefits of prokaryotic toxin-antitoxin partners. Curr Opin Struct Biol 21:109–118

Boone C, Bussey H, Greene D, Thomas DY, Vernet T (1986) Yeast killer toxin: site-directed mutations implicate the precursor protein as the immunity component. Cell 46:105–113

Bostian KA, Elliott Q, Bussey H, Burn V, Smith A, Tipper DJ (1984) Sequence of the preprotoxin dsRNA gene of type 1 killer yeast: multiple processing events produce a two-component toxin. Cell 36:741–751

Breinig F, Sendzik T, Eisfeld K, Schmitt MJ (2006) Dissecting toxin immunity in virus-infected killer yeast uncovers an intrinsic strategy of self-protection. Proc Natl Acad Sci USA 103:3810–3815

Bruenn JA (1986) The killer systems of *Saccharomyces cerevisiae* and other yeasts. In: Buck K (ed) Fungal virology. CRC Press, Boca Raton, pp 85–108
Bruenn JA (1999) The *Ustilago maydis* viruses. In: Granoff A, Webster RG (eds) Encyclopedia of virology. Academic, London, pp 1812–1817
Bruenn JA (2000) Viruses of fungi and protozoans: is everyone sick? In: Hurst CJ (ed) Viral ecology. Academic, New York, pp 297–317
Bruenn J (2008) The *Ustilago maydis* viruses. In: Mahy BWJ, Regenmortel Mv (eds) Encyclopedia of virology. Elsevier, Amsterdam, pp 214–219
Bruenn J, Taylor DJ (2012) Co-adaption of viruses and their fungal hosts. (in preparation)
Chiba S, Kondo H, Tani A, Saisho D, Sakamoto W, Kanematsu S, Suzuki N (2011) Widespread endogenization of genome sequences of non-retroviral RNA viruses into plant genomes. PLoS Pathog 7:e1002146
Crochu S, Cook S, Attoui H, Charrel RN, De Chesse R, Belhouchet M, Lemasson JJ, de Micco P, de Lamballerie X (2004) Sequences of flavivirus-related RNA viruses persist in DNA form integrated in the genome of *Aedes* spp. mosquitoes. J Gen Virol 85:1971–1980
Czech B, Hannon GJ (2011) Small RNA sorting: matchmaking for argonautes. Nat Rev Genet 12:19–31
Diamond G, Beckloff N, Weinberg A, Kisich KO (2009) The roles of antimicrobial peptides in innate host defense. Curr Pharm Des 15:2377–2392
Dignard D, Whiteway M, Germain D, Tessier D, Thomas DY (1991) Expression in yeast of a cDNA copy of the K2 killer toxin gene. Mol Gen Genet 227:127–136
Drinnenberg IA, Fink GR, Bartel DP (2011) Compatibility with killer explains the rise of RNAi-deficient fungi. Science 333:1592
Frank AC, Wolfe KH (2009) Evolutionary capture of viral and plasmid DNA by yeast nuclear chromosomes. Eukaryot Cell 8:1521–1531
Fujimura T, Esteban R, Esteban LM, Wickner RB (1990) Portable encapsidation signal of the L-A double-stranded RNA virus of *S. cerevisiae*. Cell 62:819–828
Geuking MB, Weber J, Dewannieux M, Gorelik E, Heidmann T, Hengartner H, Zinkernagel RM, Hangartner L (2009) Recombination of retrotransposon and exogenous RNA virus results in nonretroviral cDNA integration. Science 323:393–396
Gilbert C, Schaack S, Pace JK 2nd, Brindley PJ, Feschotte C (2010) A role for host-parasite interactions in the horizontal transfer of transposons across phyla. Nature 464:1347–1350
Hammond TM, Andrewski MD, Roossinck MJ, Keller NP (2008) Aspergillus mycoviruses are targets and suppressors of RNA silencing. Eukaryot Cell 7:350–357
Hoffmann JA (2003) The immune response of Drosophila. Nature 426:33–38
Holmes EC (2009) The emergence and evolution of RNA viruses, vol 1. Oxford University Press, Oxford
Horie M, Honda T, Suzuki Y, Kobayashi Y, Daito T, Oshida T, Ikuta K, Jern P, Gojobori T, Coffin JM et al (2010) Endogenous non-retroviral RNA virus elements in mammalian genomes. Nature 463:84–87
Katzourakis A, Gifford RJ (2010) Endogenous viral elements in animal genomes. PLoS Genet 6: e1001191
Koonin EV (2010) Taming of the shrewd: novel eukaryotic genes from RNA viruses. BMC Biol 8:2
Li N, Erman M, Pangborn W, Duax WL, Park C-M, Bruenn JA, Ghosh D (1999) Structure of *Ustilago maydis* killer toxin KP6 a-subunit: a multimeric assembly with a central pore. J Biol Chem 274:20425–20431
Liu H, Fu Y, Jiang D, Li G, Xie J, Cheng J, Peng Y, Ghabrial SA, Yi X (2010) Widespread horizontal gene transfer from double-stranded RNA viruses to eukaryotic nuclear genomes. J Virol 84:11876–11887
Lolle S, Skipper N, Bussey H, Thomas DY (1984) The expression of cDNA clones of yeast M1 double-stranded RNA in yeast confers both killer and immunity phenotypes. EMBO J 3:1383–1387

Mayor C, Brudno M, Schwartz JR, Poliakov A, Rubin EM, Frazer KA, Pachter LS, Dubchak I (2000) VISTA: visualizing global DNA sequence alignments of arbitrary length. Bioinformatics 16:1046–1047

Moran NA, Jarvik T (2010) Lateral transfer of genes from fungi underlies carotenoid production in aphids. Science 328:624–627

Palffy R, Gardlik R, Behuliak M, Kadasi L, Turna J, Celec P (2009) On the physiology and pathophysiology of antimicrobial peptides. Mol Med 15:51–59

Park C-M, Banerjee N, Koltin Y, Bruenn JA (1996) The *Ustilago maydis* virally encoded KP1 killer toxin. Mol Microbiol 20:957–963

Peery T, Shabat-Brand T, Steinlauf R, Koltin Y, Bruenn J (1987) The virus encoded toxin of *Ustilago maydis* – two polypeptides are essential for activity. Mol Cell Biol 7:470–477

Reimann-Philipp U (1998) Mechanisms of resistance. Expression of coat protein. Methods Mol Biol 81:521–532

Ribas JC, Fujimura T, Wickner RB (1994) Essential RNA binding and packaging domains of the Gag-Pol fusion protein of the L-A double-stranded RNA virus of *Saccharomyces cerevisiae*. J Biol Chem 269:28420–28428

Rodriguez-Cousino N, Maqueda M, Ambrona J, Zamora E, Esteban R, Ramirez M (2011) A new wine *Saccharomyces cerevisiae* killer toxin (Klus), encoded by a double-stranded RNA virus, with broad antifungal activity is evolutionarily related to a chromosomal host gene. Appl Environ Microbiol 77:1822–1832

Russell PJ, Bennett AM, Love Z, Baggott DM (1997) Cloning, sequencing and expression of a full-length cDNA copy of the M1 double-stranded RNA virus from the yeast, *Saccharomyces cerevisiae*. Yeast 13:829–836

Russell AB, Hood RD, Bui NK, LeRoux M, Vollmer W, Mougous JD (2011) Type VI secretion delivers bacteriolytic effectors to target cells. Nature 475:343–347

Schmitt MJ, Breinig F (2002) The viral killer system in yeast: from molecular biology to application. FEMS Microbiol Rev 26:257–276

Schmitt MJ, Breinig F (2006) Yeast viral killer toxins: lethality and self-protection. Nat Rev Microbiol 4:212–221

Schmitt MJ, Neuhausen F (1994) Killer toxin-secreting double-stranded RNA mycoviruses in the yeasts *Hanseniaspora uvarum* and *Zygosaccharomyces bailii*. J Virol 68:1765–1772

Schmitt MJ, Tipper DJ (1990) K28, a unique double-stranded RNA killer virus of *Saccharomyces cerevisiae*. Mol Cell Biol 10:4807–4815

Schmitt MJ, Tipper DJ (1995) Sequence of the M28 dsRNA: preprotoxin is processed to an alpha/beta heterodimeric protein toxin. Virology 213:341–351

Schmitt MJ, Poravou O, Trenz K, Rehfeldt K (1997) Unique double-stranded RNAs responsible for the anti-Candida activity of the yeast *Hanseniaspora uvarum*. J Virol 71:8852–8855

Shen Y, Bruenn JA (1993) RNA Structural requirements for RNA binding in a double-stranded RNA virus. Virology 195:481–491

Smith HA, Swaney SL, Parks TD, Wernsman EA, Dougherty WG (1994) Transgenic plant virus resistance mediated by untranslatable sense RNAs: expression, regulation, and fate of nonessential RNAs. Plant Cell 6:1441–1453

Sommer SS, Wickner RB (1982) Co-curing of plasmids affecting killer double-stranded RNAs of *Saccharomyces cerevisiae*: [HOK], [NEX], and the abundance of L are related and further evidence that M1 requires L. J Bacteriol 150:545–551

Steinlauf R, Peery T, Koltin Y, Bruenn J (1988) The *Ustilago maydis* virus encoded toxin – effect of KP6 on cells and spheroplasts. Exp Mycol 12:264–274

Sturley SL, Elliot Q, Levitre J, Tipper DJ, Bostian KA (1986) Mapping of functional domains within the *Saccharomyces cerevisiae* type 1 killer preprotoxin. EMBO J 5:3381–3389

Tao J, Ginsberg I, Banerjee N, Koltin Y, Held W, Bruenn JA (1990) The *Ustilago maydis* KP6 killer toxin: structure, expression in *Saccharomyces cerevisiae* and relationship to other cellular toxins. Mol Cell Biol 10:1373–1381

Taylor DJ, Bruenn J (2009) The evolution of novel fungal genes from non-retroviral RNA viruses. BMC Biol 7:88

Taylor DJ, Leach RW, Bruenn J (2010) Filoviruses are ancient and integrated into mammalian genomes. BMC Evol Biol 10:193–202

Tipper DJ, Schmitt MJ (1991) Yeast dsRNA viruses: replication and killer phenotypes. Mol Microbiol 5:2331–2338

Valle RP, Wickner RB (1993) Elimination of L-A double-stranded RNA virus of *Saccharomyces cerevisiae* by expression of *gag* and *gag-pol* from L-A cDNA clone. J Virol 67:2764–2771

Weiler F, Schmitt MJ (2003) Zygocin, a secreted antifungal toxin of the yeast *Zygosaccharomyces bailii*, and its effect on sensitive fungal cells. FEMS Yeast Res 3:69–76

Weiler F, Rehfeldt K, Bautz F, Schmitt MJ (2002) The *Zygosaccharomyces bailii* antifungal virus toxin zygocin: cloning and expression in a heterologous fungal host. Mol Microbiol 46:1095–1105

Wickner RB (1979) The killer double-stranded RNA plasmids of yeast. Plasmid 2:303–322

Wickner RB (1983) Killer systems in *Saccharomyces cerevisiae*: three distinct modes of exclusion of M2 double-stranded RNA by three species of double-stranded RNA, M1, L-A-E, and L-A-HN. Mol Cell Biol 3:654–661

Wickner RB (1996) Double-stranded RNA viruses of *Saccharomyces cerevisiae*. Microbiol Rev 60:250–265

Wickner RB, Fujimura T, Esteban R (1986) Overview of double-stranded RNA replication in *Saccharomyces cerevisiae*. Basic Life Sci 40:149–163

Wloch-Salamon DM, Gerla D, Hoekstra RF, de Visser JA (2008) Effect of dispersal and nutrient availability on the competitive ability of toxin-producing yeast. Proc Biol Sci 275:535–541

Yao W, Bruenn JA (1995) Interference with replication of two double-stranded RNA viruses by production of N-terminal fragments of capsid polypeptides. Virology 214:215–221

Yao W, Muqtadir K, Bruenn JA (1995) Packaging in a yeast double-stranded RNA virus. J Virol 69:1917–1919

Yao W-S, Adelman K, Bruenn JA (1997) In vitro selection of packaging sites in a double-stranded RNA virus. J Virol 71:2157–2162

Yie Y, Zhao F, Zhao SZ, Liu YZ, Liu YL, Tien P (1992) High resistance to cucumber mosaic virus conferred by satellite RNA and coat protein in transgenic commercial tobacco cultivar G-140. Mol Plant Microbe Interact 5:460–465

Yoshida S, Maruyama S, Nozaki H, Shirasu K (2010) Horizontal gene transfer by the parasitic plant *Striga hermonthica*. Science 328:1128

Zhu H, Bussey H (1991) Mutational analysis of the functional domains of yeast k1 killer toxin. Mol Cell Biol 11:175–181

Signal Transduction Pathways Regulating Switching, Mating and Biofilm Formation in *Candida albicans* and Related Species

David R. Soll

Abstract The developmental programs of *Candida albicans* are complex and intertwined. They include hypha formation, white-opaque switching, mating and biofilm formation. Here, the regulation of the latter three programs are considered. White-opaque switching is repressed by the a1-α2 corepressors complex produced in a/α but not a/a or α/α cells, and regulated in the latter by WOR1, a master regulator of switching. Mating of opaque cells is regulated by pheromone induction of a MAP kinase pathway targeting the transcription factor Cph1. Biofilm formation by a/a and α/α cells is regulated by the same pheromone-induced MAP kinase pathway, but targets a different transcription factor, Tec1. And biofilm formation by a/α cells is regulated by the Ras1/cAMP pathway, the same pathway regulating hypha formation, but targets an additional transcription factor, Bcr1. Specific overlaps suggest quite interesting scenarios for the evolution of these pathways, most notably that for a/a and α/α biofilm formation.

1 Introduction

Through most of the latter half of the twentieth century, the only developmental program that received significant attention by scientists studying the yeast pathogen *Candida albicans* was filamentation (Odds 1988). Filamentation includes the formation of both pseudohyphae and hyphae. Because *C. albicans* was diploid (Olaiya and Sogin 1979; Whelan and Magee 1981), the difficulty of generating null mutants slowed identification of the pathways that regulated hypha formation, but even so, by the early 1990s, a number of laboratories had begun to identify genes in signal transduction pathways that regulated the process (Sadhu et al. 1992; Navarro-García

D.R. Soll (✉)
Department of Biology, Developmental Studies Hybridoma Bank, University of Iowa, Iowa City, IA 52242, USA
e-mail: david-soll@uiowa.edu

G. Witzany (ed.), *Biocommunication of Fungi*, DOI 10.1007/978-94-007-4264-2_6,

et al. 1995; Clark et al. 1995; Malathi et al. 1994). The formation of pseudohyphae by *Saccharomyces cerevisiae*, one of the most established organisms for mutational analyses, provided a readily tractable model for predicting at least some of the genes in the *C. albicans* pathways, although unlike *C. albicans*, *S. cerevisiae* was not a true human pathogen and did not make true hyphae. In the past decade, however, strategies for rapidly obtaining homozygous mutants have evolved (Fonzi and Irwin 1993; Wilson et al. 1999; Reuss et al. 2004; Noble and Johnson 2009; Nobile and Mitchell 2009) that have helped identify multiple developmental pathways involved in *C. albicans* filamentation, presumably emanating from different receptors induced by different environmental stimuli (Braun and Johnson 2000; Davis et al. 2000; Lane et al. 1991; Biswas and Morschhäuser 2005; Shapiro and Cowen 2010; Hnisz et al. 2010). These pathways have been reviewed in detail elsewhere (Biswas et al. 2007) and will, therefore, not be dealt with here.

The developmental biology of *C. albicans*, however, began to get more complex in the 1980s. A second complex developmental program was identified in *C. albicans* in 1987 by Slutsky et al. (1987). They found that a clinical strain in *C. albicans* underwent a highly conserved, reversible two phase transition referred to as the white-opaque transition, or "white-opaque switching". It involved extraordinary changes in cell morphology and physiology. Between 1992 and 1993, the first phase-specific genes were identified (Morrow et al. 1992, 1993; Srikantha and Soll 1993). Subsequent studies of expression profiles revealed that this phenotypic transition involved changes in a large number of genes (Lan et al. 2002; Tsong et al. 2003), and was regulated through a master switch locus, *WOR1* (Zordan et al. 2006; Huang et al. 2006; Srikantha et al. 2006). But, even though it had become a focus of attention for the 15 years following its discovery, a clear role for white-opaque switching did not emerge until 2002. In that year Miller and Johnson discovered that white-opaque switching was regulated by the mating type locus (*MTL*) (Miller and Johnson 2002; Lockhart et al. 2002) and that the phenotypic transition from white to opaque was an essential prerequisite to mating (Miller and Johnson 2002; Lockhart et al. 2002).

The third major developmental program was identified in 1999. In that year Hull and Johnson (1999) published the first description of the **a** and α copies of the mating type locus in *C. albicans*, suggesting a complex mating program like that in *S. cerevisiae* (Fig. 1a, b). This discovery was followed by clear demonstration, both through complementation (Hull et al. 2000; Magee and Magee 2000) and microscopy (Lockhart et al. 2003a), that only **a/a** and α/α cells mated in a cell type-specific fashion. And, as mentioned, Johnson and colleagues (Miller and Johnson 2002; Lockhart et al. 2003a, b) made the surprising discovery that the unique opaque phase (Slutsky et al. 1987; Anderson and Soll 1987) represented the mating-competent *MTL*-homozygous phenotype. Members of the *Saccharomyces*, clade including *S. cerevisiae*, had no similar switching requirement (Fig. 1c, d). Nor did other members of the *Candida* clade, except for the closely related species *Candida dubliniensis* (Pujol et al. 2004).

And finally, a fourth developmental program, that of biofilm formation, slowly emerged, moving the complexity of *C. albicans* development into the realm of multicellularity (Kreft and Bonhoeffer 2005; Webb et al. 2003; Bordi and de

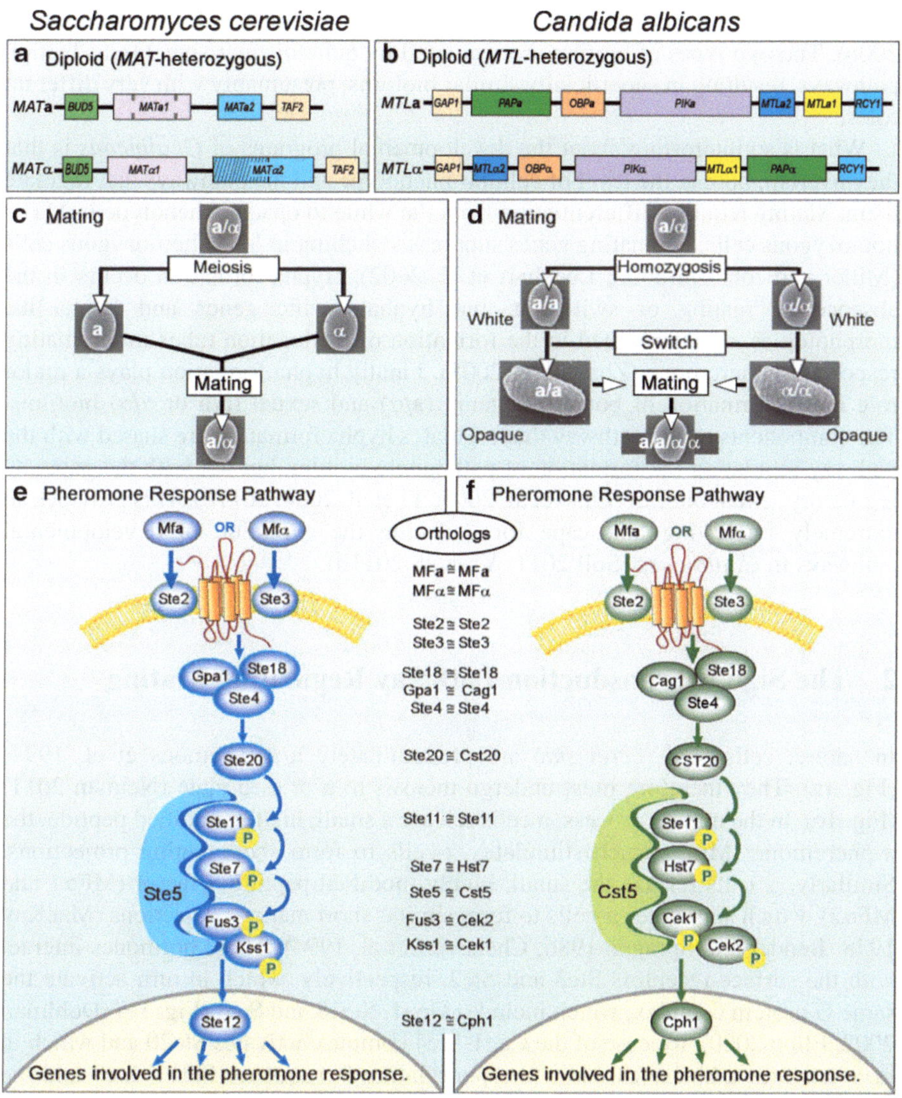

Fig. 1 A comparison of the mating process between *Saccharomyces cerevisiae* and *Candida albicans*. (**a**, **b**) The diploid mating type locus in the respective species. (**c**, **d**) The acquisition of mating competence and fusion in the respective species. (**e**, **f**) The signal transduction pathways and orthologs (noted in the *homology box*) in the respective species

Bentzmann 2011). And this story has recently become even more complex with the discovery that depending upon the configuration of the mating type locus (i.e., **a**/α versus **a**/**a** or α/α), *C. albicans* can make two different types of biofilms (Yi et al. 2011b). One (**a**/α) exhibits all of the characteristics of traditional pathogenic biofilms, including impermeability to antifungals, drug resistance and impenetrability by white blood cells. The other (**a**/**a** or α/α) exhibits none of these characteristics

(Yi et al. 2011b), but is capable of facilitating mating (Yi et al. 2011b; Daniels et al. 2006). The two types of biofilm are regulated by quite distinct signal transduction pathways, resulting in superficially similar biofilms, presumably with very different roles.

What is so interesting about the developmental programs of *C. albicans* is that they overlap, both at the level of cellular phenotype and the pathways that regulate them. Mating requires differentiation from the white to opaque phenotype in MTL-homozygous cells, but mating genes suppress switching in MTL-homozygous cells (Miller and Johnson 2002; Lockhart et al. 2002). Hypha formation occurs in the absence of mating or switching, but hypha-specific genes and hypha-like morphologies are intertwined in the formation of conjugation tubes in the mating response to pheromone (Zhao et al. 2005). Finally hypha formation plays a major role in the formation of both pathogenic (**a**/α) and sexual (**a**/**a** or α/α) biofilms. The components of the pathway that regulates hypha formation are shared with the pathway regulating the formation of pathogenic biofilm, but not with the pathway regulatory sexual biofilm (Zhao et al. 2005; Yi et al. 2008, 2011b). This provides an extremely interesting landscape for deducing the evolution of developmental pathways in *C. albicans* (Soll 2011; Yi et al. 2011b).

2 The Signal Transduction Pathway Regulating Mating

In nature, cells of *S. cerevisiae* are predominately **a**/α (Clemons et al. 1997) (Fig. 1a). They therefore must undergo meiosis to **a** or α to mate (Neiman 2011) (Fig. 1c). In the mating process, **a** cells release a small, highly modified peptide, the **a**-pheromone (Mf**a**), which stimulates α-cells to form short mating projections. Similarly, α cells release the small, highly modified peptide α-factor (MFα1 and MFα2), which stimulates **a**-cells to form similar short mating projections (MacKay 1978; Bender and Sprague 1986; Chenevert et al. 1994). These hormones interact with the surface receptors Ste3 and Ste2, respectively, which in turn activate the same G protein complex, which includes Gpa1, Ste18 and Ste4 (Fig. 1e) (Dohlman 2002; Elion 2000). Release of the Cag1-Ste4 complex activates Ste20 and which in turn activates a kinase cascade, phosphorylating in sequence Ste11, Ste7 and the terminal kinases Fus 3 and Kss1. The terminal kinases activate Ste12, a transcription factor (Fig. 1e). Ste12 up-regulates a number of genes involved in the formation of a mating evagination and fusion. The **a** and α evaginations fuse with each other at their terminal ends to form a conjugation bridge (Cross et al. 1988). Nuclei migrate into the bridge, fuse and dictate the position of the first **a**/α daughter cell (Fig. 1c). The daughter cell divides, thus initiating a new **a**/α cell population.

Both the cell biology of the mating process and the signal transduction pathway regulating it are highly similar between *S. cerevisiae* and *Candida albicans* (Yi et al. 2008; Magee et al. 2002). As is the case for **a** and α cells of *S. cerevisiae*, **a**/**a** and α/α cells form an evagination without constricting at the cell-evagination interface (Lockhart et al. 2003a, b), a final morphology referred to as a "shmoo".

In the process of fusion, the evaginations of *S. cerevisiae* undergo orientated chemotropism up pheromone gradients of opposite mating type, fusing end to end to form a conjugation bridge (Schrick et al. 1997; Segall 1993). The evaginations of *C. albicans* differ from those of *S, cerevisiae* in that they can become quite long prior to fusion, which involves upregulation of hypha-associated genes (Zhao et al. 2005). In a fashion similar to that of *S. cerevisiae*, nuclear fusion occurs along the bridge, as well as the formation of the first daughter bud (Lockhart et al. 2003a; Bennett et al. 2005).

The peptide pheromones, receptors and pheromone response pathway of C. albicans consist of orthologs of virtually all of the components in the pheromone response pathway of *S. cerevisiae* (Soll 2011) (see box between Fig. 1e and f). Even the scaffold protein, Cst5, which localizes the components of the kinase cascade, is an ortholog of the *S. cerevisiae* scaffold protein, Ste5 (Yi et al. 2011a; Côte et al. 2011), and the downstream targeted transcription factor Cph1 is an ortholog of the targeted *S. cerevisiae* transcription factor Ste12 (Liu et al. 1994). Hence, once cells are mating-competent in the two species, the same conserved pheromone response pathway and presumably many of the associated proteins involved in pheromone maturation and transport, as well as the cellular and chemotropism processes (Soll 2011; Madden and Snyder 1998; Jones and Bennett 2011), are utilized.

The two major differences, therefore, between the mating processes of *S. cerevisiae* and *C. albicans* are the unique switching requirement of *C. albicans* and the manner in which the two species reduce ploidy, after fusion, to reestablish mating competency. The latter difference arises from the ploidy of the fusion partners. In *S. cerevisiae* an **a** and α cell fuse to form an **a**/α fusant (Fig. 1c), and in *C. albicans*, an **a**/**a** and α/α cell fuse to form an a/a/α/α fusant (Fig. 1d). *S. cerevisiae* undergoes a traditional meiotic reduction to return to either **a** or α (Hartwell 1974). In contrast *C. albicans* undergoes a parasexual reduction through random loss of chromosomes, returning to the **a**/α state, then must undergo homozygosis to **a**/**a** or α/α (Bennett and Johnson 2003; Forche et al. 2008). This reduction from a tetraploid to a diploid or a near-diploid state by what may be random loss of chromosomes must be stimulated by either starvation medium or use of a medium containing sorbose (Bennett and Johnson 2003). But even though the reduction may be random, it depends upon a key gene *SPO11*, involved in meiosis in *S. cerevisiae* (Forche et al. 2008). In the host, the **a**/α progeny resulting from this apparently random process may be selected for due to the increased virulence of **a**/α over **a**/**a**/α/α cells, and, by inference, the increased capacity of **a**/α reduction products to survive in the host (Wu et al. 2007).

2.1 *The Regulation of Switching*

White-opaque switching represents a conserved, reversible, spontaneous phase transition. At the cellular level, the transition occurs between a "white" round to slightly ellipsoid budding cell phenotype and an "opaque" elongate, asymmetric cell phenotype, twice as big as a white cell and covered with pimples, each containing a channel (Anderson and Soll 1987; Soll 1992). When first discovered, it was observed that a majority of natural strains did not undergo the transition (Soll 2009). In 2002, Miller and Johnson discovered that in order to mate, cells had to undergo homozygosis from **a**/α to **a**/**a** or α/α, then switch from white to opaque, indicating that the **a**1-α2 corepressor, which represses mating, also represses switching (Miller and Johnson 2002; Lockhart et al. 2002). Hence, the phenotypically complex white-opaque transition, which involved changes in the expression of close to 5% of the genes in the genome (Lan et al. 2002; Tsong et al. 2003), finally was shown to play a basic role in the *C. albicans* life history. Now it was time to elucidate the molecular basis of the transition. Although conserved, reversible and spontaneous phase transitions had been elucidated in bacteria and shown to involve the conserved reorganization of DNA sequences (Marjan et al. 2004; van den Broek et al. 2005) or RNA translation (Smith et al. 2010), it seemed more plausible to consider an epigenetic model in which spontaneous, heritable changes in chromatin state occurred at a "master switch locus", as proposed in 1991 (Soll et al. 1991), given the beautiful example in *S. cerevisiae* of metastable changes in gene activity that occurred spontaneously in the subtelomeric region of chromosomes (Gottschling et al. 1990). These changes involved a switch between an on and off state that did not involve a change in DNA sequence. In the early 2000s and more recently (Klar et al. 2001; Srikantha et al. 2001; Stevenson and Liu 2011; Hnisz et al. 2009), mutant analyses revealed that the acetylation-deacetylation of chromatin could affect switching frequencies. But it was not until 2006 that a master switch locus, *WOR1* (also referred to earlier as *TOS9*), was identified independently in three laboratories and published in the same month (Zordan et al. 2006; Huang et al. 2006; Srikantha et al. 2006). Null mutants of *WOR1* were blocked in the white phase. The gene included an 11,000 basepair (bp) upstream regulatory region and a 4,530 bp transcribed region. The transcribed region contained an untranslated 5′ region of 2,185 bases and a translated region of 2,355 bases. The transcript possessed a polyadenosine tail. The upstream region contained several Wor1 binding sites (Zordan et al. 2006). The latter observations as well as mutant analyses led to a basic mechanism for switching that involved auto regulation (Huang et al. 2006; Zordan et al. 2006; Srikantha et al. 2006). In the simple models that evolved, it was proposed that the level of Wor1 in the white phase was below a threshold concentration (Fig. 2a). Induction of the opaque phase (a spontaneous switch from white to opaque) occurred when the intracellular level of Wor1 increased stochastically above a threshold concentration (Fig. 2b). The switch did not require a continuously high concentration of Wor1, at least for a few generations (Zordan et al. 2006). This may be due to a heritable change in chromatin structure at the *WOR1* locus (Hnisz et al. 2009; Stevenson and Liu 2011). Wor1, therefore, functioned as a positive auto inducer (Fig. 2a, b) by binding to several cis-acting regulatory sequences in the *WOR1* promoter. Wor1 binding sites have also been identified in a number of other genes distributed throughout the genome (Zordan et al. 2007).

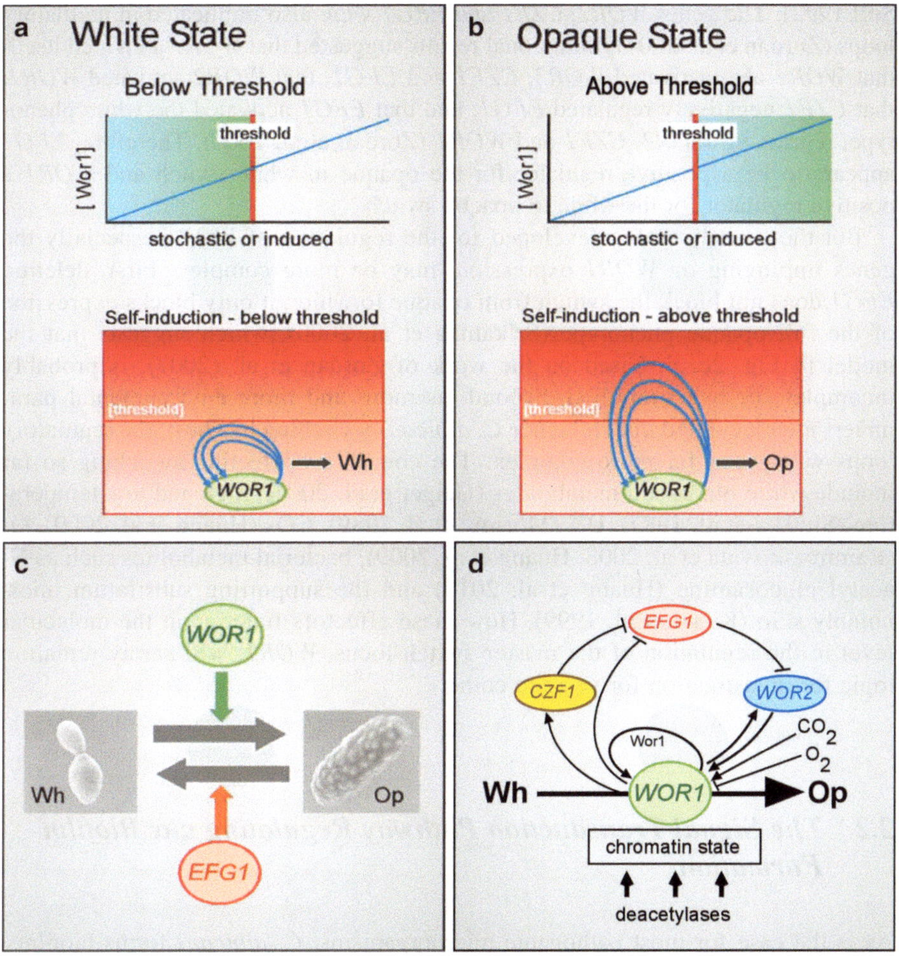

Fig. 2 The regulation of white-opaque switching in *Candida albicans* at the master switch locus, *WOR1*. (**a**, **b**) Models of the regulation of a spontaneous switch based upon threshold levels of the Wor1 protein. Below the threshold the cell is "white" and above it "opaque". Self-induction at the transcriptional level is modeled at the bottom of each panel. (**c**) A simple model of regulation based on the regulation of *white* to *opaque* by Wor1 and *opaque* to *white* by Efg1. (**d**) The regulation of the *white* to *opaque* transition by *WOR1* by genes that induce it and environmental conditions that induce it

Several secondary genes, most notably *EFG1*, play regulatory roles in the *WOR1*-based transitions between white and opaque (Zordan et al. 2007). Inhibiting or mutating select histone deacetylase genes, including *HDA1* (Klar et al. 2001; Srikantha et al. 2001), *RPD3* (Srikantha et al. 2001) and genes of the deacetylase complex Set3/Hos2 (Hnisz et al. 2009), also affected rates of switching in both directions, adding weight to the suggestion that the switch to the opaque phase involved changes in the structure of *WOR1* chromatin (Srikantha et al. 2006;

Soll 1992). The genes *WOR2, CZF1* and *EFG1* were also implicated in regulatory loops (Zordan et al. 2007). Mutational results suggested that *WOR1* activated itself, that *WOR1* also activated *WOR2*, *CZF1* and *EFG1*, that *WOR2* activated *WOR1*, that *CZF1* negatively regulated *EFG1*, and that *EFG1* activated the white phenotype, repressing *WOR2*, *CZF1* and *WOR1* (Zordan et al. 2007). Therefore, *EFG1* appears to be a positive regulator for the opaque to white switch and *WOR1* a positive regulator for the white to opaque switch.

But the models so far developed for the regulation of *WOR1* especially the genes impinging on *WOR1* expression, may be more complex. First, deleting *EFG1* does not block the switch from opaque to white; it only blocks expression of the full opaque phenotype (Srikantha et al. 2000), which suggests that the model in Fig. 2c, d, based on the work of Zordan et al. (2007), is probably incomplete or oversimplified. Second, as more and more environmental parameters are elucidated that influence *C. albicans* switching in a host, the regulatory loops will prove far more complex. The conditions affecting switching so far include white blood cell metabolites (Geiger et al. 2004), high and low temperature (Slutsky et al. 1987), UV (Morrow et al. 1989), CO_2 (Huang et al. 2009), O_2 (Ramírez-Zavala et al. 2008; Huang et al. 2009), bacterial metabolites such as N-acetyl glucosamine (Huang et al. 2010) and the supporting substratum, most notably skin (Kvaal et al. 1999). How these effectors function at the molecular level in the regulation of the master switch locus, *WOR1*, will surely remain a topic for investigation for years to come.

2.2 The Signal Transduction Pathway Regulating a/α Biofilm Formation

As is the case for most pathogenic microorganisms, *C. albicans* forms biofilms. In the case of *C. albicans*, these biofilms form on catheters, dentures and epithelia (Douglas 2003; Gendreau and Loewy 2011; Liu and Filler 2011). The majority of biofilms must be formed by strains with the mating type locus (*MTL*) configuration **a**/α, since *MTL* heterozygotes comprise the majority (~90%) of strains colonizing or infecting hosts in nature (Soll 2004). The pioneering work of Douglas and coworkers (2003, 2009) revealed that **a**/α biofilms were comprised of a basal layer of tightly packed yeast cells transitioning phenotypically to vertically oriented hyphae embedded in a polymolecular, acellular matrix in the upper three fourths of the biofilm. At the top of these biofilms, hyphae protrude with little matrix. Yeast cells are released from this top layer for dispersal (Finkel and Mitchell 2011). **a**/α biofilms have been shown to be highly drug-resistant (Mukherjee et al. 2005; Kumamoto 2002). A variety of genes had been shown to be involved in **a**/α biofilm formation, many of them also related to hypha formation (Richard et al. 2005). Although no clear regulatory pathway had been identified for **a**/α biofilm formation prior to 2011, three transcription factors were identified that appeared to play key roles, Efg1, Tec1 and Bcr1 (Ramage et al. 2002; Nobile and Mitchell 2005). Efg1 is a DNA binding protein that had also been shown to be involved in hypha formation (Leng et al. 2001), Bcr1 a C_2H_2 zinc finger protein that can bind both DNA and RNA, and Tec1 a DNA binding protein also involved in the regulation of hypha formation (Schweizer et al. 2000). A number of protein kinases, including Ire1 and Cbk1, have also been identified that play roles in **a**/α biofilm formation (Blankenship et al. 2010). A much more extensive review of proteins involved in **a**/α biofilm formation can be found in the review by Finkel and Mitchell (2011). In hypha formation, Efg1 and Tec1 had been shown to function downstream of the Ras1/cAMP pathway (Cintia et al. 2001), but Bcr1 had not been shown to be involved in that developmental process. In biofilm formation, Bcr1 has been shown to function downstream of Tec1 (Nobile and Mitchell 2005). Using a mutational approach, Yi et al. (2011b) recently demonstrated that Efg1, Tec1 and Bcr1 represented a transcription factor cascade (Efg1 → Tec1 → Bcr1) that in turn was regulated by the Ras1/cAMP pathway (Fig. 3a). Dwivedi et al. (2011) recently demonstrated that Bcr1 regulated Hwp1 and Hyr1, which play roles in oral mucosal biofilms. Although Bcr1 appears to be the last component in the transcription factor cascade (Fig. 3a), it is not the only transcription factor in the cascade that regulates biofilm genes. Overexpression of *BCR1* in an **a**/α *tec1/tec1* background does not fully rescue the *tec1/tec1* mutant phenotype (Nobile et al. 2006; Yi et al. 2011b), suggesting that Tec1 regulated biofilm genes in addition to activating Bcr1, as diagrammed in Fig. 3a.

Because the Ras1/cAMP pathway represents a major regulator of both hypha formation (Cintia et al. 2001) and the switching response to CO_2 (Huang et al. 2010), Yi et al. (2011b) tested whether it might also play a role in **a**/α biofilm formation. By mutational analyses, they demonstrated that biofilms formed on silicon elastomers were regulated by the Ras1/cAMP pathway and that this pathway was upstream of the transcription factor cascade (Fig. 3a). They also showed that of the two *PKA* kinases, CeKz played the major role in activating Efg1, the first transcription factor in the cascade. The signals and receptors activating the **a**/α biofilm pathway have not been identified, although Mitchell and coworkers (Finkel and Mitchell 2011) cogently argue that interactions between the substratum and the basal layer of cells may initiate the process. In support of this hypothesis, Yi et al. (2011b) found that the Ras1/cAMP pathway not only regulated biofilm maturation, which includes hypha and matrix formation, but also early adhesion and cohesion of yeast cells. Moreover, the pathway controlled impermeability to low and high molecular weight drugs, drug resistance and impenetrability by white blood cells. But caution should be applied when viewing the singular Ras1/cAMP regulatory pathway presented in Fig. 3a. As is evident in a regulatory model developed by Finkel and Mitchell (2011), inevitably there will be multiple signals, multiple receptors and multiple pathways, including cross talk, between pathways, as more is learned about the regulation of the pathogenic **a**/α biofilms.

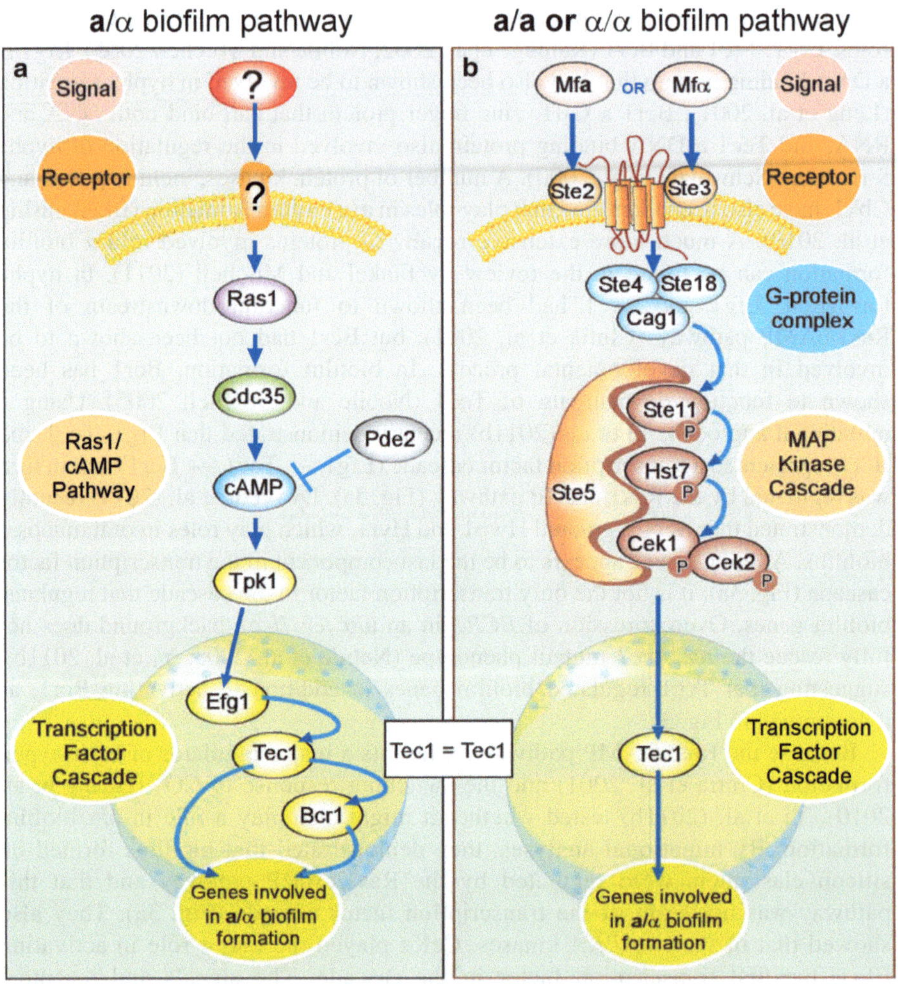

Fig. 3 The alternative pathway and the one ortholog (noted in *homology box*) that regulate **a**/α and **a**/**a** or α/α biofilm formation

2.3 *The Signal Transduction Pathways Regulating a/a and α/α Biofilm Formation*

The discovery that upon homozygosis, a cell had to switch from white to opaque to mate (Miller and Johnson 2002) was enigmatic. Given that the mating process in *C. albicans* is so similar to that of *S. cerevisiae*, and given that *S. cerevisiae* does not have to switch to mate and actually does not possess a white-opaque switching program, one must wonder why *C. albicans* and the highly related species *C. dubliniensis* are the lone species in the hemiascomycetes that have incorporated this complex developmental program into their mating systems?

Could the incorporation of this program have something to do with the success of *C. albicans* as a human commensal and pathogen? Could it have something to do with its intense relationship with vertebrate hosts and pathogenesis? A key to this puzzle was provided in 2006 when Daniels et al. (2006) discovered that **a**/**a** and α/α white, but not opaque, cells made biofilms (Soll 2009). Daniels et al. (2006) further found that a minority (as low as 1%) of opaque cells of opposite mating type increased the thickness of white **a**/**a** or α/α biofilms of opposite mating type by up to 50% and that majority white cell biofilms then facilitated mating between **a**/**a** and α/α opaque cells.

Because of the stimulation in biofilm thickness by minority opaque cells of opposite mating type, it was hypothesized that *MTL*-homozygous biofilms were induced by pheromone of opposite mating type. Through a mutational analysis, Yi et al. (2008) demonstrated that induction of an **a**/**a** or α/α biofilm involved a pathway composed of the same pheromone signals (Mfα or Mf**a**, respectively), the same receptors (Ste2 or Ste3, respectively), the same G-protein complex and the same MAP kinase cascade (Ste11, Hst7, Cek2[Cek1]) as the signal transduction pathway regulating the opaque cell pheromone response in the mating process, but that the targeted transcription factor differed. Whereas the transcription factor Cph1 is targeted in the pheromone response pathway for mating (Chen et al. 2002), the transcription factor Tec1 is targeted in white cell biofilm formation (Sahni et al. 2010). Subsequently, Yi et al. (2011b) showed that the same scaffold protein, *Cst5*, facilitated the signal-induced kinase cascade in the opaque cell mating and white cell biofilm responses. It was demonstrated that Tec1 activated genes by interacting with a cis-acting sequence WPRE, which is A-rich, whereas Cph1 activated genes by interacting with the sequence OPRE, which is G-rich (Sahni et al. 2009, 2010). The composition of the signal response pathway for **a**/**a** and α/α biofilms has interesting evolutionary implications (Soll 2011). Since only *C. albicans* (Slutsky et al. 1987) and the closely related species *Candida dubliniensis* (Pujol et al. 2004) undergo the white-opaque transition in the hemiascomycetes, it is reasonable to hypothesize that the signal transduction pathway regulating white cell biofilm formation is relatively new, evolving roughly 40 million years ago in the ancestor of the two related species (Mishra et al. 2007). In the evolution of the new pathway, all of the upstream components of the pheromone-induced signal transduction pathway in the mating process were borrowed intact, but the Tec1 transcription factor, known to play a role in hypha formation, became the target of the MAP kinase pathway in order to affect a different phenotypic outcome. The fact that the upper portion of the pathway has been borrowed intact from the conserved pheromone response pathway supports the idea that it is quite new and hasn't had time to diverge. This transference has been possible because white cells respond to the same signal as opaque cells, so the upper portion of the pathways can use common components (Soll 2011). One difference, however, did evolve in a unique fashion. The first intracellular loop of Ste2, the α-pheromone receptor, contains a 55 amino acid sequence that is absent from the homologous loop in Ste2 of *S. cerevisiae* (Yi et al. 2009). If selectively deleted in *C. albicans*, Ste2 functions in the pheromone response, but not in the biofilm response. The enigma is that the differential dependency is at the most upstream portion of the pathway, leading to the suggestion that a second pathway, essential and unidentified, for biofilm formation is induced upon activation of the receptor.

Therefore, **a**/**a** or α/α cell populations form biofilm that are thinner than **a**/α biofilms, but which can be stimulated by adding a minority of opaque cells of opposite mating type, a source of pheromone of opposite mating type (Yi et al. 2011b). These biofilms are regulated by a signal transduction pathway that shares only one component, a transcription factor, with the pathway regulating **a**/α biofilm formation (Yi et al. 2011b) (Fig. 3). Superficially, **a**/α and **a**/**a** or α/α appear to be similar in architecture. Each is composed of a basal layer of yeast cells from which emanates vertically oriented hyphae and an extracellular matrix. However, the two biofilms exhibit fundamentally different characteristics related to pathogenesis. The **a**/α biofilm is far less permeable to molecules with a molecular mass greater than 0.3 kDa, is resistant to fluconazole and is impenetrable by human polymorphonuclear leukocytes, all characteristics of a "pathogenic biofilm" (Yi et al. 2011b). The **a**/**a** or α/α biofilm, however, is permeable to low and high molecular weight molecules, susceptible to fluconazole and readily penetrated by polymorphonuclear leukocytes (Yi et al. 2011b). It, however, has been shown to facilitate mating between minority opaque **a**/**a** and α/α cells (Daniels et al. 2006; Soll 2009), suggesting it may play a role as a "sexual biofilm".

2.4 The Unexpected Role the Mating Type Locus Has on a/α Biofilm Formation

Although the close evolutionary relationship between mating and *MTL*-homozygous biofilm formation was clearly established, there's no reason to think that there was a relationship between mating and **a**/α biofilm formation, given that mating was suppressed by the **a**1-α2 corepressor complex in **a**/α cells (Tsong et al. 2003). However, a set of observations suggested that a relationship did in fact exist between pathogenesis and the mating type locus (*MTL*). *C. albicans* strains that were **a**/**a** or α/α in nature, were found to be less virulent in a mouse model for systemic infection than natural **a**/α strains (Wu et al. 2007). Likewise strains that spontaneously underwent *MTL* homozygosis were less virulent on average than the **a**/α parental strains from which they evolved (Wu et al. 2007). Srikantha et al. (2012), therefore, tested, by mutational analysis, not only if the mating type genes *MTL***a**1 and *MTL*α2 played a role in biofilm formation, but also if the three nonsex genes (NSGs) *PIK, PAP* and *OBP*, which are also embedded in the *MTL* locus, play a role. *PIK* encodes a phosphoinositol kinase (Flanagan et al. 1993), *PAP* a polyadenosine polymerace (Lingner et al. 1991) and *OBP* an oxysterol binding protein (Beh et al. 2001). The first two of the three are essential genes. Deleting the **a** or α allele of any one of the NSGs, or deleting either *MTL***a**1 or *MTL*α2, affected biofilm thickness. Since *PIK* and *PAP* are essential genes, a homozygous mutant only of

OBP, could be generated. The *obp***a**/*obp*α mutant not only exhibited major defects in all biofilm parameters, but also caused loss of impermeability and drug resistance (Srikantha et al. 2012). Therefore, genes in the mating type locus play major roles in the formation of a nonsexual developmental program, namely the formation of the pathogenic **a**/α biofilm. However, their roles involve the modulation of biofilm characteristics, rather than in the basic regulatory pathway.

3 Concluding Remark

The developmental programs of the human pathogen *C. albicans* have indeed become complex, presumably because they facilitate host–pathogen interactions and pathogenic success. Although we have focused on the primary pathways regulating these programs, there are numerous other pathways that must modulate and fine-tune these processes. Some of them may prove more important than the basic signal transduction pathways, as targets for drug development. Perhaps the most interesting aspect is the rapidity with which these developmental programs and pathways have evolved due to the recent transitions to a pathogenic lifestyle. It is for this reason that these pathways provide new general insights into how new signal transduction pathways evolve (Soll 2011).

Acknowledgements This work was funded by the Developmental Studies Hybridoma Bank, a NIH National Resource. The author is indebted to Drs. T. Srikantha, C. Pujol and K. Daniels for help in organizing the manuscript.

References

Anderson JM, Soll DR (1987) Unique phenotype of opaque cells in the white-opaque transition of *Candida albicans*. J Bacteriol 169:5579–5588

Beh CT, Cool L, Phillips J, Rine J (2001) Overlapping functions of the yeast oxysterol-binding protein homologues. Genetics 157:1117–1140

Bender A, Sprague GF Jr (1986) Yeast peptide pheromones, **a**-factor and alpha-factor, activate a common response mechanism in their target cells. Cell 47:929–937

Bennett RJ, Johnson AD (2003) Completion of a parasexual cycle in *Candida albicans* by induced chromosome loss in tetraploid strains. EMBO J 22:2505–2515

Bennett RJ, Miller MG, Chua PR, Maxon ME, Johnson AD (2005) Nuclear fusion occurs during mating in *Candida albicans* and is dependent on the *KAR3* gene. Mol Microbiol 55:1046–1059

Biswas K, Morschhäuser J (2005) The Mep2p ammonium permease controls nitrogen starvation-induced filamentous growth in *Candida albicans*. Mol Microbiol 56:649–669

Biswas S, Van Dijck P, Datta A (2007) Environmental sensing and signal transduction pathways regulating morphopathogenic determinants of *Candida albicans*. Microbiol Mol Biol Rev 71:348–376

Blankenship JR, Fanning S, Hamaker JJ, Mitchell AP (2010) An extensive circuitry for cell wall regulation in *Candida albicans*. PLoS Pathog 6:e1000752

Bordi C, de Bentzmann S (2011) Hacking into bacterial biofilms: a new therapeutic challenge. Ann Intensive Care 1:19

Braun BR, Johnson AD (2000) *TUP1*, *CPH1* and *EFG1* make independent contributions to filamentation in *Candida albicans*. Genetics 155:57–67

Chen J, Chen J, Lane S, Liu H (2002) A conserved mitogen-activated protein kinase pathway is required for mating in *Candida albicans*. Mol Microbiol 46:1335–1344

Chenevert J, Valtz N, Herskowitz I (1994) Identification of genes required for normal pheromone-induced cell polarization in *Saccharomyces cerevisiae*. Genetics 136:1287–1296

Cintia R, Rocha C, Schröppel K, Harcus D, Marcil A, Dignard D, Taylor BN, Thomas DY, Whiteway M, Leberer E (2001) Signaling through adenylyl cyclase is essential for hyphal growth and virulence in the pathogenic fungus *Candida albicans*. Mol Biol Cell 12:3631–3643

Clark KL, Feldmann PJ, Dignard D, Larocque R, Brown AJ, Lee MG, Thomas DY, Whiteway M (1995) Constitutive activation of the *Saccharomyces cerevisiae* mating response pathway by a MAP kinase kinase from *Candida albicans*. Mol Gen Genet 249:609–621

Clemons KV, Park P, McCusker JH, McCullough MJ, Davis RW, Stevens DA (1997) Application of DNA typing methods and genetic analysis to epidemiology and taxonomy of *Saccharomyces* isolates. J Clin Microbiol 35:1822–1828

Côte P, Sulea T, Dignard D, Wu C, Whiteway M (2011) Evolutionary reshaping of fungal mating pathway scaffold proteins. MBio 2:e00230-10

Cross F, Hartwell LH, Jackson C, Konopka JB (1988) Conjugation in *Saccharomyces cerevisiae*. Annu Rev Cell Biol 4:429–457

Daniels KJ, Srikantha T, Lockhart SR, Pujol C, Soll DR (2006) Opaque cells signal white cells to form biofilms in *Candida albicans*. EMBO J 25:2240–2252

Davis D, Wilson RB, Mitchell AP (2000) *RIM101*-dependent and-independent pathways govern pH responses in *Candida albicans*. Mol Cell Biol 20:971–978

Dohlman HG (2002) G proteins and pheromone signaling. Annu Rev Physiol 64:129–152

Douglas LJ (2003) Candida biofilms and their role in infection. Trends Microbiol 11:30–36

Douglas LJ (2009) Penetration of antifungal agents through *Candida* biofilms. Methods Mol Biol 499:37–44

Dwivedi P, Thompson A, Xie Z, Kashleva H, Ganguly S, Mitchell AP, Dongari-Bagtzoglou A (2011) Role of Bcr1-activated genes Hwp1 and Hyr1 in *Candida albicans* oral mucosal biofilms and neutrophil evasion. PLoS One 6:e16218

Elion EA (2000) Pheromone response, mating and cell biology. Curr Opin Microbiol 3:573–581

Finkel JS, Mitchell AP (2011) Genetic control of *Candida albicans* biofilm development. Nat Rev Microbiol 9:109–118

Flanagan CA, Schnieders EA, Emerick AW, Kunisawa R, Admon A, Thorner J (1993) Phosphatidylinositol 4-kinase: gene structure and requirement for yeast cell viability. Science 262:1444–1448

Fonzi WA, Irwin MY (1993) Isogenic strain construction and gene mapping in *Candida albicans*. Genetics 134:717–728

Forche A, Alby K, Schaefer D, Johnson AD, Berman J, Bennett RJ (2008) The parasexual cycle in *Candida albicans* provides an alternative pathway to meiosis for the formation of recombinant strains. PLoS Biol 6:e110

Geiger J, Wessels D, Lockhart SR, Soll DR (2004) Release of a potent polymorphonuclear leukocyte chemoattractant is regulated by white-opaque switching in *Candida albicans*. Infect Immun 72:667–677

Gendreau L, Loewy ZG (2011) Epidemiology and etiology of denture stomatitis. J Prosthodont 20:251–260

Gottschling DE (1992) Telomere-proximal DNA in *Saccharomyces cerevisiae* is refractory to methyltransferase activity in vivo. Proc Natl Acad Sci USA 89:4062–4065

Gottschling DE, Aparicio OM, Billington BL, Zakian VA (1990) Position effect at *S. cerevisiae* telomeres: reversible repression of Pol II transcription. Cell 63:751–762

Hartwell LH (1974) *Saccharomyces cerevisiae* cell cycle. Bacteriol Rev 38:164–198

Hnisz D, Schwarzmüller T, Kuchler K (2009) Transcriptional loops meet chromatin: a dual-layer network controls white-opaque switching in *Candida albicans*. Mol Microbiol 74:1–15

Hnisz D, Majer O, Frohner IE, Komnenovic V, Kuchler K (2010) The Set3/Hos2 histone deacetylase complex attenuates cAMP/PKA signaling to regulate morphogenesis and virulence of *Candida albicans*. PLoS Pathog 6(5):e1000889

Huang G, Wang H, Chou S, Nie X, Chen J, Liu H (2006) Bistable expression of *WOR1*, a master regulator of white-opaque switching in *Candida albicans*. Proc Natl Acad Sci USA 103:12813–12818

Huang G, Srikantha T, Sahni N, Yi S, Soll DR (2009) CO(2) regulates white-to-opaque switching in *Candida albicans*. Curr Biol 19:330–334

Huang G, Yi S, Sahni N, Daniels KJ, Srikantha T, Soll DR (2010) N-acetylglucosamine induces white to opaque switching, a mating prerequisite in *Candida albicans*. PLoS Pathog 6: e1000806

Hull CM, Johnson AD (1999) Identification of a mating type-like locus in the asexual pathogenic yeast *Candida albicans*. Science 285:1271–1275

Hull CM, Raisner RM, Johnson AD (2000) Evidence for mating of the "asexual" yeast *Candida albicans* in a mammalian host. Science 289:307–310

Jones SK Jr, Bennett RJ (2011) Fungal mating pheromones: choreographing the dating game. Fungal Genet Biol 48:668–676

Klar AJ, Srikantha T, Soll DR (2001) A histone deacetylation inhibitor and mutant promote colony-type switching of the human pathogen *Candida albicans*. Genetics 158:919–924

Kreft JU, Bonhoeffer S (2005) The evolution of groups of cooperating bacteria and the growth rate versus yield trade-off. Microbiology 151(Pt3):637–641

Kumamoto CA (2002) *Candida* biofilms. Curr Opin Microbiol 5:608–611

Kvaal C, Lachke SA, Srikantha T, Daniels K, McCoy J, Soll DR (1999) Misexpression of the opaque-phase-specific gene *PEP1* (*SAP1*) in the white phase of *Candida albicans* confers increased virulence in a mouse model of cutaneous infection. Infect Immun 67:6652–6662

Lan CY, Newport G, Murillo LA, Jones T, Scherer S, Davis RW, Agabian N (2002) Metabolic specialization associated with phenotypic switching in *Candida albicans*. Proc Natl Acad Sci USA 99:14907–14912

Lane S, Birse C, Zhou S, Matson R, Liu H (1991) DNA array studies demonstrate convergent regulation of virulence factors by Cph1, Cph2, and Efg1 in *Candida albicans*. J Biol Chem 276:48988–48996

Leng P, Lee PR, Wu H, Brown AJ (2001) Efg1, a morphogenetic regulator in *Candida albicans*, is a sequence-specific DNA binding protein. J Bacteriol 183:4090–4093

Lingner J, Kellermann J, Keller W (1991) Cloning and expression of the essential gene for poly(A) polymerase from *S. cerevisiae*. Nature 354:496–498

Liu Y, Filler SG (2011) Candida albicans Als3, a multifunctional adhesin and invasin. Eukaryot Cell 10:168–173

Liu H, Köhler J, Fink GR (1994) Suppression of hyphal formation in *Candida albicans* by mutation of a *STE12* homolog. Science 266:1723–1726, Erratum in: Science (1995) 267:271

Lockhart SR, Pujol C, Daniels KJ, Miller MG, Johnson AD, Pfaller MA, Soll DR (2002) In *Candida albicans*, white-opaque switchers are homozygous for mating type. Genetics 162:737–745

Lockhart SR, Daniels KJ, Zhao R, Wessels D, Soll DR (2003a) Cell biology of mating in *Candida albicans*. Eukaryot Cell 2:49–61

Lockhart SR, Zhao R, Daniels KJ, Soll DR (2003b) Alpha-pheromone-induced "shmooing" and gene regulation require white-opaque switching during *Candida albicans* mating. Eukaryot Cell 2:847–855

MacKay VL (1978) Mating-type specific pheromones as mediators of sexual conjugation in yeast. Symp Soc Dev Biol 35:243–259

Madden K, Snyder M (1998) Cell polarity and morphogenesis in budding yeast. Annu Rev Microbiol 52:687–744

Magee BB, Magee PT (2000) Induction of mating in *Candida albicans* by construction of *MTL*a and *MTL*alpha strains. Science 289:310–313

Magee BB, Legrand M, Alarco AM, Raymond M, Magee PT (2002) Many of the genes required for mating in *Saccharomyces cerevisiae* are also required for mating in *Candida albicans*. Mol Microbiol 46:1345–1351

Malathi K, Ganesan K, Datta A (1994) Identification of a putative transcription factor in *Candida albicans* that can complement the mating defect of *Saccharomyces cerevisiae* ste12 mutants. J Biol Chem 269:22945–22951

Marjan W, Woude V, Bäumler AJ (2004) Phase and antigenic variation in bacteria. Clin Microbiol Rev 17:581–611

Miller MG, Johnson AD (2002) White-opaque switching in *Candida albicans* is controlled by mating-type locus homeodomain proteins and allows efficient mating. Cell 110:293–302

Mishra PK, Baum M, Carbon J (2007) Centromere size and position in *Candida albicans* are evolutionarily conserved independent of DNA sequence heterogeneity. Mol Genet Genomics 278:455–465

Morrow B, Anderson J, Wilson J, Soll DR (1989) Bidirectional stimulation of the white-opaque transition of *Candida albicans* by ultraviolet irradiation. J Gen Microbiol 135:1201–1208

Morrow B, Srikantha T, Soll DR (1992) Transcription of the gene for a pepsinogen, PEP1, is regulated by white-opaque switching in *Candida albicans*. Mol Cell Biol 12:2997–3005

Morrow B, Srikantha T, Anderson J, Soll DR (1993) Coordinate regulation of two opaque-specific genes during white-opaque switching in *Candida albicans*. Infect Immun 61:1823–1828

Mukherjee PK, Zhou G, Munyon R, Ghannoum MA (2005) *Candida* biofilm: a well-designed protected environment. Med Mycol 43:191–208

Navarro-García F, Sánchez M, Pla J, Nombela C (1995) Functional characterization of the *MKC1* gene of *Candida albicans*, which encodes a mitogen-activated protein kinase homolog related to cell integrity. Mol Cell Biol 15:2197–2206

Neiman AM (2011) Sporulation in the budding yeast *Saccharomyces cerevisiae*. Genetics 189:737–765

Nobile CJ, Mitchell AP (2005) Regulation of cell-surface genes and biofilm formation by the *C. albicans* transcription factor Bcr1p. Curr Biol 15:1150–1155

Nobile CJ, Mitchell AP (2009) Large-scale gene disruption using the UAU1 cassette. Methods Mol Biol 499:175–194

Nobile CJ, Andes DR, Nett JE, Smith FJ, Yue F, Phan QT, Edwards JE, Filler SG, Mitchell AP (2006) Critical role of Bcr1-dependent adhesins in *C. albicans* biofilm formation in vitro and in vivo. PLoS Pathog 2:e63

Noble SM, Johnson AD (2009) Strains and strategies for large-scale gene deletion studies of the diploid human fungal pathogen *Candida albicans*. Eukaryot Cell 4:298–309

Odds FC (1988) Candida and candidosis, 2nd edn. Bailliere Tindall, London

Olaiya AF, Sogin SJ (1979) Ploidy determination of *Candida albicans*. J Bacteriol 140:1043–1049

Pujol C, Daniels KJ, Lockhart SR, Srikantha T, Radke JB, Geiger J, Soll DR (2004) The closely related species *Candida albicans* and *Candida dubliniensis* can mate. Eukaryot Cell 3:1015–1027

Ramage G, VandeWalle K, López-Ribot JL, Wickes BL (2002) The filamentation pathway controlled by the Efg1 regulator protein is required for normal biofilm formation and development in *Candida albicans*. FEMS Microbiol Lett 214:95–100

Ramírez-Zavala B, Reuss O, Park YN, Ohlsen K, Morschhäuser J (2008) Environmental induction of white-opaque switching in *Candida albicans*. PLoS Pathog 4:e1000089

Reuss O, Vik A, Kolter R, Morschhäuser J (2004) The SAT1 flipper, an optimized tool for gene disruption in *Candida albicans*. Gene 341:119–127

Richard ML, Nobile CJ, Bruno VM, Mitchell AP (2005) *Candida albicans* biofilm-defective mutants. Eukaryot Cell 4(8):1493–1502

Sadhu C, Hoekstra D, McEachern MJ, Reed SI, Hicks JB (1992) A G-protein alpha subunit from asexual *Candida albicans* functions in the mating signal transduction pathway of *Saccharomyces cerevisiae* and is regulated by the **a**1-alpha 2 repressor. Mol Cell Biol 12:1977–1985

Sahni N, Yi S, Daniels KJ, Srikantha T, Pujol C, Soll DR (2009) Genes selectively up-regulated by pheromone in white cells are involved in biofilm formation in *Candida albicans*. PLoS Pathog 5:e1000601

Sahni N, Yi S, Daniels KJ, Huang G, Srikantha T, Soll DR (2010) Tec1 mediates the pheromone response of the white phenotype of *Candida albicans*: insights into the evolution of new signal transduction pathways. PLoS Biol 8:e1000363

Schrick K, Garvik B, Hartwell LH (1997) Mating in *Saccharomyces cerevisiae*: the role of the pheromone signal transduction pathway in the chemotropic response to pheromone. Genetics 147:19–32

Schweizer A, Rupp S, Taylor BN, Röllinghoff M, Schröppel K (2000) The *TEA/ATTS* transcription factor CaTec1p regulates hyphal development and virulence in *Candida albicans*. Mol Microbiol 38:435–445

Segall JE (1993) Polarization of yeast cells in spatial gradients of alpha mating factor. Proc Natl Acad Sci USA 90:8332–8336

Shapiro RS, Cowen L (2010) Coupling temperature sensing and development: Hsp90 regulates morphogenetic signalling in *Candida albicans*. Virulence 1:45–48

Slutsky B, Staebell M, Anderson J, Risen L, Pfaller M, Soll DR (1987) "White-opaque transition": a second high-frequency switching system in *Candida albicans*. J Bacteriol 169:189–197

Smith AM, Fuchs RT, Grundy FJ, Henkin TM (2010) The SAM-responsive S(MK) box is a reversible riboswitch. Mol Microbiol 78:1393–1402

Soll DR (1992) High-frequency switching in *Candida albicans*. Clin Microbiol Rev 5:183–203

Soll DR (2004) Mating-type locus homozygosis, phenotypic switching and mating: a unique sequence of dependencies in *Candida albicans*. Bioessays 26:10–20

Soll DR (2009) Why does *Candida albicans* switch? FEMS Yeast Res 9:973–989

Soll DR (2011) Evolution of a new signal transduction pathway in *Candida albicans*. Trends Microbiol 19:8–13

Soll DR, Anderson J, Bergen M (1991) The developmental biology of the white-opaque transition in *Candida albicans*. In: Prasad R (ed) *Candida albicans*: cellular and molecular biology. Springer, Berlin, pp 20–45

Srikantha T, Soll DR (1993) A white-specific gene in the white-opaque switching system of *Candida albicans*. Gene 131:53–60

Srikantha T, Tsai LK, Daniels K, Soll DR (2000) EFG1 null mutants of *Candida albicans* switch but cannot express the complete phenotype of white-phase budding cells. J Bacteriol 182:1580–1591

Srikantha T, Tsai L, Daniels K, Klar AJ, Soll DR (2001) The histone deacetylase genes *HDA1* and *RPD3* play distinct roles in regulation of high-frequency phenotypic switching in *Candida albicans*. J Bacteriol 183:4614–4625

Srikantha T, Borneman AR, Daniels KJ, Pujol C, Wu W, Seringhaus MR, Gerstein M, Yi S, Snyder M, Soll DR (2006) TOS9 regulates white-opaque switching in *Candida albicans*. Eukaryot Cell 5:1674–1687

Srikantha T, Daniels KJ, Pujol C, Sahni N, Yi S, Soll DR (2012) Non-sex genes in the mating type locus (*MTL*) of *Candida albicans* play roles in **a**/α biofilm formation, permeability and drug resistance. PLoS Pathog 8:e1002476

Stevenson JS, Liu H (2011) Regulation of white and opaque cell-type formation in *Candida albicans* by Rtt109 and Hst3. Mol Microbiol 81:1078–1091

Tsong AE, Miller MG, Raisner RM, Johnson AD (2003) Evolution of a combinatorial transcriptional circuit: a case study in yeasts. Cell 115:389–399

van der Woude MW, Bäumler AJ (2004) Phase and antigenic variation in bacteria. Clin Microbiol Rev 17:581–611

Webb JS, Givskov M, Kjelleberg S (2003) Bacterial biofilms: prokaryotic adventures in multicellularity. Curr Opin Microbiol 6:578–585

Whelan WL, Magee PT (1981) Natural heterozygosity in *Candida albicans*. J Bacteriol 145:896–903

Wilson RB, Davis D, Mitchell AP (1999) Rapid hypothesis testing with *Candida albicans* through gene disruption with short homology regions. J Bacteriol 181:1868–1874

Wu W, Lockhart SR, Pujol C, Srikantha T, Soll DR (2007) Heterozygosity of genes on the sex chromosome regulates *Candida albicans* virulence. Mol Microbiol 64:1587–1604

Yi S, Sahni N, Daniels KJ, Pujol C, Srikantha T, Soll DR (2008) The same receptor, G protein, and mitogen-activated protein kinase pathway activate different downstream regulators in the alternative white and opaque pheromone responses of *Candida albicans*. Mol Biol Cell 19:957–970

Yi S, Sahni N, Pujol C, Daniels KJ, Srikantha T, Ma N, Soll DR (2009) A *Candida albicans*-specific region of the alpha-pheromone receptor plays a selective role in the white cell pheromone response. Mol Microbiol 71:925–947

Yi S, Sahni N, Daniels KJ, Lu KL, Huang G, Garnaas AM, Pujol C, Srikantha T, Soll DR (2011a) Utilization of the mating scaffold protein in the evolution of a new signal transduction pathway for biofilm development. MBio 2:e00237-10

Yi S, Sahni N, Daniels KJ, Lu KL, Srikantha T, Huang G, Garnaas AM, Soll DR (2011b) Alternative mating type configurations (**a**/α versus **a/a** or α/α) of *Candida albicans* result in alternative biofilms regulated by different pathways. PLoS Biol 9:e1001117

Zhao R, Daniels KJ, Lockhart SR, Yeater KM, Hoyer LL, Soll DR (2005) Unique aspects of gene expression during *Candida albicans* mating and possible G(1) dependency. Eukaryot Cell 4:1175–1190

Zordan RE, Galgoczy DJ, Johnson AD (2006) Epigenetic properties of white-opaque switching in *Candida albicans* are based on a self-sustaining transcriptional feedback loop. Proc Natl Acad Sci USA 103:12807–12812

Zordan RE, Miller MG, Galgoczy DJ, Tuch BB, Johnson AD (2007) Interlocking transcriptional feedback loops control white-opaque switching in *Candida albicans*. PLoS Biol 5:e256

Cell-to-Cell Communication in the Tip Growth of Mycelial Fungi

Tatiana Potapova

Abstract The capacity of fungi to explore solid substrate, invade tissues and secrete digestive enzymes are all linked to their particular mode of growth, extension of a tip. The high rate of tip growth provide by the coordinated activity dozens of cells having the septal pores allowing ions, molecules and organelles move along the hyphae. Young apical cells are deficient to generate a potential difference across the plasma membrane. For this reason, in the apical area of about 300 μm a significant electric field (100 V/m) appears and strong intercellular current flows (some nanoA). Perhaps this electrical heterogeneity plays an important role in the self-organization of interactions between cells and intracellular structures in the tip growth.

Keywords *Neurospora crassa* • Tip growth • Movement of mitochondria • Intercellular electrical currents

1 Introduction

In the biosphere fungi occupy an ecological position of reducers. They are the processors of residual organic matter and this is their principle distinction from animals and plants. In total there are about 1.5 million known species of fungi. Some of them are used in the production of dairy products, wine, bakery products, antibiotics. Other species are pests or pathogens and they are the objects of merciless struggle. According to the structure of cells the fungi are eukaryotes and have much in common with animals, making them an attractive target for laboratory experiments. The mycelial fungus *Neurospora crassa* is unique among

T. Potapova (✉)
Belozersky Institute of Physicochemical Biology,
Moscow Lomonosov State University, Moscow 119991, Russia
e-mail: potapova@belozersky.msu.ru

G. Witzany (ed.), *Biocommunication of Fungi*, DOI 10.1007/978-94-007-4264-2_7,

the fungi as the object of research. It is one of the few organisms – along with *Drosophila*, white mouse and *E. coli* – which can called "biological supermodels" (Davis 2000; Borkovich et al. 2004).

Vegetative hyphae of the *N. crassa* grow and develop on a solid substrate in the form of two-dimensional trees, continuously elongating at the apexes of the branches at a speed of 20–30 μm/min and giving occasionally new lateral branches. New plasma membrane and chitinous cell wall are formed at the apexes of the branches from the substances transported there in microvesicles from more adult cells situated at a distance of 2,000–6,000 μm from the apex (Trinci et al. 1994; Davis 2000).

The morphology of a mycelium is determined by mechanisms which regulate the polarity and the direction of hyphal growth and the frequency of branching. The growth of a fungal mycelium can be characterized by the special constant parameter, namely, the ratio of the hyphal length to the tip number. This ratio is called the hyphal growth unit (HGU) and represents the average length of hypha which support the extension of each tip in the mycelium (Caldwell and Trinci 1973). Thus, growth of a mycelium can be considered in terms of the duplication of a physiological unit of growth which consists of a tip and a segment of hypha. Useful information for understanding the self-organization of the mycelium in the growth process can be gathered from an analysis of failures in the reproduction of HGU. According to our data (Aslanidi et al. 1996; Alekseevskii et al. 1999; Potapova et al. 2001, 2003) it is held constant with enviable persistence except the mycelium incubation without glucose (Potapova et al. 2008).

From the point of views Harold (2001): "...growth and many other cellular operations depend upon energy requiring, directional processes; self-assembly in the cell is directed, in space and in time...".

Due to the high activity of plasma membrane H^+-ATPases, the fungi generate electrical currents up to several nanoamperes per cell owing to the energy of hydrolysis of its own ATP, thus stably maintaining a membrane potential (E_m) of about 180–200 mV (electronegativity inside) (Slayman 1987; Slayman et al. 1973; Chailakhyan et al. 1985). High E_m is a driving force for many transport systems in plasma membrane of the fungal cell (Slayman and Slayman 1974; Sanders et al. 1983; Rodriquez-Navarro 1986; Slayman 1987).

The growing hyphae of *N. crassa* drive longitudinal proton currents through themselves (Fig. 1b). The pattern of the current flow may be described as a spatially extended chemiosmotic system, with proton pumps and proton leaks separated in space (Harold 1986; Takeuchi et al. 1988).

In the most of animal and plant tissues as well as in multicellular microorganisms the cells are interconnected through permeable junctions (PJ) permitting adjacent cells to "socialize" components of the low-molecular pool (Loewenstein 1981; Potapova and Boitzova 1998). The ability of PJ to pass ion fluxes, comparable in values with those through the plasma membrane, makes these structures a paramount mechanism of self-organisation in a multicellular system. Realising these possibilities a quasi-homogeneous multicellular population may be transformed into a spatially organising system possessing greater functional possibilities.

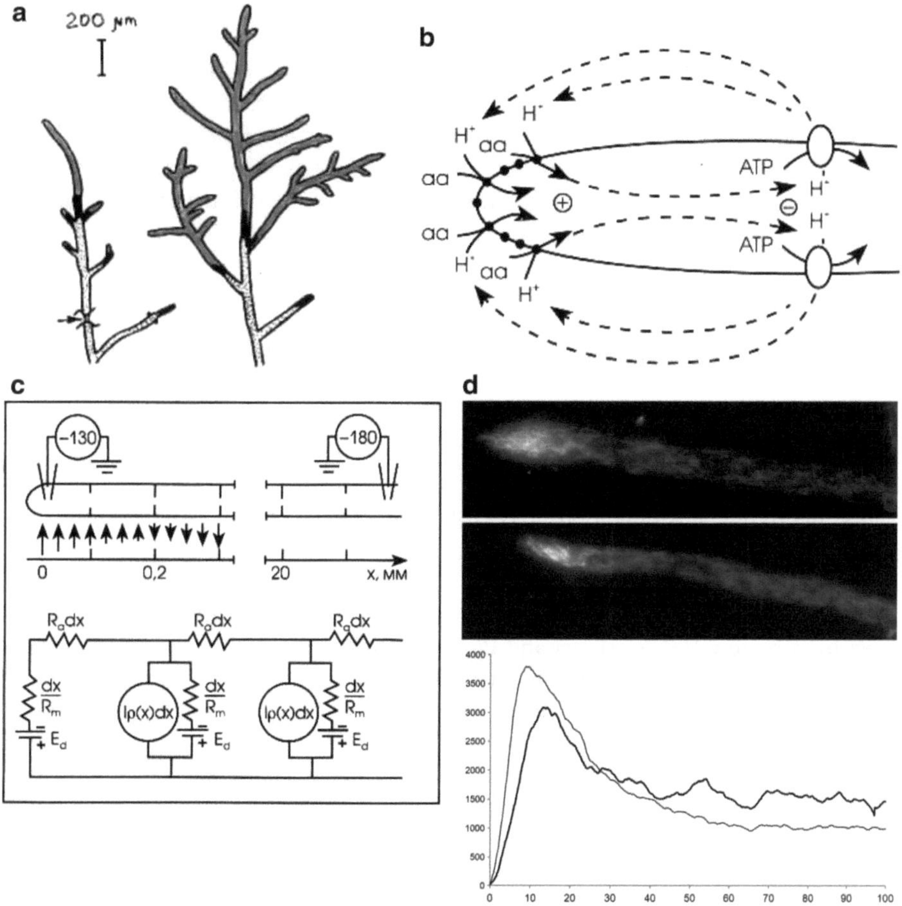

Fig. 1 (**a**) The morphology of two adjacent hyphae growing at the same V_{ext} immediately before the operation (see *dark parts* of the hyphal tips new-formed once during the 10 min growth). Then from the left hyphae the apical fragment had been isolated by one cell bursting via the micropipette puncture (the site of isolation shown). It is well seen that during the next some dozen minutes the intact hypha grew and developed more effectively, then the hyphal mini-tree. (**b**) The scheme of electrical current circulation in a spatially extended chemiosmotic system of *N. crassa* hypha, where proton pumps (H^+-ATPases) and proton leaks separated. (**c**) A schematic diagram and the electrical equivalent circuit for the apical part of a *N. crassa* hypha. The results of microelectrode measurements on the intact hypha (Slayman et al. 1973; Chailakhyan et al. 1985; Potapova et al. 1988) and of mapping the external currents near the apex (Takeuchi et al. 1988) are given. The *arrows* show the direction of the external currents. The sizes of the *arrows* are proportional to the magnitudes of the currents. The *horizontal scale* is the distance from the growing apex. (**d**) Clusters of mitochondrial filaments in the apexes of intact hypha (*upper photo*) and 700-μm fragment (*lower photo*). Objective: 100×. On the *bottom scheme* – distribution of MTR signal intensity on the front tips of intact hyphae (*thin line*) and isolated fragments (*thick line*). Along the *vertical axis* – fluorescence intensity (conditional units) along the *horizontal axis* – the distance from the apical edge, μm

The studies of *N. crassa* tip growth allow to illustrate the power of chemiosmotic approach (Mitchell 1961, 1966) to illuminate the physiology and energy supply of multicellular system possessing PJ. As Harold (2001) considers: "It is true but subtly misleading to envisage a cell as executing with instructions written down in its genome; better think of it as a spatially structured self-organizing system made up of gene-specified elements. Briefly, the genes specify What; the cell as a whole directs Where and When; and at the end of the day, it is the cell that usually supplies the best answer to the question Why".

2 Electrical Coupling in the *Neurospora* Hyphae

The *N. crassa* hyphae are multicellular chains (10–20 μm in diameter) incapsulated into a common cell wall. Every 50–100 μm, hyphal cells are separated by transverse septae with a channel in it (septal pore). Septal pore allows cytoplasmic streaming along the hypha at a rate about 4–6 cm/h as well as exchange of metabolites and even of nuclei and mitochondria between cells. Under unfavorable conditions septal pores are plugged with hexagonal crystals, which were originally identified as ergosterol, but later suggested to be composed of protein, like Voronin bodies (see for a review Belozerskaya and Potapova 1993). As well as the PJ of animals and plants, septal pores ensure electric coupling along hyphae (Gradmann et al. 1978; Chailakhyan et al. 1985; Potapova et al. 1988; Potapova and Boitzova 1998).

Intercellular channels, or PJ, of different organisms do not resemble each other: plasmadesmae of plants, animal's gap junctions, septal pores of fungi, microplasmadesma of cyanobacteria. However, the electrical properties of all multicellular systems connected via PJ are described by the equations once developed to describe the electrical processes in the telegraph cable laid along the ocean's bottom. An important characteristic of this cable is the space constant, λ. This is a length of cable's segment along which the voltage drops "e" times. The value of λ in *Neurospora* is a few hundred microns, as well as many other organisms (Gradmann et al. 1978; Chailakhyan et al. 1985).

All types of intercellular channels, providing an electrical coupling, can become uncoupled. This is a very important property, since on the random damage of one of the cell it protects the neighbors from the catastrophic loss of ions and small molecules. As in animals, the PJ conductivity of the fungus is blocked by the strong depolarization or mechanical damage one of electrically connected cells (Chailakhyan et al. 1985). Attention should be paid to the asymmetric reaction of the individual hyphal compartments to injuring agents. Mechanical injury or passing depolarizing current through the cell ($>10^{-8}$ A for 1–2 min) evoked the asymmetrical reactions in the longitudinal direction. In the direction of the colony center 4–5 adjacent cells became depolarized too. But the adjacent cell on the apical direction preserves the high E_m value via uncoupling with the damaged cell.

A unique feature of the PJ in *N. crassa* is quick repair of electrical coupling locally interrupted by the mechanical damage. After 15–20 min of such procedure from the septal pores on both sides of the empty cells small shoots appear, moving toward each other and after 1.5–2 h electrical communication became completely restored on this site of the hyphae (Potapova and Levina 1985).

In all multicellular systems the value of the cable constant λ ranges from several tens to several hundreds of microns, i.e., electrical connection is effective for a limited number of cells. What kind of problems can be solved by such a small group of cells, to socialize their ion and molecular resources? To answer this question it is needed to record simultaneously the PJ properties and functional characteristics of individual cells in the related group of cells. In the *N. crassa* hyphae a functionally related group should consist of 4–6 cells. It turned out that a number of cells at the apex of the growing hyphae are functioning effectively as a group successfully solving the problem of energy cooperation.

3 Electrical Heterogeneity of *Neurospora* Hyphal Tip and the Energy Cooperation

According to our data obtained with intracellular microelectrodes (Potapova et al. 1988; Potapova and Boitzova 1998), on the tips of *N. crassa* hyphae, there are significant gradients in E_m (Fig. 1c): at a distance $L_1 \sim 100$ μm from the apical end $E_{m1} \sim -130$ mV, and at a distance $L_2 \sim 400$ μm – $E_{m2} \sim -160$ mV. Then we obtained the apical cells to be incapable of maintaining the normal E_m value after the electrical isolation of 150–300 μm long apical fragments from the hyphae. The high values of E_m recorded in the hyphal apexes prior to their electrical isolation we explained as created by the generators in the membrane of the mature cells situated at some distance away from the growing tip (Aslanidi et al. 1997).

Near simultaneously the electrical heterogeneity in the apical cells of *N. crassa* hyphae has been described by Takeuchi et al. (1988) on the base of data, obtained by the "vibrating probe". This invention of Jaffe and Nuccitelli (1974) provided possibility to measure the strength of the external electric field in steps of 5–10 μ and discover the phenomenon of local electric currents flow in a variety of biological objects: eggs and embryos, muscles and nerves, roots, amoebae, algae and cyanobacteria trichomes. A detailed analysis made by Takeuchi et al. (1988) showed that: (i) in the apical 30 μ part of the hyphae the current is smaller than at a greater distances from apex; (ii) the current reversion point is located at a distance of about 80–180 μ from apex; (iii) the current disappearance point is located at a distance of about 300–400 μ. Takeuchi et al. (1988), analyzing the nature of electric currents within the growing *N. crassa* hyphae, expressed the firm conviction that the longitudinal intercellular electric currents in *N. crassa* hyphal tip arise due to the absence of H^+-ATPases activity at the tip of the hyphae. However, the significance of this phenomenon, they argue, is not clear.

Microelectrode studies on the growing hyphal tips of the water mold *Achlea*, allowed conclusively link the local incoming electric current with the transport of amino acids (Kropf 1986).

To explain the electrical heterogeneity of *Neurospora* hyphal tip the hypothesis of energy cooperation has been implicated. We have assumed the possibility of such cell specialisation that allows a great part of the cells to receive an "energetic subsidy" from neighbours via ionic fluxes through PJ. On the grounds of analytical calculations and experimental evaluations, we have shown (Chailakhyan et al. 1982; Aslanidi et al. 1988, 1991; Potapova and Aslanidi 1995) that inorganic ion fluxes across PJ, moving along corresponding electrochemical gradients, can maintain "energy cooperation" on a scale comparable with the total energy production (energy consumption) of an individual cell. The existence of such energy cooperation of cells is revealed in electrophysiological experiments by the presence of intercellular currents and/or electrical gradients across PJ (Chailakhyan et al. 1982; Aslanidi et al. 1986, 1988, 1991; Potapova et al. 1988, 1990; Potapova and Boitzova 1998).

Accordingly with the theoretical model analysis (Aslanidi et al. 1997; Smolyaninov and Potapova 2003) we can represent the hyphal tip as consumers of power, connected in series and localised in an area 200–300 μm (about 3–4 the youngest cells), and the electrical generators localised in the plasma membranes of more mature cells. Within the framework of the mathematical model the cell structure of hypha was fitted by a one-dimensional continuous semi-infinite electric cable, each elementary unit of which corresponded to the equivalent circuit proposed for the *N. crassa* cell in (Slayman et al. 1973). Detailed mathematical description of the model is given in (Aslanidi et al. 1997). Unlike other models this approach considers the electric current I_p generated by the proton pumps of the plasma membrane. The proton pump was modelled as a source of the current.

The analysis of the model allows to obtain the profiles of pump current densities and membrane potentials for apical fragments of various length by solving the appropriate equations. The model predicts the hyperpolarization of the apex to normal values (−180 mV) as the length of isolated fragment increases and approach cable constant value. According to the results of computer simulation (Aslanidi et al. 1997), the growing apexes of the fungal hyphae obtain more than 80% of energy from the energy cooperation with the mature parts of the mycelium. These estimates suggest that in intact hypha only 5–7% of the substances are transported to the growth zone by means of the E_m generated by the endogeneous proton pumps of the apical cell; about 45% of the substances are transported owing to the energy cooperation by means of ionic fluxes and up to 50% are transported by means of the intrahyphal vesicular transport from the trunk cells.

The model predicts also that formation of distal regions of the fragment requires more substances than that of apical regions. This can be interpreted as the accumulation of substances in the distal cells from the growth zone: the more distal a cell, the more substance it accumulates.

Of course, the cable model mentioned above is the rather simple image of the real fungal hypha. For example, it assumes the constancy of the diffusion component of E_m in the apex. However, the careful electron probe microanalysis of *Neurospora crassa* hyphae preparations revealed significant non-homogeneity of K^+ contents in the hyphae. The presence of areas with low K^+ contents in close proximity of the growing tip as well as in distal part of the hyphae in the cross-membrane area of the width by 1 m order were discovered (Aslanidi et al. 2000). It is shown that $[K^+]_{in}$ can change more then 3–5 times from apical to distal areas of the leading hyphae, the K^+ contents in the apical part of the hyphae usually being minimal. One can assume that the $[K^+]_{in}$ is defined by the volume occupied by free intracellular water, i.e. by the size of the space available for diffusion of substances (including inorganic ions). In any case, it is obvious, that the mathematical model mentioned above (Aslanidi et al. 1997) is only an approximation to the reality.

To verify the theoretically predicted link between tip growth rate and the apical hyphal fragment length, we worked out an original experimental model of hyphal mini-tree: an electrically isolated hyphal tip, comparable in length to the magnitude of the characteristic cable constant, λ (Aslanidi et al. 1996; Alekseevskii et al. 1999; Potapova et al. 2001, 2003).

An important result obtained with this experimental model is that the ability of the apical fragments to grow autonomously after separation from the hyphae is limited by a threshold length close to that at which the activity of the proton pumps appears to be absent, i.e. 200–300 μm (Potapova et al. 1988; Potapova and Boitzova 1998). Increase of the fragment length leads to the increase of extension rate (V_{ext}) of the leading tip. During 5–15 min after isolation from the hypha V_{ext} of 900–1,100 μm apical fragment leading tip does not change. Later, it decreases to about 50% of V_{ex} of the adjacent intact hypha. The delayed V_{ext} drop can be explained as a result of depletion of stocks accumulated in the apical part of hypha before isolation. Thus, the true autonomous growth of the isolated fragment begins only after 5–15 min lag-period. From this moment V_{ext} of the leading tip increases as a function of the fragment's length according to the mathematical model (Aslanidi et al. 1997) predictions. But only V_{ext} of leading tip increase, not of the side branches. The growth and development of the side branches of the isolated mini-tree were found to be influenced by the isolation procedure more drastically then the leading tip growth and development (Fig. 1a).

4 The Dynamics of Intracellular Structures During Polarized Tip Growth

Events that occur in the immediate vicinity of the tips of hyphal branches play a special role in the growth of fungal mycelium. The tip of hyphae is longitudinally polarized in a high degree. There are described the gradient of shape and wall thickness (Collinge and Trinci 1974), the distribution of mitochondria and vesicles (Collinge and Trinci 1974; Hickey et al. 2005), the concentration of Ca^{2+} and pH (Silverman-Gavrila and Lew 2003), K^+ concentrations (Aslanidi et al. 2000), respiratory rate (Lew and Levina 2004), the activity of genes (Tey et al. 2005), etc. Recently, research attentions are careful focused on describing the interactions between intracellular structures in the tip growth and the elucidation of molecular and genetic mechanisms for the implementation of this process as well as its reaction to events in the external environment (Fischer et al. 2008; Held et al. 2010; Riquelme et al. 2002; Steinberg 2007; Westermann 2008).

At the front end of the growing hyphae a series of events take place that require energy expenditure: the transmembrane transport of ions and metabolites, the exocytosis of vesicle content and endocytosis, cytoskeletal reorganization, the synthesis of chitin and cell wall formation. Not surprisingly, that clusters of mitochondria are found by the various methods in this area (Trinci et al. 1994; Davis 2000; Lew 1999; Levina and Lew 2006).

We show that in the growing *N. crassa* hyphae mitochondria are not simply concentrated in the apex but move behind elongating tip in the way that "effective concentration" of mitochondria is higher when the tip grows faster (Potapova et al. 2011) (Fig. 1d). To study the behavior of mitochondria in growing *N. crassa* hyphae we have used cell-permeable fluorescent marker for mitochondria MitoTracker Red (MTR) (Haugland 2002). Observations of the labeled mitochondria populations revealed that in the growing hyphae these organelles are concentrated at the apex as the clusters of filaments which move with the elongating tip of the growing hyphae for several hours.

Hyphal growth and development of the tree are characterized by a hierarchy of different growth velocities of the leading tip and side branches (Trinci et al. 1994). Analyzing the distribution of the MTR signal in the tips of the various branches we observed the obvious link of the efficiency of mitochondria clustering and the growth rate of a particular branch (Potapova et al. 2011).

In our experiments during initial period of novel lateral branches' formation their tips were devoid of clusters of filamentous mitochondria, but with the extension of new branches the more distinct clusters of mitochondrial filaments became visible at their tips, suggesting that the accumulation of mitochondria is not a factor in initiating branching, but the mitochondria clusters are formed in the process of hyphal growth. In some experiments, we observed the emergence of new branches by dividing the hyphal tip. In such a case, the apical cluster of filamentous mitochondria divides too and distributes between the new tips then moving with them on their elongation (Potapova et al. 2011).

Mitochondria are the intracellular organelles that produce ATP and are accumulated near the structures with the high energy consumption: the nuclei, endoplasmic reticulum or plasma membranes (Bereihter-Hahn 1990). All eukaryotic cells possess effective mechanisms for delivery of mitochondria to the places where these organelles are functionally necessary (Yaffe 1999). For example, it has been shown in chick embryo that during the active elongation of the axons mitochondria are accumulated in the active growth cone, and after completion of the extension they are uniformly distributed along the axon (Chada and Hollenbeck 2003).

In *N. crassa* mitochondria move along microtubules (Steinberg and Schliwa 1993; Westermann 2008). There have been characterized a number of proteins that facilitate this transport: in particular, several kinesin family proteins (Fuchs and Westermann 2005; Fischer et al. 2008). Unfortunately, a complete description of all the molecular components that provide vector transport of mitochondria along microtubules is absent, thus, the question of regulation of this transport remains open.

It is known that microtubules of *N. crassa* persisted in the form of short rods and randomly distributed along the hyphae, but are oriented strictly parallel to the axis of the hyphae in the tip section of a 100–150 μm (Held et al. 2010). The team of British biologists and mathematicians recently published a mathematical model that relates the rate of tip growth with vesicles originating from distant parts of the hyphae, and the number of microtubules, to ensure that transport. They came to conclusion that ten microtubules are enough to ensure receipt of organized vesicles to the site of their fusion with the plasma membrane in growth rate of 20–30 μ/min (Sugden et al. 2007).

Based on the data of microelectrode measurements (Potapova et al. 1988; Potapova and Boitzova 1998), we can estimate the electric field along the apical end of hyphae:

$$E = (E_{m2} - E_{m1})/(L_2 - L_1) = 30\,mV/300\ \mu m = 100\,V/m$$

According to Böhm et al. (2005) and Dujovne et al. (2008) isolated microtubules are oriented in the solution and change the speed under the electric field influence of 2×10^3 V/m. Thus, it can be assumed that electrical gradients in the hyphal tips are at least a part of the control system, which monitors the hyphal growth regulation, including its power supply. Perhaps further research in this field will shed at last light on the old question concerning the role of the longitudinal electric currents recorded by many authors (Jaffe and Nuccitelli 1974; Kropf et al. 1984; Gow et al. 1984) in the tip growth (Gow 1989; Harold 1994; Harold and Caldwell 1990; Jaffe and Nuccitelli 1977).

5 Conclusion

Clarification of the laws determining the organization of cell-to-cell and intracellular interactions in living systems is one of the most important problems in modern biology. Tip growth of *N. crassa* mycelium can be a suitable experimental model for studying patterns of such interactions, since tip growth is a typical system function which involves coordinated interactions of different cells as well as different intracellular structures.

The dependence of efficiency of mitochondria accumulation in the growing tips of hyphal branches on the velocity of elongation, has shown significant expenditure of ATP in these areas, despite the absence of ion pumps. The cost of ATP on the work of moving vesicles with microtubules can be estimated, suggesting that this movement is a step by step it takes to implement each step of the hydrolysis of one ATP molecule. However, microtubules are not the only consumers of ATP in the growth zone. To create a complete picture of energy consumption in tip growth the ATP costs of exocytosis and endocytosis, the work of microfilaments, biosynthesis and other important processes must be estimated. With such assessments in hands it will be possible to create a mathematical model of energy supply of tip growth considered the coordinated interaction of the complex intracellular structures in the functional active multicellular system coupled via the PJ.

References

Alekseevskii AV, Belozerskaya TA, Boitzova LJu, Lukina EN, Potapova TV, Toms KS (1999) New approaches to analysis of characteristics of self-organization of multicellular systems during polarized apical growth. Dokl Biophys 369:108–111

Aslanidi KB, Potapova TV, Shalapjenok AA, Karnauhov VN, Chailakhyan LM (1986) Photoelectric activity and spectral parameters of the unit trichome of *Phormidium uncinatum*. Dokladi Akademii Nauk SSSR 290:1504–1507 (in Russian)

Aslanidi KB, Potapova TV, Chailakhyan LM (1988) Energy transmission through the high permeable intercellular junctions. Biol Membrany 5:613–627 (in Russian)

Aslanidi KB, Boitzova LJu, Chailakhyan LM, Kublik LN, Marachova II, Moch VN, Potapova TV, Trepakova EK, Vinogradova TN (1991) Maintenance of ion-osmotic homeostasis in multicellular animal systems: the role of permeable junctions. Biol Membrany 8:837–853 (in Russian)

Aslanidi KB, Boitzova LJu, Potapova TV, Smolianinov VV (1996) Information-energy module as the functional unit of the *Neurospora crassa* polarized growth. Membr Cell Biol (Russia) 10:27–37

Aslanidi KB, Aslanidi OV, Vachadze DM, Mornev OA, Potapova TV, Chailakhyan LM, Shtemanetian EG (1997) Analysis of electrical phenomena accompanying the growth of *Neurospora crassa* hyphae: theory and experiment. Membr Cell Biol (Russia) 11:349–365

Aslanidi KB, Pogorelov FG, Aslanidi OV, Mornev OA, Potapova TV (2000) The distribution of potassium in *Neurospora crassa* hyphae. Doklady Russ Acad Sci 372:253–256 (in Russian)

Belozerskaya TA, Potapova TV (1993) Intrahyphal communication in segmented mycelium. Exp Mycol 17:157–169

Bereihter-Hahn J (1990) Behavior of mitochondria in the living cell. Int Rev Cytol 122:1–63

Böhm KJ, Nikolaos E, Mavromatos NE, Michette A, Stracke R, Unger E (2005) Movement and alignment of microtubules in electric fields and electric-dipole-moment estimates. Electromagn Biol Med 24:319–330

Borkovich KA, Alex LA, Yarden O, Freitag M et al (2004) Lessons from the genome sequence of *Neurospora crassa*: tracing the path from genomic blueprint to multicellular organism. Microbiol Mol Biol Rev 68:1–108

Caldwell JY, Trinci APJ (1973) The growth unit of the mould *Geotrichum candidum*. Arch Microbiol 88:1–10

Chada SR, Hollenbeck PJ (2003) Mitochondrial movement and positioning in axons: the role of growth factor signalling. J Exp Biol 206:1985–1992

Chailakhyan LM, Glagolev AN, Glagoleva TN, Murvanidze GM, Potapova TV, Skulachev VP (1982) Intercellular power transmission along trichomes of cyanobacteria. Biochem Biophys Acta 679:60–73

Chailakhyan LM, Potapova TV, Levina NN, Belozerskaya TA, Kritzky MS (1985) A study of intercellular communication in *Neurospora crassa* with special reference to photoelectric responses of membranes. Biol Membrany 1:76–101 (in Russian)

Collinge AJ, Trinci APJ (1974) Hyphal tips of wild-type and spreading colonial mutants of *Neurospora crassa*. Arch Microbiol 99:353–368

Davis RH (2000) *Neurospora*: contributions of a model organism. Oxford University Press, Oxford

Dujovne I, van den Heuvel M, Shen Y, de Graaff M, Dekker C (2008) Velocity modulation of microtubules in electric fields. Nano Lett 8:4217–4220

Fischer R, Zekert N, Takeshita N (2008) Polarized growth in fungi – interplay between the cytoskeleton, positional markers and membrane domains. Mol Microbiol 68:813–826

Fuchs F, Westermann B (2005) Role of *Unc104/KIF1*-related motor proteins in mitochondrial transport in *Neurospora crassa*. Mol Biol Cell 16:153–161

Gow NAR (1989) Circulating ionic currents in micro-organisms. Adv Microb Physiol 30:90–123

Gow NAR, Kropf DL, Harold FM (1984) Growing hyphae of *Achlya bisexualis* generate a longitudinal pH gradient in the surrounding media. J Gen Microbiol 130:2967–2974

Gradmann D, Hansen U-P, Long WS, Slayman CL, Warncke J (1978) Current–voltage relationships for the plasma membrane and its principal electrogenic pump in *Neurospora crassa*. J Membr Biol 39:333–367

Harold FM (1986) The vital force: a study of bioenergetics. Freeman and Co., New York

Harold FM (1994) Ionic and electrical dimensions of hyphal growth. In: Wessels JGH, Meinhardt F (eds) The mycota I. Growth, differentiation and sexuality. Springer, Berlin/Heidelberg, pp 89–109

Harold FM (2001) The way of the cell: molecules, organisms and the order of life. Oxford University Press, Oxford

Harold FM, Caldwell JH (1990) Tips and currents: electrobiology of apical growth. In: Heath IB (ed) Tip growth in plant and fungal cells. Academic Press Inc., Orlando/San Diego, pp 59–90

Haugland RP (2002) Handbook of fluorescent probes, 9th edn. Molecular Probes, Eugene

Held M, Edwards C, Nicolau DV (2010) Temporal and spatial *in vivo* optical analysis of microtubules in *Neurospora crassa*. In: Farkas DL, Nicolau DV, Leif RC (eds) Imaging, manipulation, and analysis of biomolecules, cells, and tissues VIII. SPIE, San Francisco, p 75680V

Hickey PC, Swift SR, Roca MG, Read ND (2005) Live-cell imaging of filamentous fungi using vital fluorescent dyes and confocal microscopy. Methods Microbiol 34:63–87

Jaffe LF, Nuccitelli R (1974) An ultrasensitive vibrating probe for measuring steady extracellular currents. J Cell Biol 63:614–628

Jaffe LF, Nuccitelli R (1977) Electrical control of development. Annu Rev Biophys Bioeng 6:445–476

Kropf DL (1986) Electrophysiological study of *Achlea* hyphae: ionic currents studied by intracellular potential recording. J Cell Biol 102:1209–1216

Kropf DL, Caldwell JH, Gow NAR, Harold FM (1984) Transcellular ion currents in the water mold *Achlea*: amino acid proton symport as a mechanism of current entry. J Cell Biol 99:86–96

Levina NN, Lew RR (2006) The role of tip-localized mitochondria in hyphal growth. Fungal Genet Biol 43:65–74

Lew RR (1999) Comparative analysis of Ca and H flux magnitude and location along growing hyphae of *Saprolegnia ferax* and *Neurospora crassa*. Eur J Cell Biol 78:892–902

Lew RR, Levina NN (2004) Oxygen flux magnitude and location along growing hyphae of *Neurospora crassa*. FEMS Microbiol Lett 233:125–130

Loewenstein WR (1981) Junctional intercellular communication: the cell-to-cell membrane channel. Physiol Rev 61:829–913

Mitchell P (1961) Coupling of phosphorylation to electron and hydrogen transfer by a chemiosmotic type of mechanism. Nature 191:144–148

Mitchell P (1966) Chemiosmotic coupling in oxidative and photosynthetic phosphorylation. Biol Rev Camb Philos Soc 41:445–502

Potapova TV, Aslanidi KB (1995) Energy coupling of adjacent cells as an universal function of cell-to-cell permeable junctions. In: Kanno Y (ed) Intercellular communication through gap junctions, Progress in cell res 4. Elsevier Science B.V., Amsterdam, pp 53–56

Potapova TV, Boitzova LJu (1998) Structure, function, regulation: experimental analysis in groups of non-excitable cells coupled via permeable junctions. Membr Cell Biol (Russia) 11:817–829

Potapova TV, Levina NN (1985) Reactions of *Neurospora* to a local lesion of the cell membrane. Biol Membrany 2:123–127 (in Russian)

Potapova TV, Aslanidi KB, Belozerskaya TA, Levina NN (1988) Transcellular ionic currents studied by intracellular potential recordings in *Neurospora crassa* hyphae. (Transfer of energy from proximal to apical cells). FEBS Lett 241:173–176

Potapova TV, Aslanidi KB, Boitzova LJu (1990) Energy transfer via cell-to-cell junctions: ouabain-resistant cells maintain a membrane potential in ouabain-sensitive cells. FEBS Lett 262:69–71

Potapova TV, Boytzova LJu, Alexeevskii AV, Smoljaninov VV (2001) The functional role of intercellular interactions at the *Neurospora crassa* hyphae: influence of distant mycelium part on the tip growth. Biol Membrany 18:364–369 (in Russian)

Potapova TV, Boytzova LJu, Alexeevskii AV, Smoljaninov VV, Belozerskaya TA (2003) Tip growth of the genetically different *Neurospora crassa* strains: effects of electrical isolation and nitrogen depletion. Biol Membrany 20:395–400 (in Russian)

Potapova TV, Alexeevskii TA, Boitzova LJu (2008) Tip growth of *Neurospora* during glucose deprivation. Membr Cell Biol (Russia) 25:171–177

Potapova TV, Boitzova LJu, Golyshev SA, Popinako AV (2011) Dynamics of mitochondria during *Neurospora crassa* tip growth. Membr Cell Biol (Russia) 28:345–353

Riquelme M, Robetson RW, McDaniel DP, Bartnicki-Garcia S (2002) The effect of *ropy-1* mutation on cytoplasmic organization and intracellular motility in mature hyphae of *Neurospora crassa*. Fungal Genet Biol 37:171–179

Rodriquez-Navarro A (1986) A potassium-proton symport in *Neurospora crassa*. J Gen Physiol 87:649–674

Sanders D, Slayman CL, Pall ML (1983) Stoichiometry of H^{+}/amino acid cotransport in *Neurospora crassa* revealed by current–voltage analysis. Biochim Biophys Acta 735:67–76

Silverman-Gavrila LB, Lew RR (2003) Calcium gradient dependence of *Neurospora crassa* hyphal growth. Microbiology 149:2475–2485

Slayman CL (1987) The plasma membrane potential of *Neurospora crassa*: a proton pumping electroenzyme. J Bioenerg Biomembr 19:1–20

Slayman CL, Slayman CW (1974) Depolarization of the plasma membrane of *Neurospora crassa* during active transport of glucose. Proc Natl Acad Sci USA 71:1931–1939

Slayman CL, Long WS, Lu CY-H (1973) The relationship between ATP and an electrogenic pump in the plasma membrane of *Neurospora crassa*. J Membr Biol 14:305–338

Smolyaninov VV, Potapova TV (2003) Evaluation of the critical fragment length of *Neurospora crassa* hyphae. Biol Membrany 20:314–321 (in Russian)

Steinberg G (2007) Hyphal growth: a tale of motors, lipids, and spitzenkorper. Eukaryot Cell 6:351–360

Steinberg G, Schliwa M (1993) Organelle movements in the wild type and wall-less fz; sg; os-1 mutants of *Neurospora crassa* are mediated by cytoplasmic microtubules. J Cell Sci 106:555–564

Sugden KEP, Evans MR, Poon WCK, Read ND (2007) Model of hyphal tip growth involving microtubule-based transport. Phys Rev E 75:031909

Takeuchi G, Schmidt J, Caldwell JH, Harold FM (1988) Transcellular ion currents and extension of *Neurospora crassa* hyphae. J Membr Biol 101:33–41

Tey WK, North AJ, Reyes JL, Lu YF, Jedd G (2005) Polarized gene expression determines Woronian body formation at the leading edge of the fungal colony. Mol Biol Cell 16:2651–2659

Trinci APJ, Wiebe MG, Robson GD (1994) The mycelium as an integrated entity. In: Wessels JGH, Meinhart F (eds) The mycota I. Growth, differentiation and sexuality. Springer, Berlin/Heidelberg, pp 175–193

Westermann B (2008) Molecular machinery of mitochondrial fusion and fission. J Biol Chem 283:13501–13505

Yaffe MP (1999) The machinery of mitochondrial inheritance and behavior. Science 283:1493–1497

Programmed Cell Death and Heterokaryon Incompatibility in Filamentous Fungi

Elizabeth A. Hutchison and N. Louise Glass

Abstract Filamentous fungal hyphal networks can be formed by intercellular interactions within and between fungal colonies. However, intercellular interactions can be restricted by a self/nonself recognition mechanism, such that networks can only be formed between colonies that are identical at all self/nonself recognition (*het*) loci. Hyphal fusion between fungal colonies that are nonself results in the rapid induction of programmed cell death (PCD) in the fusion compartment and subtending cells. Although PCD in fungi exhibits some morphological similarities with apoptotic cell death, the genetic regulation of PCD appears to be unique. Self/nonself recognition loci have been cloned in the model fungal systems *Podospora anserina* and *Neurospora crassa* and have revealed that self/nonself recognition loci often involve both allelic and nonallelic interactions, show evolutionary features consistent with their selection for recognition and have implicated a filamentous fungal-specific protein domain (termed HET) in the induction of PCD.

1 Introduction to Self/Nonself Genetic Recognition Mechanisms

Mechanisms to distinguish self from nonself are prevalent across a wide range of species, including eukaryotes and prokaryotes, and the types of nonself recognition mechanisms that have evolved are equally diverse. One of the best-studied systems

E.A. Hutchison
Department of Plant and Microbial Biology, University of California, Berkeley, 111 Koshland Hall, Berkeley, CA 94720-3102, USA

Microbiology Department, Cornell University, 157 Wing Hall, Ithaca, NY 14853-8101, USA
e-mail: eah259@cornell.edu

N.L. Glass (✉)
Department of Plant and Microbial Biology, University of California, Berkeley, 111 Koshland Hall, Berkeley, CA 94720-3102, USA
e-mail: Lglass@berkeley.edu

G. Witzany (ed.), *Biocommunication of Fungi*, DOI 10.1007/978-94-007-4264-2_8,

is the mammalian immune system, which consists of innate and adaptive responses that control self/nonself recognition and mediate pathogen defense. Somatic recombination events produce polymorphic major histocompatibility (MHC) proteins that mediate recognition (Chaplin 2010). Self/nonself recognition systems also occur in basal eukaryotic invertebrates (Burnet 1971; Kvell et al. 2007) such as *Botryllus schlosseri* (McKitrick and De Tomaso 2010), a colonial ascidian, and *Hydractinia symbiolongicarpus*, a cnidarian (Cadavid 2004). In these systems, polymorphic transmembrane proteins regulate allorecognition responses (De Tomaso et al. 2005; Nicotra et al. 2009; Rosa et al. 2010). The ability of plants to recognize self from nonself has also been well characterized, and shares some broad mechanistic similarities to that of animals (Ronald and Beutler 2010). Plant self/nonself recognition is important for both pathogen defense and the prevention of self-fertilization for outcrossing species (Sanabria et al. 2008; Dodds and Rathjen 2010). In the protists, the amoeba *Dictyostelium discoideum* exhibits kin recognition behaviors, and self/nonself recognition in this organism limits only "self" genotypes to the sporulating fruiting bodies (Strassmann et al. 2000; Li and Purugganan 2010). In the multinucleate amoeba *Physarum polycephalum,* incompatibility reactions following vegetative fusion events have been reported (Lane and Carlile 1979). More recently, mechanisms for prokaryotic self/nonself have begun to be characterized; though variable in molecular mechanism, they usually function to inhibit colony interactions (Gibbs et al. 2008; Be'er et al. 2009, 2010; MacIntyre et al. 2010) or to prevent nonself cells from reaching sporulation structures in multicellular bacterial fruiting bodies (Velicer and Vos 2009). In addition, a form of programmed cell death has been observed in the bacterium *Streptomyces antibioticus*, which undergoes complex differentiation processes and forms filamentous, multi-nucleate hyphae (Miguelez et al. 1999, 2000). Finally, filamentous fungi also possess a complex, sophisticated mechanism for distinguishing genetic identity called heterokaryon incompatibility (HI; also termed vegetative incompatibility, VI) (Glass and Dementhon 2006; Lu 2006; Pinan-Lucarre et al. 2007).

An important aspect of the filamentous fungal lifestyle is the ability to undergo fusion within and between colonies in order to form a network, which is thought to aid in nutrient transport and resource utilization (Fleissner et al. 2008). If two individual colonies come into contact, they can fuse and form a heterokaryotic mycelium that contains nuclei and cellular contents from both individuals (Fig. 1). If the two strains are genetically identical at all heterokaryon (*het* or *vic*) loci they are compatible, meaning that formation of a viable, stable heterokaryon can occur. However, if the two individuals are genetically different at any one of their *het* loci, the fusion cell is rapidly compartmentalized and programmed cell death (PCD) ensues (Fig. 1). In this case, the strains are considered incompatible (Glass and Dementhon 2006; Pinan-Lucarre et al. 2007). HI has been characterized primarily in the model fungal systems *Neurospora crassa*, *Podospora anserina* and *Cryphonectria parasitica*, but also in basidiomycetes (Worrall 1997) and arbuscular mycorrhizal (AM) fungi (Giovannetti et al. 2006). Pre-recognition of an incompatible strain has also been reported in *Tuber borchii* (Sbrana et al. 2007),

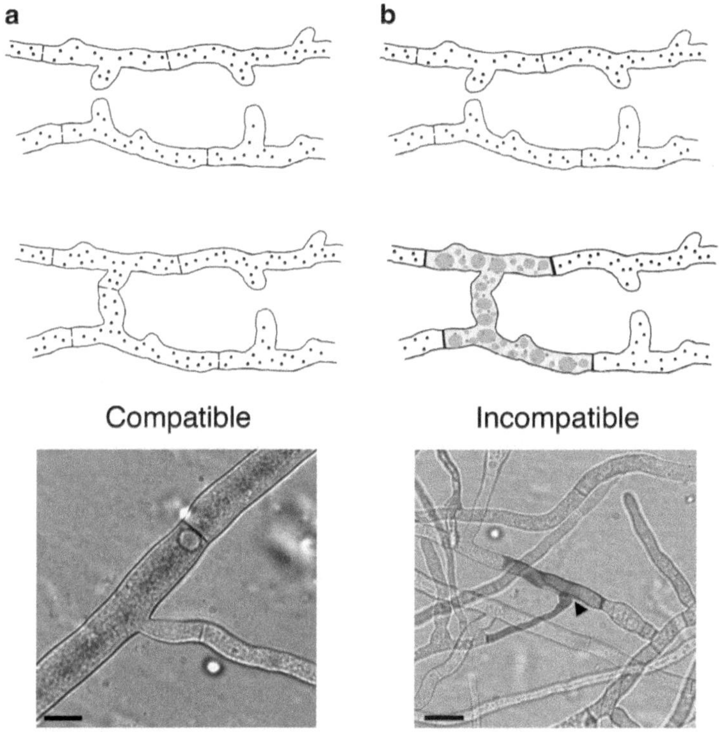

Fig. 1 Schematic of compatible and incompatible hyphal interactions. (**a**) If two individual hyphae are genetically identical, they can successfully interact chemotropically, and fusions generate a viable heterokaryon. (**b**) If individuals are genetically different (at any *het* loci), fusion can still occur, yet the fusion cell is compartmentalized and cell death rapidly occurs. DIC images below the diagram show examples of wild type compatible and incompatible heterokaryons that were stained with a vital dye (methylene blue) such that dead cells appear dark in color, as seen in the image from (**b**)

Rhizoctonia solani (Hyakumachi and Ui 1987), *Glomus mosseae* (Giovannetti et al. 2003) and *Gibberella fujikuroi* (Correll et al. 1989), where incompatible hyphae actually avoid fusion (anastomosis). These observations suggest a link between hyphal fusion signaling and nonself recognition. In *N. crassa*, hyphal and germling fusion is mediated by the oscillation of signaling proteins to the tips of cells undergoing chemotropic interactions prior to cell fusion (Fleissner et al. 2009). However, when incompatible fusions do occur, the fusion cell is compartmentalized by occluding septal pores and undergoes a rapid programmed cell death (Glass and Kaneko 2003). Thus, HI can be separated into two parts, the first involving mechanisms for distinguishing self versus nonself, while the second aspect involves mechanisms mediating cell death. Thus, self/nonself recognition mechanisms work in tandem with a PCD pathway in order to restrict fusion to genetically identical individuals.

2 Biological Functions of HI

HI has significant biological implications for fungi, primarily in preventing the transfer of deleterious genetic elements between colonies, including mycoviruses and senescence plasmids. It has also been hypothesized that HI functions as a fungal innate immune system (Paoletti and Saupe 2009). The effect of HI on viral transmission between colonies has been studied extensively in the causative agent of chestnut blight, *C. parasitica* (Cortesi et al. 2001; Milgroom and Cortesi 2004). Viral infection of *C. parasitica* causes a hypovirulence phenotype, and thus mycoviruses have been used as a biological control agent (Milgroom and Cortesi 2004). Fungal viruses generally do not have an extracellular phase to their infection cycle, and depend on cytoplasmic transfer via hyphal fusion rather than external infection for transmission to new hosts (Nuss 2010). Fungal viral transmission can be controlled by exclusion mechanisms (often, viruses are not transferred 100% of the time into sexual or asexual spores), by RNA silencing, and by differences at *het* loci (Milgroom 1999; Nuss 2010). Differences at *het* loci cause a significant reduction in the viral transmission rate by restricting cytoplasmic transfer between *C. parasitica* colonies (Cortesi et al. 2001; Biella et al. 2002). HI in several *Aspergillus* species has also been associated with restriction of the transfer of deleterious cytoplasmic elements (Caten 1972; van Diepeningen et al. 1997; Cortesi et al. 2001; Milgroom and Cortesi 2004). Unlike *C. parasitica*, differences at *het* loci are sufficient to completely block viral transfer in the black *Aspergilli* (van Diepeningen et al. 1997). Genetic differences at *het* loci also restrict the transfer of hypovirulence-associated dsRNAs in the pathogen *Sclerotinia homoeocarpa* (Deng et al. 2002). Where *het* locus diversity is high in a fungal population, biological control using viruses is likely to be less effective (Milgroom and Cortesi 2004).

The Dutch elm disease pathogen *Ophiostoma novo-ulmi* (Brasier 1991) typically infects a new area as a clonal population (Brasier 1988; Paoletti et al. 2006) and is thus highly susceptible to rampant viral infection. Like *C. parasitica*, this fungal pathogen is rendered hypovirulent upon viral infection (Paoletti et al. 2006). However, it has been observed that *O. novo-ulmi* acquires genetic information for the opposite mating type, as well as alternate alleles at *het* loci, from rare crossing events with *O. ulmi* species. Genetic diversity at *het* loci can restrict fusion and viral transfer between strains of *O. novo-ulmi*, and the ability to undergo sexual crosses may provide a selective advantage to progeny with genetic differences at multiple *het* loci (Paoletti et al. 2006). Thus, this study provides very strong evidence that genetic differences at *het* loci can inhibit viral transmission as well as influence fungal population structure.

In addition to affecting viral transmission, HI was shown to restrict the transfer of a mitochondrial senescence phenotype in *P. anserina* (Aanen et al. 2010) and of mitochondrial senescence plasmids in *N. crassa* (Debets et al. 1994), though in *C. parasitica* HI had little to no effect on mitochondrial plasmid transfer (Baidyaroy et al. 2000). Mitochondrial senescence plasmids are circular DNA molecules that integrate into fungal mitochondrial DNA, thereby causing eventual lethal mutations; the defective mitochondria divide faster than functional mitochondria, and eventually cause death of the colony (Maheshwari and Navaraj 2008).

Another case in which HI confers an evolutionary advantage is the production of aflatoxin by *Aspergilli*. Aflatoxin production is significantly higher in compatible heterokaryon pairings and lower in incompatible heterokaryons, suggesting that mycelial networks formed by compatible hyphae offer a selective advantage (Wicklow and Horn 2007). Finally, Debets and Griffiths (1998) investigated the role of HI with respect to mitochondrial plasmids and nuclear genes during the *N. crassa* sexual cycle and purport that HI prevents "resource plundering" of maternal tissues by genetically different strains, thereby only allowing genetically identical strains to access nutrients and resources from maternal tissues. Thus, differences at *het* loci are at least a partial barrier to transmission of deleterious elements between hyphae, particularly cytoplasmic elements (Nauta and Hoekstra 1994), and also prevent parasitism of one genotype on another in filamentous fungi.

3 Characteristics of Cell Death Programs in Filamentous Fungi

In filamentous fungi, HI is linked to cell death, such that fusion and resource sharing are restricted to genetically identical strains. Cell death programs, like self/nonself recognition programs, are also present in a wide range of organisms and fall into two very broad categories: apoptotic (type I) and non-apoptotic (Degterev and Yuan 2008). More recently, it has become appreciated that many alternative, non-apoptotic cell death pathways, including but not limited to autophagy (type II) and necroptosis, are important for cell death and development (Degterev and Yuan 2008; Yuan and Kroemer 2010). In addition to HI, filamentous fungi employ PCD mechanisms throughout their life cycle and development, including basidiocarp development, pathogenesis, and fungal ageing (Umar and Van Griensven 1997, 1998; Lu et al. 2003; Mousavi and Robson 2003; Emri et al. 2005; Hamann et al. 2008). In this section, we review similarities and differences of PCD associated with HI and with that induced by cell exposure to fungal chemical signals and ageing.

3.1 *Cell Death and Chemical Signals*

Filamentous fungal PCD has been observed in response to exposure to a variety of chemical stresses (Robson 2006; Castro et al. 2008; Hamann et al. 2008; Ramsdale 2008; Sharon et al. 2009). Cell death can also be induced by chemical signals produced during fungal–fungal interactions. For example, farnesol, a well-characterized *Candida albicans* quorum signaling molecule involved in the yeast to hyphal transition (Langford et al. 2009), was found to induce an apoptotic-like phenotype in *A. nidulans*, *Fusarium graminearum*, and *Pencillium expansum*

(Semighini et al. 2006b, 2008; Liu et al. 2010). In many fungi, reactive oxygen species (ROS) have been implicated in either signaling or cell death. Upon contact with each other, *Coprinus cinereus* and *P. anserina* exhibit peroxide accumulation, and in some cases cell death. ROS in this instance appears to be a signaling molecule between species, not necessarily a toxic death effector (Silar 2005). ROS have been implicated in HI interactions in several species of grape disease-causing ascomycetes (Freitas et al. 2009), and ROS also increased during cell death due to HI in *N. crassa* (Hutchison et al. 2009). In *Colletotrichum trifolii*, dominant active Ras mutants cause an increase in intracellular ROS and subsequent cell death (Chen and Dickman 2004), and the authors later showed that the cell death phenotype could be suppressed using the antioxidant proline (Chen and Dickman 2005). Similarly, an activated Ras pathway was also shown to cause programmed cell death in *C. albicans* and *Saccharomyces cerevisiae* (Phillips et al. 2006; Hlavata et al. 2008). In some cases the chemical signal is protein-based, such as the small, secreted, anti-fungal protein (PAF) of *Pencillium chrysogenum*. This protein has been found to inhibit the growth of several other fungal species (Kaiserer et al. 2003; Marx et al. 2008), and causes a cell death phenotype in *Aspergillus nidulans* (Leiter et al. 2005; Marx et al. 2008). Thus, fungal cell death can be induced via exogenous addition of stress-inducing compounds or drugs, via chemical signaling between fungi and also by signals produced during HI.

3.2 Conservation of Classical Cell Death Genes

Though filamentous fungal cell death has phenotypic characteristics of apoptosis, fungal species often lack the complete complement of apoptotic genes that have been identified in higher eukaryotic species. In a comparative genomics study, Fedorova et al. (2005) characterized the programmed cell death machinery of the *Aspergilli* and related fungi. Filamentous fungi lack key upstream PCD components, such as members of the *BAX* and *BCL* gene families (Fedorova et al. 2005), but have conserved downstream components such as PARP (poly (ADP-ribose) polymerase), AIF1 (apoptosis inducing factor 1), AMID (AIF-homologous mitochondrion-associated inducer of death), APAF1 (apoptotic protease activating factor 1), EndoG (endonuclease G), and metacaspases (Fedorova et al. 2005; Sharon et al. 2009). Some filamentous fungal apoptosis homologs have an affect on cell death when mutated, while others do not, suggesting that cell death pathways may be regulated differently between filamentous fungi and other eukaryotic organisms, and that death programs can vary even among fungi. For example, *N. crassa* strains carrying a deletion of an AIF1 homolog were more resistant to ROS, while mutants in an AMID homolog were actually more sensitive to ROS (Castro et al. 2008). However, strains carrying deletions of the AIF1 homolog or metacaspases had no effect on cell death due to HI in *N. crassa* (Hutchison et al. 2009). In addition, metacaspase activity has been observed in *Aspergillus fumigatus* upon entry into stationary phase, but deletion of both metacaspases in *A. fumigatus* did not affect induction of PCD by a variety of cell death inducers (Mousavi and Robson 2004; Richie et al. 2007). In *A. nidulans,* exposure to sphingolipids results in cell death that shares phenotypic characteristics with apoptosis, however, the metacaspase *casA* was not necessary for cell death (Cheng et al. 2003). Interestingly, the *Aspergillus* metacaspase mutants were only sensitive to chemicals that induced endoplasmic reticulum (ER) stress and the unfolded protein response (UPR) (Richie et al. 2007). Farnesol-induced cell death in *A. nidulans*, unlike *A. fumigatus*, is affected by deletion mutations in the metacaspase genes *casA* and *casB* (Colabardini et al. 2010). In addition, mutants in an *A. nidulans* protein kinase C (*pkcA*) homolog were more resistant to farnesol-induced cell death (Savoldi et al. 2008). However, over-expression of *pkcA* increased cell death during farnesol exposure, as well as metacaspase activity and induction of the UPR (Colabardini et al. 2010). Homologs of additional apoptosis pathway genes were observed to affect farnesol-induced cell death, including a PARP homolog (Semighini et al. 2006a) and an AIF1 homolog (Savoldi et al. 2008; Dinamarco et al. 2010). In addition, increased degradation of the PARP protein, a known caspase target, correlated with an increase in metacaspase activity observed during asexual sporulation (Thrane et al. 2004). However, mutations in an *A. nidulans endoG* homolog had no effect on farnesol-induced cell death, and were instead involved in the DNA damage response (Pimentel Figueiredo et al. 2010). Thus, it is likely that farnesol induces a PCD pathway in *A. nidulans* that uses components of the classical apoptosis pathway, but that its regulation and function may be different than previously characterized systems.

3.3 Cell Death and Lifespan

In studies on longevity and lifespan in *P. anserina*, it was found that deletion of both metacaspases as well as an AMID homolog led to a significant lifespan increase (Hamann et al. 2007). A more recent study identified two AIF1 and two AMID homologs (*PaAif1-2*, *PaAmid1-2*) in the *P. anserina* genome (Brust et al. 2010a). Both PaAIF2 and PaAMID2 proteins localized to mitochondria and mutations in these genes increased resistance to hydrogen peroxide and increased lifespan (Brust et al. 2010a). Additionally, *P. anserina* homologs of cyclophilin D, another player in apoptosis, and PARP were shown to have a role in regulation of lifespan (Brust et al. 2010b; Muller-Ohldach et al. 2010). Although it is clear that apoptosis homologs play a role in fungal PCD and ageing, it is not clear whether fungi are using different cell death genes for different developmental pathways and under different death inducing conditions. Many of the differences likely arise from the fact that death is being induced in different ways in fungal PCD models, and a more standardized, comprehensive look at the effect of these genes on PCD will be useful in creating a more cohesive cell death model for fungi.

3.4 Ultrastructure of HI-Induced Cell Death

Macroscopically, the phenotype of HI-induced cell death in the interaction zone between incompatible fungal strains often consists of the formation of a barrage at the contact interface. Barrages are typically raised, pigmented or clear zones where cell death occurs (Micali and Smith 2003; Pinan-Lucarre et al. 2007). However, barrage formation is not always associated with HI and the barrage appearance can vary (Micali and Smith 2003; Aanen et al. 2010; Ikeda et al. 2011). In addition to cell death, incompatible heterokaryons typically exhibit decreased growth and lack of conidiation (Aanen et al. 2010). The ultrastructure and cellular phenotype of PCD has been characterized in detail for several filamentous fungi. In *P. anserina,* cells become highly vacuolized and re-localization of known autophagy proteins to the vacuole occurs (Pinan-Lucarre et al. 2003). Induction of autophagy by rapamycin exposure elicits a phenotype identical to that of HI (Dementhon et al. 2003). However, mutations in genes required for autophagy increase cell death, suggesting that autophagy is a response to HI and is not involved in causing cell death (Pinan-Lucarre et al. 2005). The root rot ascomycete fungus *Rosellinia necatrix* also exhibits a vacuole-based PCD phenotype, where initial steps include vacuolar membrane collapse, followed by destruction of the plasma membrane and a change in nuclear structure (Inoue et al. 2011). In *N. crassa,* HI-induced cell death consists of organelle degradation, plasmolysis, formation of membrane-bound structures in the cytoplasm (Jacobson et al. 1998), and DNA fragmentation (Marek et al. 2003). A similar cell death phenotype was observed in *C. parasitica* and included breakdown of cell structure and cytoplasmic degeneration, organelle degradation, and vacuolar membrane breakdown (Newhouse and Macdonald 1991). It is currently unclear if the phenotype observed during HI-induced cell death in different filamentous fungal species results from similar or different mechanisms of dismantling cells upon nonself recognition.

4 Molecular Mechanisms of HI

Genes involved in nonself recognition and HI have been cloned and characterized in several species, namely *P. anserina* and *N. crassa* (Table 1). In general, fungal species have approximately ten *het* loci, and these loci typically have 2–3 allelic specificities (Aanen et al. 2010). In addition, all molecularly characterized *het* loci, with the exception of *het-S*, have a non-allelic partner that is required for HI (Table 1; Aanen et al. 2010). In some cases, the non-allelic partners are maintained in linkage disequilibrium via close physical linkage (*het-c*, *pin-c*), or via recombination suppression due to inversions (*het-6*, *un-24*). However, in other cases, the non-allelic partners are not linked and self-incompatible (SI) meiotic progeny can be produced, as in the *het-R*/*het-V* system in *P. anserina.* Another feature of HI genes shared among *N. crassa* and *P. anserina* is that in almost all cases, one of the

Table 1 Summary of molecularly characterized *het* loci

Gene	Features	No. of alleles	Genetic interactions	References
Neurospora crassa				
het-c	Glycine-rich membrane protein; proteins encoded by alternate allele form heterocomplexes	3	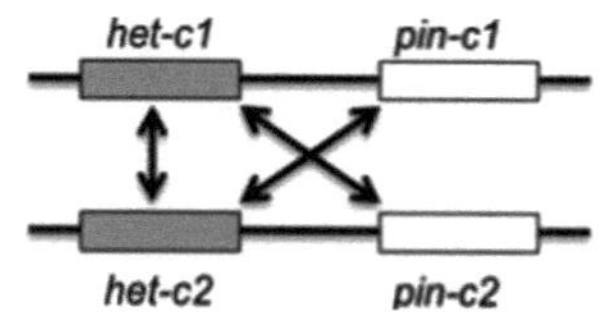	Saupe et al. (1996), Saupe and Glass (1997, Sarkar et al. (2002), Kaneko et al. (2006), and Hall et al. (2010)
pin-c	HET domain	3		
het-6	HET domain	2	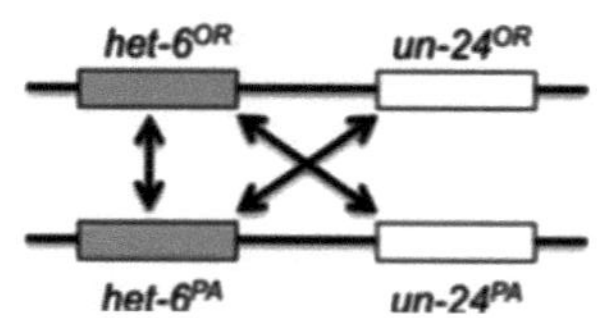	Mir-Rashed et al. (2000), Smith et al. (2000), and Micali and Smith (2006)
un-24	Ribonucleotide reductase	2		
tol	HET domain	1	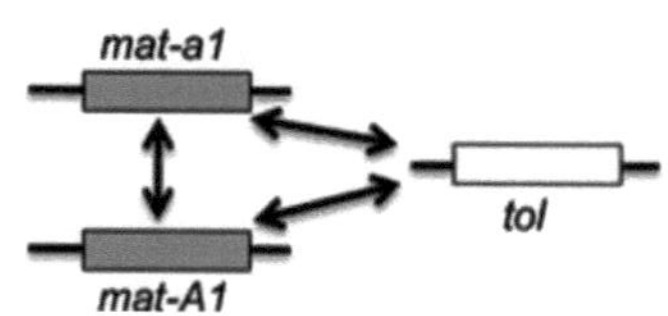	Glass et al. (1988), Jacobson (1992), and Shiu and Glass (1999)
mat	Mating type genes	2		

(continued)

Table 1 (continued)

Gene	Features	No. of alleles	Genetic interactions	References
Podospora anserina				
het-s	Prion; forms infectious amyloid structures	2	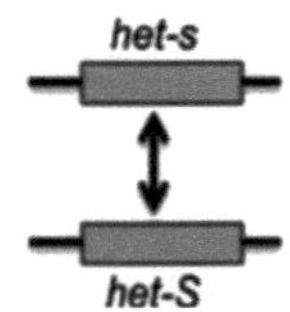	Saupe (2007)
het-r	HNWD[a]	2	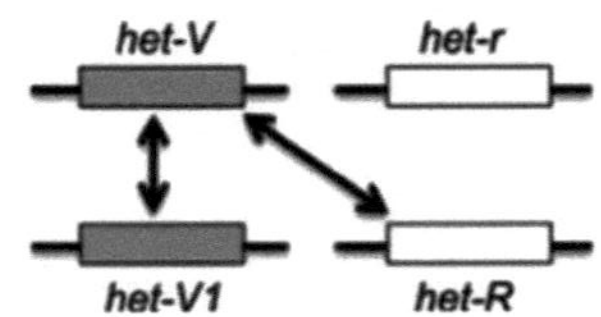	Labarere (1973, 1974) and Chevanne et al. (2009)
het-c	Glycolipid transfer protein (GLTP)	2	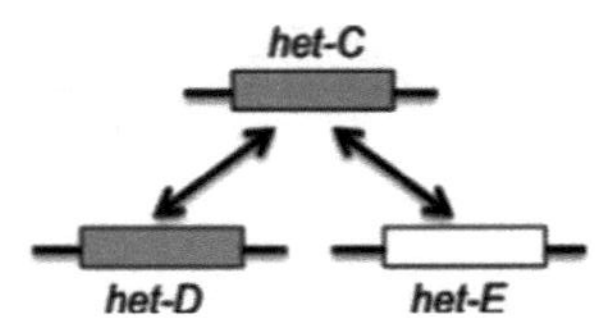	Saupe et al. (1994, 1995a, b) and Espagne et al. (1997, 2002)
het-d	HNWD	3		
het-e	HNWD	4		

[a]HNWD (contains HET, NACHT, and WD protein domains) (Paoletti et al. 2007)

non-allelic partners encodes a protein containing a HET domain (Fedorova et al. 2005; Glass and Dementhon 2006). HET domains (Pfam PF06985) are approximately 150 amino acids long, and are specific to filamentous ascomycete species. Importantly, over-expression of just the HET domain has been shown to cause cell death in *P. anserina* (Paoletti and Clave 2007).

4.1 Neurospora crassa

Neurospora crassa is a filamentous, heterothallic ascomycete that grows on burnt vegetation following forest fires. *N. crassa* has been a model system for studying HI for decades (Garnjobst 1953), and 11 *het* loci have been mapped in this fungus (Garnjobst 1953; Mylyk 1975; Perkins 1975). Three *het* loci have been molecularly analyzed, and will be discussed below.

One of the best-characterized HI loci in *N. crassa* is the *het-c/pin-c* system. The *het-c* locus encodes a glycine-rich plasma membrane protein (Saupe et al. 1996) and interacts genetically with its partner for incompatibility, *pin-c* (Kaneko et al. 2006). *pin-c* is a highly polymorphic gene that encodes a HET domain (Kaneko et al. 2006). The *het-c* and *pin-c* loci are closely linked (within 1 kbp) and exist in three different haplotypes: *het-c1 pin-c1*, *het-c2 pin-c2*, and *het-c3 pin-c3* (Wu et al. 1998; Hall et al. 2010). *het-c* and *pin-c* alleles are under balancing selection: individuals carrying different *het-c pin-c* haplotypes are equally frequent in populations, show severe linkage disequilibrium (no recombination between *het-c pin-c* haplotypes) and exhibit trans-species polymorphisms (allelic polymorphisms are maintained across speciation events, Charlesworth 2006) (Wu et al. 1998; Hall et al. 2010). *het-c* specificity is conferred by a relatively small region of ~30 amino acids (Saupe and Glass 1997; Wu and Glass 2001). While the specificity domain of *pin-c* is as yet unknown, recent population data and computational analysis have identified several candidate regions (Hall et al. 2010). Physical interaction between HET-C proteins of differing specificity has been shown (Sarkar et al. 2002), but it is as yet unclear whether HET-C and PIN-C proteins interact. While *het-c* and *pin-c* homologs are present in many filamentous fungi, these two genes have been recruited to function in HI only in *Neurospora* and closely-related genera (Hall et al. 2010). *het-c* homologs in *P. anserina, Fusarium proliferatum,* and *Aspergillus niger* do not show allelic variability (Saupe et al. 2000; Kerenyi et al. 2006; van Diepeningen et al. 2009). However, *het-c* polymorphism has been observed in *Botrytis cinerea* (Fournier et al. 2003), and in some cases transformation with an incompatible *het-c* allele from *N. crassa* triggers PCD, even if the recipient fungus does not exhibit multiple alleles for its *het-c* homolog (Saupe et al. 2000; van Diepeningen et al. 2009). Interestingly, a *het-c* homolog that may have been transferred horizontally from a filamentous fungus was identified in the bacterium *Pseudomonas syringae*, and this allele causes a PCD reaction when introduced into *N. crassa* (Wichmann et al. 2008). These observations suggest that recruitment of *het-c* in *P. syringae* offers a selective advantage when competing with filamentous fungi in the environment.

The *het* loci *het-6* and *un-24* share some similarities with *het-c* and *pin-c* in that they are present in multiple alleles (*het-6*^{PA} *un-24*^{PA} and *het-6*^{OR} *un-24*^{OR}), require allelic and non-allelic interactions and exhibit severe linkage disequilibrium (Mir-Rashed et al. 2000; Smith et al. 2000; Micali and Smith 2006; Powell et al. 2007). Linkage disequilibrium between *un-24* and *het-6* is maintained by a paracentric inversion (Micali and Smith 2006). As with most *het* loci, one of the genetic partners at this locus, *het-6*, encodes a HET domain protein. *un-24* encodes ribonucleotide reductase, a protein necessary for DNA replication. This *het* locus illustrates an interesting quality of many *het* loci, that often they encode proteins that have important cellular functions in addition to HI.

Another example of a *het* locus with dual cellular functions is the mating type locus. There are two idiomorphs of the *mat* locus, *a* and *A*, and individuals must be of opposite mating type to cross (Glass et al. 1988). However, during vegetative growth *a* and *A* function as a *het* locus, such that fusion between strains of opposite mating type induces HI and cell death (Glass et al. 1988). The non-allelic interaction partner of *mat* is *tol*, an unlinked HET domain-containing gene. Mutations in *tol* suppress mating type incompatibility (Newmeyer 1970; Shiu and Glass 1999). Interestingly, a pseudohomothallic relative of *N. crassa*, *Neurospora tetrasperma*, has a mutated, non-functional *tol* gene, such that it can exist as a natural heterokaryon containing both mating types (Jacobson 1992). In fact, introgressing a functional *N. crassa tol* gene into *N. tetrasperma* can cause mating type incompatibility, while the introgression of the *N. tetrasperma tol* allele into *N. crassa* suppresses mating type incompatibility (Jacobson 1992). Thus, in *N. crassa* HI is mediated both by filamentous fungal-specific genes (such as the HET domain containing genes), as well as conserved genes with critical developmental roles, such as *un-24* and *mat*.

4.2 Podospora anserina

P. anserina is a pseudohomothallic species with nine mapped *het* loci, five of which have been molecularly characterized, with five allelic and three non-allelic *het* interactions among the nine loci (Pinan-Lucarre et al. 2007) (Table 1). Several of the genes that mediate HI in *P. anserina* have a specific domain structure and are part of a gene family termed HNWD; these genes contain a HET domain, a NACHT domain, and WD repeats (Paoletti et al. 2007). NACHT domains are members of the STAND class of protein domains, which have NTPase activity and have been shown to be involved in programmed cell death in both animal and plant species (Leipe et al. 2004). In *P. anserina*, NTPase activity was shown to be essential for HI function (Espagne et al. 1997). WD repeats form a β-propeller structure, and typically facilitate protein–protein interactions (Smith 2008). Variation in WD repeat number has been shown to contribute to allelic specificity for all three of the HNWD genes that have been molecularly characterized (*het-d*, *het-e*, and *het-r*) (Saupe et al. 1995b; Espagne et al. 1997, 2002; Chevanne et al. 2009).

Recently, it was shown that identifying polymorphisms in genes encoding HNWD proteins at a population level could facilitate identification of *het* loci, specifically, the *het-R/r* gene (Chevanne et al. 2009). *het-R/het-V* non-allelic as well as *het-V/het-V1* allelic interactions cause HI, and this interaction is temperature sensitive (Labarere 1973, 1974). Unlike several of the *N. crassa het* loci, *het-R* and *het-V* are unlinked, which creates the possibility of creating self-incompatible meiotic (SI) progeny; these progeny germinate but exhibit growth and cell death, and can greatly facilitate mutant screens (Pinan-Lucarre et al. 2007).

het-d and *het-e* are two additional HNWD genes that have been molecularly characterized in *P. anserina*, and these genes interact with *het-c*, which encodes a glycolipid transfer protein (GLTP) (Saupe et al. 1994). *het-c* in *P. anserina* (not to be confused with *N. crassa het-c*, which encodes a glycine-rich plasma membrane protein) is also required for normal sexual spore morphogenesis (Saupe et al. 1994). There are two alleles of *het-c* in *P. anserina*, and unlike many *het* loci, a single amino acid change is sufficient to confer allelic specificity (Saupe et al. 1995a). For *het-d* and *het-e*, allelic specificity and HI function is determined by NTPase activity in the NACHT domain, and by the number of WD repeats; for both proteins, a minimum number of WD repeats is required for functionality (11 for *het-d*, 10 for *het-e*) (Espagne et al. 2002). The WD repeat regions are subject to a variety of evolutionary selection mechanisms, including repeat induced point mutation (RIP, a fungal silencing mechanism; see (Galagan and Selker 2004) for review), concerted evolution (WD repeats are exchanged between alleles and undergo sequence homogenization), and diversifying selection (Paoletti et al. 2007). The WD repeat region is highly mutable, with incompatible heterokaryons "escaping" by somatic deletions or rearrangements within this region (Chevanne et al. 2010). Thus, *P. anserina het* loci containing WD repeats have the potential to create new alleles via diversifying selection, rearrangement, or RIP, and can further modify these alleles through concerted evolution mechanisms (Paoletti et al. 2007; Chevanne et al. 2010).

The *het-s* locus in *P. anserina* is an exception to other characterized *het* loci. *het-s* incompatibility is mediated only by allelic interactions between *het-s* and *het-S* alleles (Rizet 1952), and does not involve a gene encoding a protein with a HET domain. There are two alleles at this locus, *het-s* and *het-S*, and the *het-s* allele encodes a protein that can form infectious prions (Coustou et al. 1997). Strains with *het-s* alleles exist as two phenotypes: [Het-s*], the non-prion state, and [Het-s], the infectious prion state (Saupe 2007, 2011). [Het-s*] and [Het-S] strains form compatible heterokaryons, whereas [Het-s] strains are incompatible with [Het-S] strains (Saupe 2007, 2011). The prion phenotype [Het-s] can be spontaneously acquired in a [Het-s*] strain during growth or during cytoplasmic contact, but can be lost during formation of structures that contain little cytoplasm (Saupe 2007, 2011). Structural analysis revealed that the [Het-s] prion forms highly ordered amyloid structures, but is unstructured in the non-prion [Het-s*] state (Dos Reis et al. 2002; Maddelein et al. 2002; Balguerie et al. 2003). In addition to HI, the [Het-s] prion state exhibits meiotic drive, and when crossed as a female causes abortion of some of the *het-S* containing ascospores (Aanen et al. 2010).

Through a comparison of *het* loci in *P. anserina* and *N. crassa*, certain patterns or signatures of selection are apparent. Genes encoding proteins with HET domains are key regulators of HI, and almost always form genetic interactions with other loci. These other loci encode a diverse set of proteins with diverse cellular functions, including but not limited to membrane proteins, glycolipid transfer proteins, enzymes involved in DNA synthesis, transcriptional regulators, and even prions. Strong balancing selection maintains multiple alleles for many *het* loci, yet the degree of sequence variability and the number of alleles can vary. Future experiments to clone the remaining *het* loci will provide insight into the mechanisms of genetic recognition, and will determine the degree of conservation of filamentous fungal *het* loci among different species.

5 Regulators of HI in *N. crassa* and *P. anserina*

In addition to *het* loci themselves, there are many loci that have been identified in *N. crassa* and *P. anserina* that can modify or suppress the cell death response during HI. In *P. anserina*, screens for mutations that suppress cell death in self-incompatible (SI) strains identified several modifier or "*mod*" genes (Belcour and Bernet 1969; Bernet 1971; Bernet et al. 1973). The *mod-A1* and *mod-B1* mutations exhibit no vegetative phenotypes, but the single and double deletion strains are suppressed for cell death due to non-allelic *het* interactions (cell death due to *het-s*/*het-S* allelic interactions is unaffected) (Belcour and Bernet 1969; Bernet 1971; Bernet et al. 1973; Labarere 1973). Interestingly, as for several of the *P. anserina het* loci, *mod-A1 mod-B1* mutant strains have a sexual reproduction defect, and are sterile under certain growth conditions (Barreau et al. 1998). The *mod-B* locus is not yet cloned, but *mod-A1* was found to encode a proline/leucine rich protein, with little homology to other known genes (Barreau et al. 1998). Two additional *mod* genes have been cloned and they also affect additional cellular process such as growth, pigmentation, and the sexual cycle (Loubradou et al. 1997, 1999); *mod-E* encodes an *HSP90* homolog (Loubradou et al. 1997), and *mod-D* encodes a Gα subunit of a heterotrimeric G protein (Loubradou et al. 1999). Thus, characterization of the *mod* genes suggests that in *P. anserina*, genes involved in HI are also involved in various aspects of development, particularly sexual reproduction.

For both *P. anserina* and *N. crassa*, dramatic transcriptional changes occur along with the phenotypic changes associated with HI (Boucherie et al. 1981; Bourges et al. 1998; Paoletti et al. 2001; Hutchison et al. 2009). In *P. anserina*, genes induced during incompatibility (termed *idi* genes) include a protease (*idi-6*) (Paoletti et al. 2001), an *ATG8* homolog (*idi-7*) (Pinan-Lucarre et al. 2003), and a bZIP transcription factor (*idi-4*) (Dementhon et al. 2004). All three of these genes are either homologous to a known autophagy gene, or have been shown to be involved in *P. anserina* autophagic cell death. Autophagy mutants show accelerated cell death in *P. anserina*, indicating that (i) autophagy may actually be functioning in a protective role during HI and (ii) that the autophagy-like cell death observed in *P. anserina* may actually be an alternative form of vacuolar cell death (Pinan-Lucarre et al. 2005).

In *N. crassa*, mutations in the transcription factor *vib-1* suppress cell death due to differences at *het-c pin-c*, *het-e*, *het-8*, and *mat* (Xiang and Glass 2002, 2004). In addition, *vib-1* is necessary for the expression of at least some of the *N. crassa* HET domain genes, including *het-6*, *tol*, and *pin-c* (Dementhon et al. 2006). Interestingly, like several of the *mod* genes, *vib-1* is also involved in sexual development and positively regulates protoperithecial formation (Hutchison and Glass 2010). Further, *vib-1* mutants do not produce extracellular proteases upon nitrogen or carbon starvation (Dementhon et al. 2006). Homologs of *vib-1* are present throughout the filamentous fungi (Hutchison and Glass 2010), and it is possible that its function in HI is conserved.

6 Evolutionary Implications of Genetic Recognition Programs

Allelic diversity is required to enable nonself recognition, and thus strong evolutionary pressures are predicted to function in the restriction of fusion events to genetically identical strains. Filamentous fungi whose *het* loci have been mapped typically have 8–11 (ascomycetes) or 1–4 (basidiomycetes) loci, and 2–3 alleles at each locus (Aanen et al. 2010). *het* loci are not clustered, but are scattered throughout the genome in filamentous ascomycete fungi. Thus, for a species with eight *het* loci and two alleles at each locus, there would be 2^8 or 256 different allele combinations; thus, successful fusions between strains that are self or at least very closely related are likely to be almost nonexistent in nature. When characterized phylogenetically, self/nonself recognition loci often exhibit balancing selection and occasionally exhibit trans-species polymorphism (Charlesworth 2006). Though not always the case in fungi (Milgroom and Cortesi 1999; Kauserud et al. 2006), many genetic recognition loci have been shown to exhibit this type of evolutionary signature. In *N. crassa*, both *het-c* and *pin-c* loci are under balancing selection and exhibit trans-species polymorphism (Wu et al. 1998; Hall et al. 2010). A recent study on the *het-c pin-c* locus in *Neurospora* found that this *het* locus likely arose through duplication and rearrangements that resulted in a change in gene order such that the *pin-c* became located directly adjacent to *het-c* (Hall et al. 2010). More distantly-related fungi contain a clear *het-c* ortholog and *pin-c* homologs, but it seems that HI function is restricted to those species where *het-c* and *pin-c* are adjacent and under linkage disequilibrium (Hall et al. 2010). In addition, the authors found that the *het-c pin-c* locus was originally bi-allelic, and that a third allele was created by a rare recombination event between alleles (Hall et al. 2010).

Alleles of *het-6* and *un 24*, like the *het-c pin-c* locus, are present in equal frequencies in a population, and these alleles have been maintained during speciation (Smith et al. 2000; Powell et al. 2007), as has the pericentric inversion associated with the *un-24-het-6* genomic region containing the alternative *un-24-het-6* alleles. Evolutionary simulations estimate that more than two alleles at a *het* locus would occur only if there was a constraint or limit to forming new *het* loci (Muirhead et al. 2002). In addition, the authors' model predicts that trans-species polymorphism would only be observed when the mutation rate is rather low, consistent with the fact that many *het* loci polymorphisms consists of insertions/deletions (indels) or inversions, which are relatively rare mutational events (Muirhead et al. 2002). Thus, it is likely that alleles for at least some of the *het* loci have been maintained for long periods of time, and that they are under strong evolutionary constraints.

In *P. anserina, the* family of genes that contain NACHT/WD domains (some of which also contain HET domains), undergo concerted evolution of the WD repeat domain, which vary widely in repeat number (Paoletti et al. 2007). Even pseudogenes that have been inactivated by RIP contain functional WD repeat sequences (Paoletti et al. 2007). Four residues in the WD repeat region are under positive diversifying selection, and these residues are at a predicted protein interaction surface (Paoletti et al. 2007). In this case, it seems that new alleles could arise by repeat variation or by mutations in the WD region, and these mutations could also spread to other WD repeat-containing proteins through concerted evolution. Thus, there are multiple ways to create new *het* loci, as well as to create new alleles for those loci. Further evolutionary analyses are likely to uncover additional mechanisms for creating new *het* loci and alternate *het* alleles.

7 Concluding Remarks

The ability of an individual to maintain and perceive genetic identity is a vital process for a wide variety of organisms. Research in filamentous fungi has shown both similarities and differences between fungal mechanisms of recognition with those in other eukaryotic species. Though several *het* loci have been cloned and characterized, molecular data is only available from two different species. Identification of *het* genes from other species will aid in identifying additional common genetic themes for fungal nonself recognition loci. In particular, although regions required for allelic specificity have been identified, the molecular mechanisms associated with nonself recognition have not been elucidated for any system. Genes encoding HET domain proteins have been identified as important players for both nonself recognition and PCD, but the molecular role of these proteins in either function remains elusive. Alleles at characterized *het* loci often show characteristic evolutionary signatures; further characterization of new het loci in a variety of filamentous fungi will reveal whether this characteristic extends to these nonself recognition loci. Finally, further identification and characterization of fungal HI regulatory genes will likely provide clues as to additional components of the cell death machinery as well as downstream cellular responses to PCD.

References

Aanen DK, Debets AJ, Glass NL, Saupe SJ (2010) Biology and genetics of vegetative incompatibility in fungi. In: Borkovich KA, Ebbole DJ (eds) Cellular and molecular biology of filamentous fungi. ASM Press, Washington, DC, pp 274–288

Baidyaroy D, Glynn JM, Bertrand H (2000) Dynamics of asexual transmission of a mitochondrial plasmid in *Cryphonectria parasitica*. Curr Genet 37:257–267

Balguerie A, Dos Reis S, Ritter C, Chaignepain S, Coulary-Salin B, Forge V, Bathany K, Lascu I, Schmitter JM, Riek R, Saupe SJ (2003) Domain organization and structure-function relationship of the HET-s prion protein of *Podospora anserina*. EMBO J 22:2071–2081

Barreau C, Iskandar M, Loubradou G, Levallois V, Begueret J (1998) The *mod-A* suppressor of nonallelic heterokaryon incompatibility in *Podospora anserina* encodes a proline-rich polypeptide involved in female organ formation. Genetics 149:915–926

Be'er A, Zhang HP, Florin EL, Payne SM, Ben-Jacob E, Swinney HL (2009) Deadly competition between sibling bacterial colonies. Proc Natl Acad Sci U S A 106:428–433

Be'er A, Ariel G, Kalisman O, Helman Y, Sirota-Madi A, Zhang HP, Florin EL, Payne SM, Ben-Jacob E, Swinney HL (2010) Lethal protein produced in response to competition between sibling bacterial colonies. Proc Natl Acad Sci U S A 107:6258–6263

Belcour L, Bernet J (1969) Sur la mise en evidence du'un gene dont la mtuation supprime sepcifiquement certaines manifestations d'incompatibilite chez le *Podospora anserina*. CR Acad Sci Paris 269:712–714

Bernet J (1971) Sur un cas de suppression de l'incompatibilite cellulaire chez le champignon *Podospora anserina*. CR Acad Sci Paris 273:1120–1122

Bernet J, Begueret J, Labarere J (1973) Incompatibility in the fungus *Podospora anserina*. Are the mutations abolishing the incompatibility reaction ribosomal mutations? Mol Gen Genet 124:35–50

Biella S, Smith ML, Aist JR, Cortesi P, Milgroom MG (2002) Programmed cell death correlates with virus transmission in a filamentous fungus. Proc Biol Sci 269:2269–2276

Boucherie H, Dupont CH, Bernet J (1981) Polypeptide synthesis during protoplasmic incompatibility in the fungus *Podospora anserina*. Biochim Biophys Acta 653:18–26

Bourges N, Groppi A, Barreau C, Clave C, Begueret J (1998) Regulation of gene expression during the vegetative incompatibility reaction in *Podospora anserina*. Characterization of three induced genes. Genetics 150:633–641

Brasier CM (1988) Rapid changes in genetic structure of epidemic populations of *Ophiostoma ulmi*. Nature 332:538–541

Brasier CM (1991) *Ophiostoma novo-ulmi* sp. nov., causative agent of the current Dutch elm disease pandemics. Mycopathologia 15:151–161

Brust D, Hamann A, Osiewacz HD (2010a) Deletion of *PaAif2* and *PaAmid2*, two genes encoding mitochondrial AIF-like oxidoreductases of *Podospora anserina*, leads to increased stress tolerance and lifespan extension. Curr Genet 56:225–235

Brust D, Daum B, Breunig C, Hamann A, Kuhlbrandt W, Osiewacz HD (2010b) Cyclophilin D links programmed cell death and organismal aging in *Podospora anserina*. Aging Cell 9:761–775

Burnet FM (1971) "Self-recognition" in colonial marine forms and flowering plants in relation to the evolution of immunity. Nature 232:230–235

Cadavid LF (2004) Self-discrimination in colonial invertebrates: genetic control of allorecognition in the hydroid *Hydractinia*. Dev Comp Immunol 28:871–879

Castro A, Lemos C, Falcao A, Glass NL, Videira A (2008) Increased resistance of complex I mutants to phytosphingosine-induced programmed cell death. J Biol Chem 283:19314–19321

Caten CE (1972) Vegetative incompatibility and cytoplasmic infection in fungi. J Gen Microbiol 72:221–229

Chaplin DD (2010) Overview of the immune response. J Allergy Clin Immunol 125:S3–S23

Charlesworth D (2006) Balancing selection and its effects on sequences in nearby genome regions. PLoS Genet 2:e64

Chen C, Dickman MB (2004) Dominant active Rac and dominant negative Rac revert the dominant active Ras phenotype in *Colletotrichum trifolii* by distinct signalling pathways. Mol Microbiol 51:1493–1507

Chen C, Dickman MB (2005) Proline suppresses apoptosis in the fungal pathogen *Colletotrichum trifolii*. Proc Natl Acad Sci U S A 102:3459–3464

Cheng J, Park T-S, Chio L-C, Fischl AS, Ye XS (2003) Induction of apoptosis by sphingoid long-chain bases in *Aspergillus nidulans*. Mol Cell Biol 23:163–177

Chevanne D, Bastiaans E, Debets A, Saupe SJ, Clave C, Paoletti M (2009) Identification of the *het-r* vegetative incompatibility gene of *Podospora anserina* as a member of the fast evolving HNWD gene family. Curr Genet 55:93–102

Chevanne D, Saupe SJ, Clave C, Paoletti M (2010) WD-repeat instability and diversification of the *Podospora anserina hnwd* non-self recognition gene family. BMC Evol Biol 10:134

Colabardini AC, De Castro PA, De Gouvea PF, Savoldi M, Malavazi I, Goldman MH, Goldman GH (2010) Involvement of the *Aspergillus nidulans* protein kinase C with farnesol tolerance is related to the unfolded protein response. Mol Microbiol 78:1259–1279

Correll JC, Klittich CJR, Leslie JF (1989) Heterokaryon self-incompatibility in *Gibberella fujikuroi* (*Fusarium moniliforme*). Mycol Res 93:21–27

Cortesi P, McCulloch CE, Song H, Lin H, Milgroom MG (2001) Genetic control of horizontal virus transmission in the chestnut blight fungus, *Cryphonectria parasitica*. Genetics 159:107–118

Coustou V, Deleu C, Saupe S, Begueret J (1997) The protein product of the *het-s* heterokaryon incompatibility gene of the fungus *Podospora anserina* behaves as a prion analog. Proc Natl Acad Sci USA 94:9773–9778

Debets AJM, Griffiths AJF (1998) Polymorphism of het-genes prevents resource plundering in *Neurospora crassa*. Mycol Res 102:1343–1349

Debets F, Yang X, Griffiths AJ (1994) Vegetative incompatibility in *Neurospora*: its effect on horizontal transfer of mitochondrial plasmids and senescence in natural populations. Curr Genet 26:113–119

Degterev A, Yuan J (2008) Expansion and evolution of cell death programmes. Nat Rev Mol Cell Biol 9:378–390

Dementhon K, Paoletti M, Pinan-Lucarre B, Loubradou-Bourges N, Sabourin M, Saupe SJ, Clave C (2003) Rapamycin mimics the incompatibility reaction in the fungus *Podospora anserina*. Eukaryot Cell 2:238–246

Dementhon K, Saupe SJ, Clave C (2004) Characterization of IDI-4, a bZIP transcription factor inducing autophagy and cell death in the fungus *Podospora anserina*. Mol Microbiol 53:1625–1640

Dementhon K, Iyer G, Glass NL (2006) VIB-1 is required for expression of genes necessary for programmed cell death in *Neurospora crassa*. Eukaryot Cell 5:2161–2173

Deng F, Melzer MS, Boland GJ (2002) Vegetative compatibility and transmission of hypovirulence-associated dsRNA in *Sclerotinia homoeocarpa*. Can J Plant Pathol 24:481–488

De Tomaso AW, Nyholm SV, Palmeri KJ, Ishizuka KJ, Ludington WB, Mitchel K, Weissman IL (2005) Isolation and characterization of a protochordate histocompatibility locus. Nature 438:454–459

Dinamarco TM, Pimentel Bde C, Savoldi M, Malavazi I, Soriani FM, Uyemura SA, Ludovico P, Goldman MH, Goldman GH (2010) The roles played by *Aspergillus nidulans* apoptosis-inducing factor (AIF)-like mitochondrial oxidoreductase (*AifA*) and NADH-ubiquinone oxidoreductases (*NdeA-B* and *NdiA*) in farnesol resistance. Fungal Genet Biol 47:1055–1069

Dodds PN, Rathjen JP (2010) Plant immunity: towards an integrated view of plant-pathogen interactions. Nat Rev Genet 11:539–548

Dos Reis S, Coulary-Salin B, Forge V, Lascu I, Begueret J, Saupe SJ (2002) The HET-s prion protein of the filamentous fungus *Podospora anserina* aggregates in vitro into amyloid-like fibrils. J Biol Chem 277:5703–5706

Emri T, Malnar Z, Pocsi I (2005) The appearances of autolytic and apoptotic markers are concomitant but differently regulated in carbon-starving *Aspergillus nidulans* cultures. FEMS Microbiol Lett 251:297–303

Espagne E, Balhadere P, Begueret J, Turcq B (1997) Reactivity in vegetative incompatibility of the HET-E protein of the fungus *Podospora anserina* is dependent on GTP-binding activity and a WD40 repeated domain. Mol Gen Genet 256:620–627

Espagne E, Balhadere P, Penin ML, Barreau C, Turcq B (2002) HET-E and HET-D belong to a new subfamily of WD40 proteins involved in vegetative incompatibility specificity in the fungus *Podospora anserina*. Genetics 161:71–81

Fedorova ND, Badger JH, Robson GD, Wortman JR, Nierman WC (2005) Comparative analysis of programmed cell death pathways in filamentous fungi. BMC Genomics 6:177

Fleissner A, Simonin AR, Glass NL (2008) Cell fusion in the filamentous fungus, *Neurospora crassa*. Methods Mol Biol 475:21–38

Fleissner A, Leeder AC, Roca MG, Read ND, Glass NL (2009) Oscillatory recruitment of signaling proteins to cell tips promotes coordinated behavior during cell fusion. Proc Natl Acad Sci USA 106:19387–19392

Fournier E, Levis C, Fortini D, Leroux P, Giraud T, Brygoo Y (2003) Characterization of Bc-*hch*, the *Botrytis cinerea* homolog of the *Neurospora crassa het-c* vegetative incompatibility locus, and its use as a population marker. Mycologia 95:251–261

Freitas R, Rego C, Oliveira H, Ferreira RB (2009) Interactions among grapevine disease-causing fungi. The role of reactive oxygen species. Phytopathol Mediterr 48:117–127

Galagan JE, Selker EU (2004) RIP: the evolutionary cost of genome defense. Trends Genet 20:417–423

Garnjobst L (1953) Genetic control of heterokaryosis in *Neurospora crassa*. Am J Bot 40:607–614

Gibbs KA, Urbanowski ML, Greenberg EP (2008) Genetic determinants of self identity and social recognition in bacteria. Science 321:256–259

Giovannetti M, Sbrana C, Strani P, Agnolucci M, Rinaudo V, Avio L (2003) Genetic diversity of isolates of *Glomus mosseae* from different geographic areas detected by vegetative compatibility testing and biochemical and molecular analysis. Appl Environ Microbiol 69:616–624

Giovannetti M, Avio L, Fortuna P, Pellegrino E, Sbrana C, Strani P (2006) At the root of the wood wide web: self recognition and non-self incompatibility in mycorrhizal networks. Plant Signal Behav 1:1–5

Glass NL, Dementhon K (2006) Non-self recognition and programmed cell death in filamentous fungi. Curr Opin Microbiol 9:553–558

Glass NL, Kaneko I (2003) Fatal attraction: nonself recognition and heterokaryon incompatibility in filamentous fungi. Eukaryot Cell 2:1–8

Glass NL, Vollmer SJ, Staben C, Grotelueschen J, Metzenberg RL, Yanofsky C (1988) DNAs of the two mating-type alleles of *Neurospora crassa* are highly dissimilar. Science 241:570–573

Hall C, Welch J, Kowbel DJ, Glass NL (2010) Evolution and diversity of a fungal self/nonself recognition locus. PLoS One 5:e14055

Hamann A, Brust D, Osiewacz HD (2007) Deletion of putative apoptosis factors leads to lifespan extension in the fungal ageing model *Podospora anserina*. Mol Microbiol 65:948–958

Hamann A, Brust D, Osiewacz HD (2008) Apoptosis pathways in fungal growth, development and ageing. Trends Microbiol 16:276–283

Hlavata L, Nachin L, Jezek P, Nystrom T (2008) Elevated Ras/protein kinase A activity in *Saccharomyces cerevisiae* reduces proliferation rate and lifespan by two different reactive oxygen species-dependent routes. Aging Cell 7:148–157

Hutchison EA, Glass NL (2010) Meiotic regulators Ndt80 and Ime2 have different roles in Saccharomyces and Neurospora. Genetics 185:1271–1282

Hutchison E, Brown S, Tian CG, Glass NL (2009) Transcriptional profiling and functional analysis of heterokaryon incompatibility in *Neurospora crassa* reveals that reactive oxygen species, but not metacaspases, are associated with programmed cell death. Microbiology 155:3957–3970

Hyakumachi M, Ui T (1987) Non-self-anastomosing isolates of *Rhizoctonia solani* obtained from fields of sugarbeet monoculture. Trans Br Mycol Soc 89:155–159

Ikeda K, Inoue K, Nakamura H, Hamanaka T, Ohta T, Kitazawa H, Kida C, Kanematsu S, Park P (2011) Genetic analysis of barrage line formation during mycelial incompatibility in *Rosellinia necatrix*. Fungal Biol 115:80–86

Inoue K, Kanematsu S, Park P, Ikeda K (2011) Cytological analysis of mycelial incompatibility in *Rosellinia necatrix*. Fungal Biol 115:87–95

Jacobson DJ (1992) Control of mating type heterokaryon incompatibility by the tol gene in *Neurospora crassa* and *N. tetrasperma*. Genome 35:347–353

Jacobson D, Beurkens K, Klomparens K (1998) Microscopic and ultrastructural examination of vegetative incompatibility in partial diploids heterozygous at *het* loci in *Neurospora crassa*. Fungal Genet Biol 23:45–56

Kaiserer L, Oberparleiter C, Weiler-Gorz R, Burgstaller W, Leiter E, Marx F (2003) Characterization of the *Penicillium chrysogenum* antifungal protein PAF. Arch Microbiol 180:204–210

Kaneko I, Dementhon K, Xiang Q, Glass NL (2006) Nonallelic interactions between *het-c* and a polymorphic locus, *pin-c*, are essential for nonself recognition and programmed cell death in *Neurospora crassa*. Genetics 172:1545–1555

Kauserud H, Saetre GP, Schmidt O, Decock C, Schumacher T (2006) Genetics of self/nonself recognition in *Serpula lacrymans*. Fungal Genet Biol 43:503–510

Kerenyi Z, Olah B, Jeney A, Hornok L, Leslie JF (2006) The homologue of *het-c* of *Neurospora crassa* lacks vegetative compatibility function in *Fusarium proliferatum*. Appl Environ Microbiol 72:6527–6532

Kvell K, Cooper EL, Engelmann P, Bovari J, Nemeth P (2007) Blurring borders: innate immunity with adaptive features. Clin Dev Immunol 2007:83671

Labarere J (1973) Properties of an incompatibility system in *Podospora anserina* fungus and value of this system for the study of incompatibility. C R Acad Sci Hebd Seances Acad Sci D 276:1301–1304

Labarere J (1974) Incompatibility in *Podospora anserina*: comparative properties of the antagonistic cytoplasmic factors of a nonallelic system. J Bacteriol 120:854–860

Lane EB, Carlile MJ (1979) Post-fusion somatic incompatibility in plasmodia of *Physarum polycephalum*. J Cell Sci 35:339–354

Langford ML, Atkin AL, Nickerson KW (2009) Cellular interactions of farnesol, a quorum-sensing molecule produced by *Candida albicans*. Future Microbiol 4:1353–1362

Leipe DD, Koonin EV, Aravind L (2004) STAND, a class of P-loop NTPases including animal and plant regulators of programmed cell death: multiple, complex domain architectures, unusual phyletic patterns, and evolution by horizontal gene transfer. J Mol Biol 343:1–28

Leiter E, Szappanos H, Oberparleiter C, Kaiserer L, Csernoch L, Pusztahelyi T, Emri T, Pocsi I, Salvenmoser W, Marx F (2005) Antifungal protein PAF severely affects the integrity of the plasma membrane of *Aspergillus nidulans* and induces an apoptosis-like phenotype. Antimicrob Agents Chemother 49:2445–2453

Li SI, Purugganan MD (2010) The cooperative amoeba: *Dictyostelium* as a model for social evolution. Trends Genet 27:48–54

Liu P, Luo L, Guo JH, Liu HM, Wang BQ, Deng BX, Long CA, Cheng YJ (2010) Farnesol induces apoptosis and oxidative stress in the fungal pathogen *Penicillium expansum*. Mycologia 102:311–318

Loubradou G, Begueret J, Turcq B (1997) A mutation in an HSP90 gene affects the sexual cycle and suppresses vegetative incompatibility in the fungus *Podospora anserina*. Genetics 147:581–588

Loubradou G, Begueret J, Turcq B (1999) MOD-D, a Galpha subunit of the fungus *Podospora anserina*, is involved in both regulation of development and vegetative incompatibility. Genetics 152:519–528

Lu BCK (2006) Programmed cell death in fungi. In: Kues U, Fischer R (eds) The mycota: growth, differentiation, and sexuality, I. Springer, New York, pp 167–187

Lu BCK, Gallo N, Kues U (2003) White-cap mutants and meiotic apoptosis in the basidiomycete *Coprinus cinereus*. Fungal Genet Biol 39:82–93

MacIntyre DL, Miyata ST, Kitaoka M, Pukatzki S (2010) The *Vibrio cholerae type* VI secretion system displays antimicrobial properties. Proc Natl Acad Sci USA 107:19520–19524

Maddelein ML, Dos Reis S, Duvezin-Caubet S, Coulary-Salin B, Saupe SJ (2002) Amyloid aggregates of the HET-s prion protein are infectious. Proc Natl Acad Sci USA 99:7402–7407

Maheshwari R, Navaraj A (2008) Senescence in fungi: the view from *Neurospora*. FEMS Microbiol Lett 280:135–143

Marek SM, Wu J, Glass NL, Gilchrist DG, Bostock RM (2003) Nuclear DNA degradation during heterokaryon incompatibility in *Neurospora crassa*. Fungal Genet Biol 40:126–137

Marx F, Binder U, Leiter E, Pocsi I (2008) The *Penicillium chrysogenum* antifungal protein PAF, a promising tool for the development of new antifungal therapies and fungal cell biology studies. Cell Mol Life Sci 65:445–454

McKitrick TR, De Tomaso AW (2010) Molecular mechanisms of allorecognition in a basal chordate. Semin Immunol 22:34–38

Micali CO, Smith ML (2003) On the independence of barrage formation and heterokaryon incompatibility in *Neurospora crassa*. Fungal Genet Biol 38:209–219

Micali CO, Smith ML (2006) A nonself recognition gene complex in *Neurospora crassa*. Genetics 173:1991–2004

Miguelez EM, Hardisson C, Manzanal MB (1999) Hyphal death during colony development in *Streptomyces antibioticus*: morphological evidence for the existence of a process of cell deletion in a multicellular prokaryote. J Cell Biol 145:515–525

Miguelez EM, Hardisson C, Manzanal MB (2000) Streptomycetes: a new model to study cell death. Int Microbiol 3:153–158

Milgroom MG (1999) Viruses in fungal populations. In: Worrall JJ (ed) Structure and dynamics of fungal populations. Kluwer Academic Publishers, Dordrecht, pp 283–305

Milgroom MG, Cortesi P (1999) Analysis of population structure of the chestnut blight fungus based on vegetative incompatibility genotypes. Proc Natl Acad Sci USA 96:10518–10523

Milgroom MG, Cortesi P (2004) Biological control of chestnut blight with hypovirulence: a critical analysis. Annu Rev Phytopathol 42:311–338

Mir-Rashed N, Jacobson DJ, Dehghany MR, Micali CO, Smith ML (2000) Molecular and functional analyses of incompatibility genes at *het-6* in a population of *Neurospora crassa*. Fungal Genet Biol 30:197–205

Mousavi SAA, Robson GD (2003) Entry into the stationary phase is associated with a rapid loss of viability and an apoptotic-like phenotype in the opportunistic pathogen *Aspergillus fumigatus*. Fungal Genet Biol 39:221–229

Mousavi SAA, Robson GD (2004) Oxidative and amphotericin B-mediated cell death in the opportunistic pathogen *Aspergillus fumigatus* is associated with an apoptotic-like phenotype. Microbiology 150:1937–1945

Muirhead CA, Glass NL, Slatkin M (2002) Multilocus self-recognition systems in fungi as a cause of trans-species polymorphism. Genetics 161:633–641

Muller-Ohldach M, Brust D, Hamann A, Osiewacz HD (2010) Overexpression of *PaParp* encoding the poly(ADP-ribose) polymerase of *Podospora anserina* affects organismal aging. Mech Ageing Dev 132:33–42

Mylyk OM (1975) Heterokaryon incompatibility genes in *Neurospora crassa* detected using duplication-producing chromosome rearrangements. Genetics 80:107–124

Nauta M, Hoekstra RF (1994) Evolution of vegetative incompatibility in filamentous ascomycetes I. Deterministic models. Evolution 48:979–995

Newhouse JR, Macdonald WL (1991) The ultrastructure of hyphal anastomoses between vegetatively compatible and incompatible virulent and hypovirulent strains of *Cryphonectria parasitica*. Can J Bot 69:602–614

Newmeyer D (1970) A suppressor of the heterokaryon-incompatibility associated with mating type in *Neurospora crassa*. Can J Genet Cytol 12:914–926

Nicotra ML, Powell AE, Rosengarten RD, Moreno M, Grimwood J, Lakkis FG, Dellaporta SL, Buss LW (2009) A hypervariable invertebrate allodeterminant. Curr Biol 19:583–589

Nuss DL (2010) Mycoviruses. In: Borkovich KA, Ebbole DJ (eds) Cellular and molecular biology of filamentous fungi. ASM Press, Washington, DC, pp 145–152

Paoletti M, Clave C (2007) The fungus-specific HET domain mediates programmed cell death in *Podospora anserina*. Eukaryot Cell 6:2001–2008

Paoletti M, Saupe SJ (2009) Fungal incompatibility: evolutionary origin in pathogen defense? Bioessays 31:1201–1210
Paoletti M, Castroviejo M, Begueret J, Clave C (2001) Identification and characterization of a gene encoding a subtilisin-like serine protease induced during the vegetative incompatibility reaction in *Podospora anserina*. Curr Genet 39:244–252
Paoletti M, Buck KW, Brasier CM (2006) Selective acquisition of novel mating type and vegetative incompatibility genes via interspecies gene transfer in the globally invading eukaryote *Ophiostoma novo-ulmi*. Mol Ecol 15:249–262
Paoletti M, Saupe SJ, Clave C (2007) Genesis of a fungal non-self recognition repertoire. PLoS One 2:e283
Perkins DD (1975) The use of duplication-generating rearrangements for studying heterokaryon incompatibility genes in *Neurospora*. Genetics 80:87–105
Phillips AJ, Crowe JD, Ramsdale M (2006) Ras pathway signaling accelerates programmed cell death in the pathogenic fungus *Candida albicans*. Proc Natl Acad Sci USA 103:726–731
Pimentel Figueiredo BD, de Castro PA, Dinamarco TM, Goldman MH, Goldman GH (2010) The *Aspergillus nidulans* $nucA^{EndoG}$ homologue is not involved in cell death. Eukaryot Cell 10:276–283
Pinan-Lucarre B, Paoletti M, Dementhon K, Coulary-Salin B, Clave C (2003) Autophagy is induced during cell death by incompatibility and is essential for differentiation in the filamentous fungus *Podospora anserina*. Mol Microbiol 47:321–333
Pinan-Lucarre B, Balguerie A, Clave C (2005) Accelerated cell death in *Podospora* autophagy mutants. Eukaryot Cell 4:1765–1774
Pinan-Lucarre B, Paoletti M, Clave C (2007) Cell death by incompatibility in the fungus *Podospora*. Semin Cancer Biol 17:101–111
Powell AJ, Jacobson DJ, Natvig DO (2007) Ancestral polymorphism and linkage disequilibrium at the *het-6* region in pseudohomothallic *Neurospora tetrasperma*. Fungal Genet Biol 44:896–904
Ramsdale M (2008) Programmed cell death in pathogenic fungi. Biochim Biophys Acta 1783:1369–1380
Richie DL, Miley MD, Bhabhra R, Robson GD, Rhodes JC, Askew DS (2007) The *Aspergillus fumigatus* metacaspases *CasA* and *CasB* facilitate growth under conditions of endoplasmic reticulum stress. Mol Microbiol 63:591–604
Rizet G (1952) Les phenomenes de barrage chez *Podospora anserina*. I. Analyse genetique des barrages entre les souches S et s. Rev Cytol Biol Veg 13:51–92
Robson GD (2006) Programmed cell death in the aspergilli and other filamentous fungi. Med Mycol 44:S109–S114
Ronald PC, Beutler B (2010) Plant and animal sensors of conserved microbial signatures. Science 330:1061–1064
Rosa SF, Powell AE, Rosengarten RD, Nicotra ML, Moreno MA, Grimwood J, Lakkis FG, Dellaporta SL, Buss LW (2010) *Hydractinia* allodeterminant *alr1* resides in an immunoglobulin superfamily-like gene complex. Curr Biol 20:1122–1127
Sanabria N, Goring D, Nurnberger T, Dubery I (2008) Self/nonself perception and recognition mechanisms in plants: a comparison of self-incompatibility and innate immunity. New Phytol 178:503–514
Sarkar S, Iyer G, Wu J, Glass NL (2002) Nonself recognition is mediated by HET-C heterocomplex formation during vegetative incompatibility. EMBO J 21:4841–4850
Saupe SJ (2007) A short history of small s: a prion of the fungus *Podospora anserina*. Prion 1:110–115
Saupe SJ (2011) The [Het-s] prion of *Podospora anserina* and its role in heterokaryon incompatibility. Semin Cell Dev Biol. doi:10.1016/j.semcdb.2011.02.019
Saupe SJ, Glass NL (1997) Allelic specificity at the *het-c* heterokaryon incompatibility locus of *Neurospora crassa* is determined by a highly variable domain. Genetics 146:1299–1309

Saupe SJ, Descamps C, Turcq B, Begueret J (1994) Inactivation of the *Podospora anserina* vegetative incompatibility locus *het-c*, whose product resembles a glycolipid transfer protein, drastically impairs ascospore production. Proc Natl Acad Sci USA 91:5927–5931

Saupe SJ, Turcq B, Begueret J (1995a) Sequence diversity and unusual variability at the *het-c* locus involved in vegetative incompatibility in the fungus *Podospora anserina*. Curr Genet 27:466–471

Saupe SJ, Turcq B, Begueret J (1995b) A gene responsible for vegetative incompatibility in the fungus *Podospora anserina* encodes a protein with a GTP-binding motif and G beta homologous domain. Gene 162:135–139

Saupe SJ, Kuldau GA, Smith ML, Glass NL (1996) The product of the *het-c* heterokaryon incompatibility gene of *Neurospora crassa* has characteristics of a glycine-rich cell wall protein. Genetics 143:1589–1600

Saupe SJ, Clave C, Sabourin M, Begueret J (2000) Characterization of *hch*, the *Podospora anserina* homolog of the *het-c* heterokaryon incompatibility gene of *Neurospora crassa*. Curr Genet 38:39–47

Savoldi M, Malavazi I, Soriani FM, Capellaro JL, Kitamoto K, da Silva Ferreira ME, Goldman MHS, Goldman GH (2008) Farnesol induces the transcriptional accumulation of the *Aspergillus nidulans* Apoptosis-Inducing Factor (AIF)-like mitochondrial oxidoreductase. Mol Microbiol 70:44–59

Sbrana C, Nuti MP, Giovannetti M (2007) Self-anastomosing ability and vegetative incompatibility of *Tuber borchii* isolates. Mycorrhiza 17:667–675

Semighini CP, Savoldi M, Goldman GH, Harris SD (2006a) Functional characterization of the putative *Aspergillus nidulans* poly(ADP-ribose) polymerase homolog *PrpA*. Genetics 173:87–98

Semighini CP, Hornby JM, Dumitru R, Nickerson KW, Harris SD (2006b) Farnesol-induced apoptosis in *Aspergillus nidulans* reveals a possible mechanism for antagonistic interactions between fungi. Mol Microbiol 59:753–764

Semighini CP, Murray N, Harris SD (2008) Inhibition of *Fusarium graminearum* growth and development by farnesol. FEMS Microbiol Lett 279:259–264

Sharon A, Finkelstein A, Shlezinger N, Hatam I (2009) Fungal apoptosis: function, genes and gene function. FEMS Microbiol Rev 33:833–854

Shiu PK, Glass NL (1999) Molecular characterization of *tol*, a mediator of mating-type-associated vegetative incompatibility in *Neurospora crassa*. Genetics 151:545–555

Silar P (2005) Peroxide accumulation and cell death in filamentous fungi induced by contact with a contestant. Mycol Res 109:137–149

Smith TF (2008) Diversity of WD-repeat proteins. Subcell Biochem 48:20–30

Smith ML, Micali OC, Hubbard SP, Mir-Rashed N, Jacobson DJ, Glass NL (2000) Vegetative incompatibility in the *het-6* region of *Neurospora crassa* is mediated by two linked genes. Genetics 155:1095–1104

Strassmann JE, Zhu Y, Queller DC (2000) Altruism and social cheating in the social amoeba *Dictyostelium discoideum*. Nature 408:965–967

Thrane C, Kaufmann U, Stummann BM, Olsson S (2004) Activation of caspase-like activity and poly (ADP-ribose) polymerase degradation during sporulation in *Aspergillus nidulans*. Fungal Genet Biol 41:361–368

Umar MH, Van Griensven LJLD (1997) Morphogenetic cell death in developing primordia of *Agaricus bisporus*. Mycologia 89:274–277

Umar MH, Van Griensven LJLD (1998) The role of morphogenetic cell death in the histogenesis of the mycelial cord of *Agaricus bisporus* and in the development of macrofungi. Mycol Res 102:719–735

van Diepeningen AD, Debets AJ, Hoekstra RF (1997) Heterokaryon incompatibility blocks virus transfer among natural isolates of black *Aspergilli*. Curr Genet 32:209–217

van Diepeningen AD, Pal K, van der Lee TA, Hoekstra RF, Debets AJ (2009) The *het-c* heterokaryon incompatibility gene in *Aspergillus niger*. Mycol Res 113:222–229

Velicer GJ, Vos M (2009) Sociobiology of the myxobacteria. Annu Rev Microbiol 63:599–623

Wichmann G, Sun J, Dementhon K, Glass NL, Lindow SE (2008) A novel gene, *phcA* from *Pseudomonas syringae* induces programmed cell death in the filamentous fungus *Neurospora crassa*. Mol Microbiol 68:672–689

Wicklow DT, Horn BW (2007) Association between vegetative compatibility and aflatoxin production by *Aspergillus* species during intraspecific competition. Mycoscience 48:267–273

Worrall JJ (1997) Somatic incompatibility in basidiomycetes. Mycologia 89:24–36

Wu J, Glass NL (2001) Identification of specificity determinants and generation of alleles with novel specificity at the *het-c* heterokaryon incompatibility locus of *Neurospora crassa*. Mol Cell Biol 21:1045–1057

Wu J, Saupe SJ, Glass NL (1998) Evidence for balancing selection operating at the *het-c* heterokaryon incompatibility locus in a group of filamentous fungi. Proc Natl Acad Sci USA 95:12398–12403

Xiang Q, Glass NL (2002) Identification of *vib-1*, a locus involved in vegetative incompatibility mediated by *het-c* in *Neurospora crassa*. Genetics 162:89–8101

Xiang Q, Glass NL (2004) The control of mating type heterokaryon incompatibility by *vib-1*, a locus involved in *het-c* heterokaryon incompatibility in *Neurospora crassa*. Fungal Genet Biol 41:1063–1076

Yuan JY, Kroemer G (2010) Alternative cell death mechanisms in development and beyond. Genes Dev 24:2592–2602

Part II
Interorganismic Communication

Communication and Differentiation in the Development of Yeast Colonies

Zdena Palková and Libuse Váchová

Abstract In addition to complex natural biofilms, colonies of different yeast species represent multicellular communities that possess a specific internal organization. Cells within colonies are able to differentiate to specialized cell types that perform specific functions at specific positions. Primitive cell-tissues are thus created, the formation of which is dependent on cell–cell interactions and the transmission of signals within the colony. In addition, colonies can behave as independent multicellular entities, producing signals that enable them to mutually synchronize their development when it occurs within the same territory. As a consequence, colonies synchronously adapt to changing environments and they gain a greater capacity to exploit remnant nutrients. In this review, we summarize the current knowledge concerning cell specialization and signaling within different kinds of yeast colonies and the known aspects of communication among individual colonies.

1 Introduction

When growing under natural environmental conditions, microorganisms, including yeast, prefer to exist within multicellular communities that are attached to various surfaces and that possess specific organization. Examples are colonies and biofilms growing on solid supports, pellicles (flors) of flor yeast formed at the liquid-air

Z. Palková (✉)
Department of Genetics and Microbiology, Faculty of Science,
Charles University in Prague, Viničná 5, 128 44 Prague 2, Czech Republic
e-mail: zdenap@natur.cuni.cz

L. Váchová
Division of Cell and Molecular Microbiology, Institute of Microbiology of the ASCR,
v.v.i., Vídeňská 1083, 142 20 Prague 4, Czech Republic

Department of Genetics and Microbiology, Faculty of Science,
Charles University in Prague, Viničná 5, 128 44 Prague 2, Czech Republic
e-mail: vachova@biomed.cas.cz

G. Witzany (ed.), *Biocommunication of Fungi*, DOI 10.1007/978-94-007-4264-2_9,

boundary and flocks of flocculating yeasts formed in a liquid environment. Each of these structures can be found in natural environments, such as in fruits (yeast colonies), sherry wines (flors of top-fermenting yeast), lager beers (flocks of bottom-fermenting yeasts) and catheters (pathogenic yeast biofilms). In addition, yeast mats growing on low-concentration agar and yeast stalks growing upward from microcavities in hard agar have been investigated (Reynolds et al. 2008; Scherz et al. 2001). Although they differ in behavior, all of these communities gain multiple specific properties that are not present in individual yeast. These include the capacity for cell–cell coordination when, for example, the population synchronously changes behavior in response to changing surroundings. Cells within communities are better protected than are individual yeast cells and they differentiate into subpopulations, which fulfill specific functions. This is particularly important when the population encounters nutrient limitations (a common situation in natural settings) and the survival of individual yeast cells becomes secondary to the survival of a population.

In contrast to liquid cultivations, oriented gradients of low-molecular-weight compounds and gases can exist and may be involved in cell–cell signaling and regulation of the development of surface-attached yeast communities. Yeasts are able to produce a variety of either volatile compounds (e.g., volatile esters, ammonia, CO_2 and H_2S) or other diffusible compounds. Of these compounds, fusel and aromatic alcohols function as regulators of pseudohyphae formation in *Saccharomyces cerevisiae* (Dickinson 1996; Chen and Fink 2006); farnesol, farnesoic acid and tyrosol function in quorum sensing and the regulation of cell dimorphic transition (i.e., the transition between oval budding cells, pseudohyphae or true hyphae) (reviewed in Kruppa 2009); and CO_2 triggers filamentous growth and the white-opaque switch of *Candida albicans* (Hall et al. 2010; Huang et al. 2009). Only few pieces of information are currently available regarding the involvement of diffusible molecules in regulating the development of multicellular structures. These include the effects of alcohols (mainly tyrosol and farnesol) on biofilm development and of CO_2 and NH_3 on yeast colony development (Hall et al. 2010; Palkova and Vachova 2003).

2 Colony Morphology and Development Are Affected by Environmental Stimuli

Organized colonies can originate either from a single cell (microcolony) or from a drop of a cell suspension (giant colony). In both cases, colony formation includes an initial phase of exponential division of cells (Meunier and Choder 1999) that do not differ markedly from each other. After this phase, a colony morphology that is characteristic for a particular strain, but that varies with a varying environment, is formed. There are two possible ways of how colony morphology is affected by the environment: (i) a transient environmental impact causes the colony morphology to change when the environment is changed; however, cells from altered colonies adopt the original phenotype when they are then grown again under the original conditions (see Sect. 2.1). (ii) An environmental stimulus supports a stable change in some colonial cells that subsequently form colonies with a different morphology even under the original growth conditions. Such a morphological transition (i.e., phenotypic colony switching) is stable and is transferred to the yeast progeny (see Sects. 2.2, 2.3 and 2.4). The switch back toward the original colony phenotype is possible, but it usually requires a distinct environmental stimulus from the forward switch.

2.1 The Environment Transiently Affects S. cerevisiae Colony Morphology and Properties

Although it exhibits complex internal organization, the typical colony morphology of most *S. cerevisiae* laboratory strains is smooth. However, strains freshly isolated from natural settings form colonies with remarkable morphology even in the laboratory. They are referred to as fluffy (Kuthan et al. 2003), wrinkled (Suzuki et al. 1991), complex (Granek and Magwene 2010) or biofilm (Vachova et al. 2011) colonies. The biofilm designation refers to the fact that the colonies possess attributes common to fungal biofilms, including adhesion to substrates and the production of extracellular matrix (ECM) (Kuthan et al. 2003; Stovicek et al. 2010; Vachova et al. 2011). The term "biofilm colony" is used in this review. Formation of a structured colony phenotype is often associated with the formation of pseudohyphae; however, biofilm colonies composed exclusively of oval yeast cells also exist (Stovicek et al. 2010).

Colony architecture is influenced by the environment (i.e., nutrients, temperature, humidity, osmotic pressure, etc.). For example, both smooth and biofilm colonies grown on soft agar are much flatter than those on hard agar, which indicates changes in colony organization. The RAS-cAMP-Protein kinase A (PKA) pathway is important for this agar-concentration response (Scherz et al. 2001), which supports the prediction that the response of colonies to the environment is genetically regulated and not only caused by physical conditions.

The formation of a biofilm colony architecture that is generally more sensitive to environmental impacts is markedly influenced by nutrient sources. Low carbon levels in the presence of abundant nitrogen support the formation of a structured morphology, while high concentrations of fermentable sugars result in a smooth or less-structured morphology (Granek and Magwene 2010). The participating signaling pathways include the MAP kinase and RAS-PKA pathways, both of which respond to glucose levels. As the Flo11p cell surface adhesin is regulated by these pathways (Robertson and Fink 1998) and the absence of Flo11p abolishes the formation of structured colonies (Stovicek et al. 2010), nutrients can affect the Flo11p level and thus the colony architecture. In addition, a role for the Rim15p protein kinase in integrating signals from at least three major nutrient signaling pathways (RAS-PKA, Sch9p and TOR) in biofilm colony development has also been proposed (Granek and Magwene 2010).

2.2 The Domestication of *S. cerevisiae*

In contrast to the transient formation of the smooth colony morphology induced by nutrient availability, the stable domestication of natural *S. cerevisiae* strains can occur, after which some cells start to form smooth colonies. This domestication is accompanied by a complex reprogramming of the gene expression profile (Kuthan et al. 2003), indicating a complex change of the yeast lifestyle. What is the reason for such colony domestication? Biofilm colonies are well equipped to survive in unfavorable natural environments (see Sect. 4.1), as they produce abundant protective ECM that is rich in polysaccharides (Vachova et al. 2011). However, polysaccharide production is energetically demanding. Therefore, it is advantageous to switch production off when not needed, such as when colonies are protected against drying and toxic compounds (i.e., under conditions that are met in laboratory cultivation). Signals that induce and regulate the process of domestication remain a mystery. A small amount of available data suggest that epigenetic regulation mechanisms linked to chromatin remodeling may be involved (Kuthan et al. 2003).

2.3 White-Opaque Switching in *C. albicans*

The most studied example of phenotypic switching is the transition between the white and opaque colony-forming phenotypes of the human fungal pathogen *C. albicans* (Slutsky et al. 1987). The two phenotypes differ not only in their colony morphology but also in their cell morphology, metabolic state, mating capacity and virulence. The "white" cells, similar to common laboratory strain cells, form hemispherical white colonies. The significantly larger "opaque" bean-shaped cells form large, flat, grey colonies that can be selectively stained red with phloxine B (Anderson and Soll 1987). White-opaque switching is epigenetic (Zordan et al. 2007) and its frequency is significantly affected by environmental signals. Only white cells that are homozygous at the mating type locus (a or α) can switch to opaque cells and only opaque cells of opposite mating types can mate (Soll 2004).

The signals influencing the white-opaque and opaque-white switching involve environmental factors, such as the temperature and the concentrations of CO_2 and other compounds. The opaque phenotype is relatively stable at 24°C, while temperatures above 30°C induce the switch to the white phenotype in more than 90% of cells (Slutsky et al. 1987). White cells predominate in the blood stream, which has a physiological temperature of 37°C, while opaque cells more efficiently colonize the skin surface, an environment with lower temperature. The skin also facilitates mating (Lachke et al. 2003).

Oxidants, CO_2 and N-acetylglucosamine are inducers of white-opaque switching. CO_2, at concentrations of 5% or 20% (the physiological concentration in animal hosts), represents a signal that induces white-opaque switching at both 24°C and 37°C independent of the presence of O_2 and stabilizes the mating-competent opaque phenotype (Huang et al. 2009). N-acetylglucosamine strongly induces switching at 25°C and even more so at 37°C. In the gastrointestinal tract, *C. albicans* cells encounter high CO_2 concentrations and N-acetylglucosamine, which are products of bacterial microbiota (Huang et al. 2010). These signals can thus promote mating in this tract.

A model of the network regulating white-opaque switching includes the *WOR1*, *WOR2*, *EFG1* and *CZF1* genes, with *WOR1* (a positive regulator of the opaque state) being the main player. In rare cells of the white population, *WOR1* expression begins to increase, which then activates a positive-feedback loop. The net effect is increased expression levels of *CZF1*, *WOR2* and *WOR1* and a parallel decreased expression level of *EFG1* (a negative regulator of *WOR1* and a positive regulator of the white state) (Zordan et al. 2007). This model anticipates that the switching is stochastic and does not require an environmental signal. Wor1p, however, can also function as the main player in environmental sensing (Huang et al. 2010). It can be phosphorylated by PKA via the Ras1p/cAMP pathway, which is involved in N-acetylglucosamine induction and partially involved in low level CO_2 induction. Each of the inducers, however, also functions through pathways that have not yet been identified. In addition to transcriptional regulation, chromatin-modifying enzymes can modulate white-opaque switching (Hnisz et al. 2009).

2.4 Phenotypic Switching in Different Yeasts

Strains capable of switching among different colony phenotypes have been described in *S. cerevisiae*, *C. albicans*, *C. parapsilosis*, *C. tropicalis* and *Cryptococcus neoformans*. Some strains of *C. albicans* spontaneously switch between at least seven colony phenotypes with a frequency of approximately 10^{-4} (Slutsky et al. 1985). The switch to a more structured colony phenotype is usually connected to a dimorphic transition and increases in cell adherence and pathogenicity when colonizing the host. CO_2 has been reported to induce cell filamentation and, thus, the formation of structured colonies in *C. albicans*. This CO_2-mediated signaling involves Cyr1p adenylyl cyclase, which contains a catalytic domain with a CO_2 receptor site that is critical for CO_2 sensing (Hall et al. 2010).

3 Signaling and Interaction Between Individual Colonies

Various data indicate that the development of yeast colonies is guided and/or affected by signals that are spread among populations that occur in the same territory. In general, these signals should first somehow distinguish the particular population from its neighbors (see Sect. 3.1) and then guide the development of individual colonies with respect to external impacts (both from the neighbors and the environment) (see Sect. 3.2). In addition, these signals can participate in synchronizing colony development, as documented by the synchronous activation of gene expression and protein production in colonies growing in a group (Minarikova et al. 2001; Vachova et al. 2009a). Thus, complex interactions may occur in nature where unrelated species are nearby and when both universal and specific communications can take effect among the colonies.

3.1 Ammonia, an Attractant Involved in the Unification of Young Microcolonies

A mature colony behaves as a primitive multicellular organism composed of cells with specific functions localized at specific positions (see Sect. 4). This evokes questions regarding from which developmental point the colony starts to behave as one entity and how this process is regulated. Some hints come from the study on Σ1278-derived laboratory strains of *S. cerevisiae* that form structured colonies similar to biofilm colonies. When young microcolonies of these strains grow in near proximity, they gain the capacity to merge into more numerous colony (Vopalenska et al. 2010). The microcolony unification starts with a dimorphic transition of cells located at the colony margins adjacent to neighboring colonies and proceeds with subsequent oriented growth of these cells toward their neighbors (Fig. 1Ba). Oriented pseudohyphal growth can be induced by a gradient of volatile ammonia, which is produced by young microcolonies. When microcolony unification occurs, a joint colony (composed of two or more original microcolonies) then develops as one entity regarding its morphology and organization. However, if adjacent colonies come into physical contact after forming the first wrinkles, they remain separated and each colony develops as its own entity (Fig. 1Bb). Although this is difficult to prove because of the absence of pseudohyphal growth, the existence of early phase ammonia production in laboratory strain colonies (Palkova et al. 1997) suggests that the same colony demarcation principle may also be valid in smooth colonies. Development of either joined or separated Ato1p-GFP colonies (Vachova et al. 2009a) synchronizing Ato1p-GFP production according to their mutual distances supports this prediction (Fig. 1Cc).

3.2 Ammonia, a Quorum-Sensing Signal for Metabolic Reprogramming in Aging Colonies

In addition to a low-dosage ammonia gradient playing a role in young microcolonies, volatile ammonia is an important signaling molecule in aging colonies of various species (Gori et al. 2007; Palkova et al. 1997). The high-level ammonia production in colonies is preceded by an acidic period of development that lasts several days (Fig. 1A). During this period, most of the nutrients from the

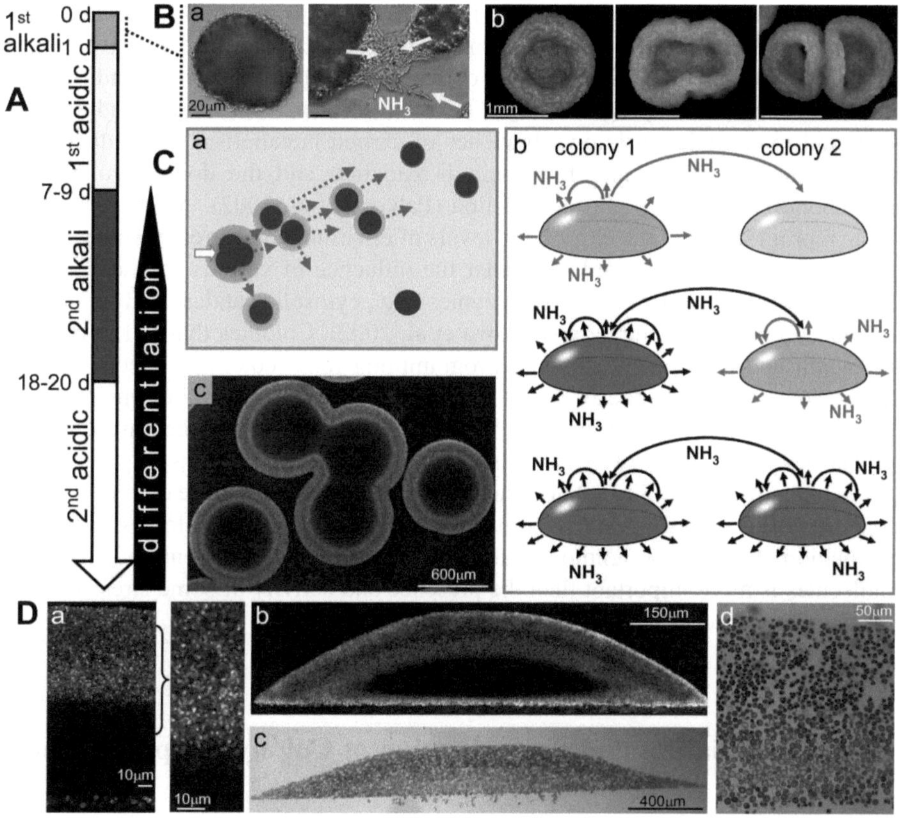

Fig. 1 Signaling and differentiation of *S. cerevisiae* colonies. (**A**) Timeline of smooth giant colony development on a complex respiratory medium. (**B**) Unification of young Σ1267 microcolonies evoked by ammonia. (*a*) Oriented pseudohyphae among neighboring microcolonies (*right*) are not visible in a solitary microcolony (*left*) (20 h after the cell inoculation). (*b*) Microcolonies unify according to their distance: a colony arising from a solitary microcolony (*left*), from two close microcolonies (*middle*) or from two more distant microcolonies (*right*). (**C**) Ammonia signaling in aged colonies. (*a*) Synchronization of ammonia production. The *black & white arrow marks* colonies starting the production of ammonia, which induces (*dotted arrows*) surrounding colonies (*black circles*) to begin producing ammonia. (*b*) Quorum-sensing mechanism of ammonia production; NH_3 acts as an auto-inducer of its own production. (*c*) Synchronized production of Ato1p-GFP in neighboring colonies; a bottom view of the colony is shown by confocal microscopy (Adapted from Vachova et al. 2009a, Fig. 4b). (**D**) Smooth colony differentiation. Only thick-surface and thin-bottom cell layers produce Ato1p (Adapted from Vachova et al. 2009a, Fig. 2c) (*b*) or sporulate (Figures courtesy of S. Honigberg, University of Missouri-Kansas City, USA; 6-day-old colony of wild *S. cerevisiae* YPS128 strain was grown on rich acetate medium as described in Piccirillo et al. 2010) (c); higher magnifications of central parts of particular colonies (*a* and *d*), respectively. A side colony view is shown

surroundings are consumed and the colonies start to starve; simultaneously, the level of intracellular stress (e.g., the level of ROS, reactive oxygen species) increases. Then, the first ammonia molecules are released and volatilized; ammonia production then quickly intensifies via a “ping-pong” auto-induction mechanism resembling quorum sensing (Fig. 1Cb) (Palkova and Forstova 2000). In this manner, all colonies in the same territory synchronously produce relatively high amounts of ammonia. In parallel, the colonies enter the alkali phase and perform extensive metabolic reprogramming. This includes changes in amino acid metabolism and in some lesser-known branches of carbon metabolism, the induction of peroxisome biogenesis and fatty acid β-oxidation and the down-regulation of mitochondrial oxidative phosphorylation (Palkova et al. 2002). Along with metabolic reprogramming, the expression levels of environmental stress response genes (i.e., genes that are up-regulated under the influence of various stresses) and the activities of various stress defense enzymes (e.g., cytosolic catalase and superoxide dismutase) gradually decrease (Palkova et al. 2002). Colonies that fail to produce ammonia and activate metabolic reprogramming (e.g., colonies defective in the pleiotropic regulator Sok2p) cannot escape increasing stress and die prematurely (Vachova et al. 2004). However, the absence of some stress defense enzymes crucial for the survival of yeast cells in liquid culture cultivations (e.g., cytosolic superoxide dismutase Sod1p) does not cause any problems for the colonies. Taken together, these data suggest that for the proper development of colonies, metabolic adaptation evoked by ammonia and very likely other unidentified signaling molecules is more important than the presence and activity of some stress-defense mechanisms for directly removing ROS (Cap et al. 2009).

4 Differentiation and Specialization of Colony Subpopulations

Long ago, microscopy studies showed that colonies are not formed from a disordered mound of more or less uniform cells; rather, they are composed of layers of cells that differ in morphology (Lindegren and Hamilton 1944; Pisova 1934). These observations suggested that both short-distance (cell–cell) and long-distance (gradients of signaling molecules) intra-colonial signals should exist that regulate colony differentiation. However, only recently implemented techniques have allowed researchers to obtain a detailed picture of the 3-D architecture of *S. cerevisiae* colonies and have started to reveal differences in the differentiation and the internal cell organizations of smooth and biofilm colonies (Piccirillo et al. 2010; Vachova et al. 2009a, 2011).

4.1 Differentiation and Protection of Biofilm Colonies of Wild S. cerevisiae Strains

From the beginning, a biofilm colony develops as a 3-D structure (Fig. 2) with elongated cells forming the cell chains at the colony bottom (Vachova et al. 2011). These cells function as roots that attach the colony to the substratum. Starting at

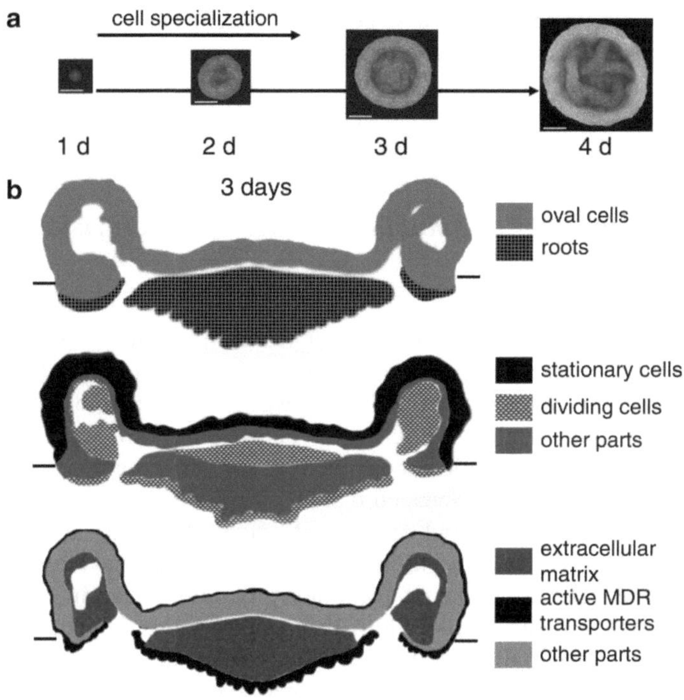

Fig. 2 Biofilm colony development and cell specialization. (**a**) Timeline of colony morphology development. Bar, 500 μm. (**b**) Schematic of colony differentiation (*side view*, based on the data from confocal microscopy) (Vachova et al. 2011). Roots are formed by pseudohyphae; *ECM* protective extracellular matrix, *MDR* multicellular drug resistance pumps. *Black line*, agar surface

approximately 35 h, a cavity free of cells is formed within the colony. The entire structure is strengthened by long, thin fibers interconnecting the cells both in the aerial colony parts and in the roots. These fibers, the presence of which is dependent on Flo11p adhesin, seem to contribute to structure formation and stability. Moreover, specialized cell subpopulations playing roles in colony protection develop at specific positions (Fig. 2). First, early in the development, stationary cells are formed near the colony surface and are oriented toward the free air. These cells are more resistant to stresses and environmental impact than are young dividing cells. Second, the entire colony is covered by cell layers with active multidrug resistance transporters Pdr5p and Snq2p, which are able to expel toxic compounds out of yeast cells. These cells are present in the uppermost layers of the stationary air-oriented cells but are also in the dividing tips of the pseudohyphae invading the agar. Third, the colony interior is protected by low permeable ECM, which begins to be produced in parallel with the appearance of the internal cavity. ECM production by the internal cells could contribute to the formation of the cavity. Well-protected new cell generations are then formed by the division of internal cells, which can later partially fill the cavity. This internal colony organization is typical for biofilm colonies formed by a variety of different non-isogenic *S. cerevisiae* natural strains, which indicates the universality of the described principles (Vachova et al. 2011).

4.2 Differentiation of Smooth Colonies of *S. cerevisiae* Laboratory Strains

From the beginning, smooth colonies are flatter than biofilm colonies, their cells are closely packed together and they contain neither intercellular fibers nor ECM (Kuthan et al. 2003; Scherz et al. 2001). However, smooth colonies also exhibit clear internal organization. Studies of center-outer colony differentiation (Vachova et al. 2009b) and microscopy analyses of vertical colony cross-sections (Vachova et al. 2009a) have revealed that at least three main cell subpopulations differing in the production and activity of specific proteins and in their fate are formed during colony transition to the second alkali developmental phase (Fig. 1A). The cells localized at the colony margin ("margin" cells) exhibit features of adapted vital cells, while the central cell subpopulation includes dying cells and cells harboring high levels of ROS and stress-related enzyme activities (Vachova et al. 2009b; Vachova and Palkova 2005). However, these central cells are not homogeneous; they diversify into cells that produce specific proteins that are localized in the upper and lower colony regions. Thus, only upper cells produce the Ato1p ammonium transporter, similarly to "margin" cells (Fig. 1Da, b) (Vachova et al. 2009a).

Another example demonstrating the differentiation in smooth colonies is the sporulation pattern observed in colonies formed by diploid *S. cerevisiae* strains (Lindegren and Hamilton 1944; Piccirillo et al. 2010). In colonies grown on acetate agar, sporulation begins within two cell layers, one inside the colony and the second near the agar at the colony bottom. Later, sporulation expands up to the colony surface, with a sharp boundary between the thick upper layer of sporulating cells and the lower layer of nonsporulating cells (Fig. 1Dc, d) (Piccirillo et al. 2010). The sporulation pattern has been shown to be approximately the same in colonies of different laboratory strains, including more structured colonies of the Σ1267-derived strain and in colonies of *S. cerevisiae* and *S. paradoxus* natural isolates. The layering is similar to that identified for Ato1p (Vachova et al. 2009a) (Fig. 1D). In more structured colonies, sporulation also proceeds efficiently in the subsurface roots of colonies invading the agar (Piccirillo and Honigberg 2010).

As in a biofilm colony, subpopulations within a smooth colony contribute to the protection of the entire colony. However, the particular mechanisms are different. Thus, a protective layer of the uppermost cells covers the colony like a "skin", in which cells are connected via their cell wall proteins. This skin blocks the inward penetration of some compounds that could be harmful to the colony (Vachova et al. 2009a).

4.3 How Is Colony Differentiation Regulated?

As shown above, different kinds of specialized cells are formed either dependently or independently of the environment within colonies. Thus, diverse differentiation programs that colonies can enter are both predetermined genetically/epigenetically (e.g., biofilm versus domesticated colonies) and affected by the current environment (e.g., the available nutrients). Usually, these programs begin after colonies exit the initial "exponential" growth phase (Meunier and Choder 1999). In addition, internal cell diversification is coordinated in neighboring colonies (Minarikova et al. 2001; Vachova et al. 2009a) (Fig. 1Cc) and the resulting pattern is typical for the particular strain and conditions. These observations imply that multiple regulatory mechanisms exist and participate in multiple differentiation programs. Currently, the mechanisms regulating internal colony organization are mostly unknown. Examples of data from some individual processes are summarized below.

The timing of ammonia signaling correlates with the time when distinct cell subpopulations are demarcated within the smooth colony (Cap et al. 2009; Vachova et al. 2009a). The inability of ammonia signaling leads to the absence of this kind of colony differentiation (e.g., colonies of *sok2*Δ or *sod2*Δ strains). These findings suggest a relationship between the extracellular ammonia signal and the activation of a particular differentiation program; however, the mechanism is unknown. Importantly, unprotonated ammonia (but not ammonium) functions as a signaling molecule in colonies (Palkova and Forstova 2000). This implies the existence of an intracellular ammonia "receptor" because ammonia can penetrate into cells by diffusion (Palkova and Vachova 2003). The initiation of ammonia production correlates with nutrient depletion and an increase in ROS and the absence of the mitochondrial superoxide dismutase Sod2p abrogates ammonia signaling and colony differentiation. Thus, changes in the homeostasis of ROS and/or Sod2p may participate in regulatory events leading to ammonia signaling (Cap et al. 2009). A potential role for ammonia in colony differentiation is additionally supported by expressive morphological changes during the transition of *C. mogii* colonies to the alkali phase. When induced by volatile ammonia, a smooth *C. mogii* colony composed of pseudohyphae changes to an aerial structure, in which pseudohyphae decompose into individual oval cells that form wrinkles (Palkova and Forstova 2000).

Studies on chimeric colonies consisting of two mixed strains have revealed a role for alkali signals (and/or perhaps CO_2) sensed through the Rim101 pathway in the regulation of colony sporulation patterns (Piccirillo et al. 2010). In particular, this pathway is involved in extending sporulation away from internal colony cells to the colony surface (see Sect. 4.2). In addition, the transmission of signals among the two cell types in chimeric colony has been observed, which demonstrates the existence of cell–cell communication among neighboring cells (Piccirillo et al. 2010).

Significant changes in the distribution of dead cells in colonies of a *whi2*-deleted strain (*WHI2* encodes a regulatory protein involved in cell proliferation and the stress response) and the suppression of this phenotype by deletion of the *TPK3* gene encoding a catalytic component of PKA suggest a role for cAMP-PKA signaling in the localization of dying cells within these colonies (Leadsham et al. 2009).

5 Conclusions and Further Perspectives

As demonstrated above, different colony developmental programs exist that lead to different colony organizations and behaviors. In addition, these programs can switch from one to the other and back under the guidance of key regulators that have not yet been identified. Some, but not all, of the programs also reflect environmental conditions. While nothing is yet known about the molecular mechanisms regulating wild-to-domesticated transitions, studies of switching systems in pathogenic yeast (e.g., white-opaque switching) have already revealed some potential regulators. Nevertheless, the current knowledge is far from complete and known regulatory pathways are still fragmentary. Thus, yeast colonies provide challenges and opportunities for the future. In fact, they could become good models for studies of cell differentiation and other "multicellular processes" (including cell–cell interactions and signaling), just as individual yeast cells long ago became indispensable for investigations of principles of some basic cellular processes. From a more practical point of view, uncovering the molecular principles of yeast multicellular behavior is crucial for formulating new ideas of how to protect against microbial communities, such as fungal biofilms occupying wet surfaces and pathogenic yeasts invading host organisms. This can then aid in the improvement of prevention strategies and therapies for mycoses in immunocompromised patients.

Acknowledgements The work was supported by grants from the Grant Agency of the Czech Republic 204/08/0718 and from the Ministry of Education LC531, Research Concepts MSM0021620858 and AV0Z50200510.

References

Anderson JM, Soll DR (1987) Unique phenotype of opaque cells in the white-opaque transition of *Candida albicans*. J Bacteriol 169:5579–5588

Cap M, Vachova L, Palkova Z (2009) Yeast colony survival depends on metabolic adaptation and cell differentiation rather than on stress defense. J Biol Chem 284:32572–32581

Chen H, Fink GR (2006) Feedback control of morphogenesis in fungi by aromatic alcohols. Genes Dev 20:1150–1161

Dickinson JR (1996) 'Fusel' alcohols induce hyphal-like extensions and pseudohyphal formation in yeasts. Microbiology 142:1391–1397

Gori K, Mortensen HD, Arneborg N, Jespersen L (2007) Ammonia production and its possible role as a mediator of communication for *Debaryomyces hansenii* and other cheese-relevant yeast species. J Dairy Sci 90:5032–5041

Granek JA, Magwene PM (2010) Environmental and genetic determinants of colony morphology in yeast. PLoS Genet 6:e1000823

Hall RA, De Sordi L, Maccallum DM, Topal H, Eaton R, Bloor JW, Robinson GK, Levin LR, Buck J, Wang Y, Gow NA, Steegborn C, Muhlschlegel FA (2010) CO(2) acts as a signalling molecule in populations of the fungal pathogen *Candida albicans*. PLoS Pathog 6:e1001193

Hnisz D, Schwarzmuller T, Kuchler K (2009) Transcriptional loops meet chromatin: a dual-layer network controls white-opaque switching in *Candida albicans*. Mol Microbiol 74:1–15

Huang G, Srikantha T, Sahni N, Yi S, Soll DR (2009) CO(2) regulates white-to-opaque switching in *Candida albicans*. Curr Biol 19:330–334
Huang G, Yi S, Sahni N, Daniels KJ, Srikantha T, Soll DR (2010) N-acetylglucosamine induces white to opaque switching, a mating prerequisite in *Candida albicans*. PLoS Pathog 6: e1000806
Kruppa M (2009) Quorum sensing and *Candida albicans*. Mycoses 52:1–10
Kuthan M, Devaux F, Janderova B, Slaninova I, Jacq C, Palkova Z (2003) Domestication of wild *Saccharomyces cerevisiae* is accompanied by changes in gene expression and colony morphology. Mol Microbiol 47:745–754
Lachke SA, Lockhart SR, Daniels KJ, Soll DR (2003) Skin facilitates *Candida albicans* mating. Infect Immun 71:4970–4976
Leadsham JE, Miller K, Ayscough KR, Colombo S, Martegani E, Sudbery P, Gourlay CW (2009) Whi2p links nutritional sensing to actin-dependent Ras-cAMP-PKA regulation and apoptosis in yeast. J Cell Sci 122:706–715
Lindegren CC, Hamilton E (1944) Autolysis and sporulation in the yeast colony. Bot Gaz 105:316–321
Meunier JR, Choder M (1999) *Saccharomyces cerevisiae* colony growth and ageing: biphasic growth accompanied by changes in gene expression. Yeast 15:1159–1169
Minarikova L, Kuthan M, Ricicova M, Forstova J, Palkova Z (2001) Differentiated gene expression in cells within yeast colonies. Exp Cell Res 271:296–304
Palkova Z, Forstova J (2000) Yeast colonies synchronise their growth and development. J Cell Sci 113:1923–1928
Palkova Z, Vachova L (2003) Ammonia signaling in yeast colony formation. Int Rev Cytol 225:229–272
Palkova Z, Janderova B, Gabriel J, Zikanova B, Pospisek M, Forstova J (1997) Ammonia mediates communication between yeast colonies. Nature 390:532–536
Palkova Z, Devaux F, Ricicova M, Minarikova L, Le Crom S, Jacq C (2002) Ammonia pulses and metabolic oscillations guide yeast colony development. Mol Biol Cell 13:3901–3914
Piccirillo S, Honigberg SM (2010) Sporulation patterning and invasive growth in wild and domesticated yeast colonies. Res Microbiol 161:390–398
Piccirillo S, White MG, Murphy JC, Law DJ, Honigberg SM (2010) The Rim101p/PacC pathway and alkaline pH regulate pattern formation in yeast colonies. Genetics 184:707–716
Pisova M (1934) Anatomy of yeast colonies. Rozpravy II Tridy Ceske Akad 154:1–13 (in Czech)
Reynolds TB, Jansen A, Peng X, Fink GR (2008) Mat formation in *Saccharomyces cerevisiae* requires nutrient and pH gradients. Eukaryot Cell 7:122–130
Robertson LS, Fink GR (1998) The three yeast A kinases have specific signaling functions in pseudohyphal growth. Proc Natl Acad Sci USA 95:13783–13787
Scherz R, Shinder V, Engelberg D (2001) Anatomical analysis of *Saccharomyces cerevisiae* stalk-like structures reveals spatial organization and cell specialization. J Bacteriol 183:5402–5413
Slutsky B, Buffo J, Soll DR (1985) High-frequency switching of colony morphology in *Candida albicans*. Science 230:666–669
Slutsky B, Staebell M, Anderson J, Risen L, Pfaller M, Soll DR (1987) "White-opaque transition": a second high-frequency switching system in *Candida albicans*. J Bacteriol 169:189–197
Soll DR (2004) Mating-type locus homozygosis, phenotypic switching and mating: a unique sequence of dependencies in *Candida albicans*. Bioessays 26:10–20
Stovicek V, Vachova L, Kuthan M, Palkova Z (2010) General factors important for the formation of structured biofilm-like yeast colonies. Fungal Genet Biol 47:1012–1022
Suzuki T, Miyamae Y, Ishida I (1991) Variation of colony morphology and chromosomal rearrangement in *Candida tropicalis* pK233. J Gen Microbiol 137:161–167
Vachova L, Palkova Z (2005) Physiological regulation of yeast cell death in multicellular colonies is triggered by ammonia. J Cell Biol 169:711–717
Vachova L, Devaux F, Kucerova H, Ricicova M, Jacq C, Palkova Z (2004) Sok2p transcription factor is involved in adaptive program relevant for long term survival of *Saccharomyces cerevisiae* colonies. J Biol Chem 279:37973–37981

Vachova L, Chernyavskiy O, Strachotova D, Bianchini P, Burdikova Z, Fercikova I, Kubinova L, Palkova Z (2009a) Architecture of developing multicellular yeast colony: spatio-temporal expression of Ato1p ammonium exporter. Environ Microbiol 11:1866–1877
Vachova L, Kucerova H, Devaux F, Ulehlova M, Palkova Z (2009b) Metabolic diversification of cells during the development of yeast colonies. Environ Microbiol 11:494–504
Vachova L, Stovicek V, Hlavacek O, Chernyavskiy O, Stepanek L, Kubinova L, Palkova Z (2011) Flo11p, drug efflux pumps, and the extracellular matrix cooperate to form biofilm yeast colonies. J Cell Biol 194:679–687
Vopalenska I, Stovicek V, Janderova B, Vachova L, Palkova Z (2010) Role of distinct dimorphic transitions in territory colonizing and formation of yeast colony architecture. Environ Microbiol 12:264–277
Zordan RE, Miller MG, Galgoczy DJ, Tuch BB, Johnson AD (2007) Interlocking transcriptional feedback loops control white-opaque switching in *Candida albicans*. PLoS Biol 5:e256

Hyphal Interference: Self Versus Non-self Fungal Recognition and Hyphal Death

Philippe Silar

Abstract Hyphal interference has been described as a mechanism that triggers hyphal death when two mycelia from different species meet. Although this phenomenon is encountered in a large array of species, it is little studied and many questions remain unanswered. Still mysterious is how fungi are able to differentiate self from non-self, what is(are) the toxic substance(s) that promote hyphal death and whether Hyphal Interference is an ancient phenomenon conserved in fungi or is, as many other fungal traits, the result of convergent evolution. Here, I review what is known about Hyphal Interference and propose that the genetically tractable model fungi *Podospora anserina* and *Coprinopsis cinerea* could be used to gain further insight into the phenomenon.

1 Introduction

Fungi are found in nearly all inhabitable biotopes on earth. In some of these, such as topsoils, they may constitute the dominant life form in term of living biomass and metabolic activity. As all the other organisms on earth, they live in constant competition and must secure their niche. Fungal hyphae have the ability to efficiently invade the growth substrate and it is thus frequent that mycelia from different individuals come into contact. When these mycelia belong to the same species they may fuse (anasotomose) and, depending upon the genetic constitution of the two partners, the fusion may give rise to an heterokaryotic mycelium or fail to persist due to some incompatibility. Although anastomoses between closely related species have been

P. Silar (✉)
UFR des Sciences du Vivant, Université Paris Diderot, Sorbonne Paris Cité,
75205 Paris Cedex 13, France

Institut de Génétique et Microbiologie, CNRS UMR 8621,
Université Paris-Sud, Bat. 400 UPS, 91405 Orsay Cedex, France
e-mail: philippe.silar@igmors.u-psud.fr

G. Witzany (ed.), *Biocommunication of Fungi*, DOI 10.1007/978-94-007-4264-2_10,

described (Ainsworth and Rayner 1989; Melzer et al. 2002; Roca et al. 2004), usually when mycelia from two different species meet they do not fuse (see for example Giovannetti et al. 1999). However, they often interact in a complex manner to keep a foothold on their substrate. Hyphal Interference (HI) that has been described as the death of hyphae from one species promoted by the meeting of those from another species (Ikediugwu and Webster 1970a) is one such interaction.

2 What Is Hyphal Interference?

2.1 *Discovery and Early Studies of Hyphal Interference*

HI was discovered when Ikediugwu and Webster (1970a) undertake to account for the decrease of sporulation of *Ascobolus crenulatus* and *Pilobolus crystallinus* in the presence of *Coprinus heptemerus*, which had been previously observed (Harper and Webster 1964). These three fungi are frequently encountered on herbivore dung (coprophilous fungi). Ikediugwu and Webster showed that abnormal sporulation was not due to the presence of an antagonistic substance that would persist in the growth medium (antibiosis) or the absence of essential nutrients that would be taken up more efficiently by *C. heptemerus* (exploitation competition), but could rather be explained by the fact that hyphae from *A. crenulatus* and *P. crystallinus* died a few minutes after meeting those of *C. heptemerus*. They then coined the term of Hyphal Interference to describe this kind of cell death. When observed under the microscope, HI required the contact between the hyphae of the two fungi. By overlaying cultures from one species by disks of cellophane onto which cultures of the second species had grown, they could obtain some hyphal death. They argued that a diffusible chemical was responsible for HI, the HI substance. HI would thus not require direct contact *per se*. However, they were unable to identify the substance. In addition, when HI involved six other basidiomycetes species, no diffusible compound could be detected by the same cellophane test (Ikediugwu and Webster 1970b). We now know that fungal hyphae may penetrate and cross through cellophane (Brun et al. 2009). It thus cannot be excluded that *C. heptemerus* breached cellophane and that HI requires actual hyphal contact to be triggered. Death would then be caused by toxic HI substance(s) released only after hyphal contact. This is supported by the fact that extension of *A. crenulatus* hyphae was not diminished until after a contact was made with *C. heptemerus* ones.

In their first report on HI, Ikediugwu and Webster (1970a) also reported that living hyphae of *C. heptemerus* were necessary for HI to proceed, as those killed by propylene oxide vapour did not trigger HI. Finally, they showed that HI between *A. crenulatus* and *C. heptemerus* occurred on a wide range of substrate and was insensitive to the carbon/nitrogen ratio, to the contrary of mycoparasitism, whose efficiency is often affected by this ratio. Further work showed that HI was triggered by the basidiomycete *A. heptemerus* on a variety of ascomycetes and that the ascomycete *A. crenulatus* was sensitive to all tested basidiomycetes, including coprophiles and wood decomposers (Ikediugwu and Webster 1970b). HI was also observed between basidiomycetes (Ikediugwu and Webster 1970b). This enabled to range fungi according to their HI ability and sensitivity (Ikediugwu and Webster 1970b). Both were inversely correlated, i.e., those able to kill the greatest number of species were the most resistant to HI. Interestingly, the two fungi having the strongest HI ability were those having the slowest growth rates. Ikediugwu and Webster proposed then that HI was one of the tools that some fungi may possess to win the competition when encountering a contestant, especially the slow growing ones that would this way be able to expand their territory and possibly access the nutrients released from the dying hyphae (Ikediugwu and Webster 1970a, b).

In their studies, Ikediugwu and Webster reported that only basidiomycetes appeared able to trigger HI and further early works evidenced additional basidiomycetes as HI promoting fungi (Ainsworth and Rayner 1991; Ikediugwu and Webster 1970b; Traquair and McKeen 1977). The modality of hyphal death after contact seems conserved in the reported cases. Interestingly, in the case of HI of *Datronia mollis* caused by *Phanerochaete magnoliae*, the hyphae of both species at first died, even if *P. magnoliae* is able in the long run to invade *D. mollis* fruiting bodies (Ainsworth and Rayner 1991). The hyphae of *D. mollis* may also coil around those of *P. magnoliae*, as observed at early stages in some mycoparasitic interactions.

2.2 HI in the Ascomycete *Podospora anserina*

More recent works suggest that ascomycetes are also able to trigger HI or HI-like phenomena on several fungi including both ascomycetes and basidiomycetes. (Cox and Scherm 2006; Perello et al. 2002; Silar 2005). Among them *Podospora anserina*, is able to both trigger HI on *Penicillium chrysogenum* and suffer HI by *Coprinopsis cinerea* (Silar 2005). *P. anserina* is a coprophilous fungus that is used as a model in several laboratories. Because of its efficient forward and reverse genetics, it can be easily used to understand the molecular basis of HI. Note that we have not yet been able to detect HI associated with other popular ascomycete models such as *Neurospora crassa, Sordaria macrospora* and *Aspergillus nidulans* (P. Silar and S. Brun, unpublished data).

The discovery of HI in *P. anserina* originated from the observation that the genomes of several filamentous fungi contain genes coding the catalytic subunit of NADPH oxidases (Nox, Lalucque and Silar 2003). These enzymes are well known as being involved in defence against pathogens in both metazoan and plants (Bedard et al. 2007). In both groups, the encounter with pathogens triggers an oxidative burst, whose origin is in part due to Nox. In fungi, Nox are known to be involved in development (Lara-Ortiz et al. 2003; Malagnac et al. 2004). Several ascomycete and basidiomycete fungi were evaluated for their ability to present an oxidative burst when encountering other microorganisms, i.e., bacteria, yeasts and moulds (Silar 2005). Two species, the basidiomycete *Coprinopsis cinerea* and

P. anserina, were shown to present localized "oxidative burst" when touching another filamentous fungus. The "burst" was visualized by accumulation of oxidized diaminobenzidine (DAB) in the presence of horseradish peroxidase, which precipitates as a reddish pigment. In *P. anserina*, it was triggered only by living hyphae and was elicited by all tested fungi. On the contrary, neutral barrier such as glass or plastic, bacteria and yeasts were unable to elicit any DAB precipitation above the level presented by an unchallenged mycelium. In *C. cinerea*, a similar "burst" was elicited not only by filamentous fungi, but also by soil-dwelling actinobacteria (*Streptomyces* and *Corynebacterium*).

Hyphal death at the contact point between fungal contestants was assayed by a trypan blue assay, as trypan blue stains dead cells. Although the oxidative burst was always present when *P. anserina* and *C. cinerea* encountered another fungus, death was observed only in some combinations of contestants. No death was observed with *Fusarium oxysporum* as challenger against all tested fungi. Death of *P. chrysogenum* hyphae was detected when in contact with those of *P. anserina* but not with those of *C. cinerea*. When confronted with *C. cinerea*, death of *P. anserina* hyphae was observed. Therefore, *P. anserina* may both promote HI or be subjected to HI, depending upon the contestant that it encounters. As reported for *P. magnoliae* (Ainsworth and Rayner 1991), coiling of *P. anserina* hyphae around the contestant ones can be observed when confronted with several species, but not all. Figure 1a illustrates the HI on *Chateomium globosum* exerted by *P. anserina*, which is associated with coiling. Coiling may exist but is not clearly evidenced during interaction between *P. anserina* and *P. chrysogenum* (S. Brun 2011, personal communication). Because of its striking similarity with HI, the hyphal death caused by *P. anserina* was assimilated to some form of HI. Other ascomycetes, including *Xylaria hypoxylon* and *Penicillium lilacinus*, trigger similar reaction (Cox and Scherm 2006; Perello et al. 2002).

As it will be developed in the following sections, HI appears to present several characteristics conserved in both basidiomycetes and ascomycetes, suggesting that it is an ancient characteristic. However, we cannot exclude that, like many fungal traits, it has evolved repeatedly by convergence or that it may differ in substantial ways in the various species. Of interest is the recent report of an HI-like mechanism in *Roselinia necatrix*, whereby incompatible strain die upon contact without cell fusion unlike what is observed in classical vegetative incompatibility reactions (Inoue et al. 2011). This is the first example, whereby hyphal death is exhibited by the meeting of strains from the same species, which may broaden the occurrence of HI in nature. It remains to be seen whether this is an exception or a more general phenomenon.

2.3 What Distinguishes HI from Other Fungal Antagonism Mechanisms?

Antagonistic behaviours between fungal mycelia have been known for a long time (DeVay 1956) and are described for numerous species inhabiting various substrata

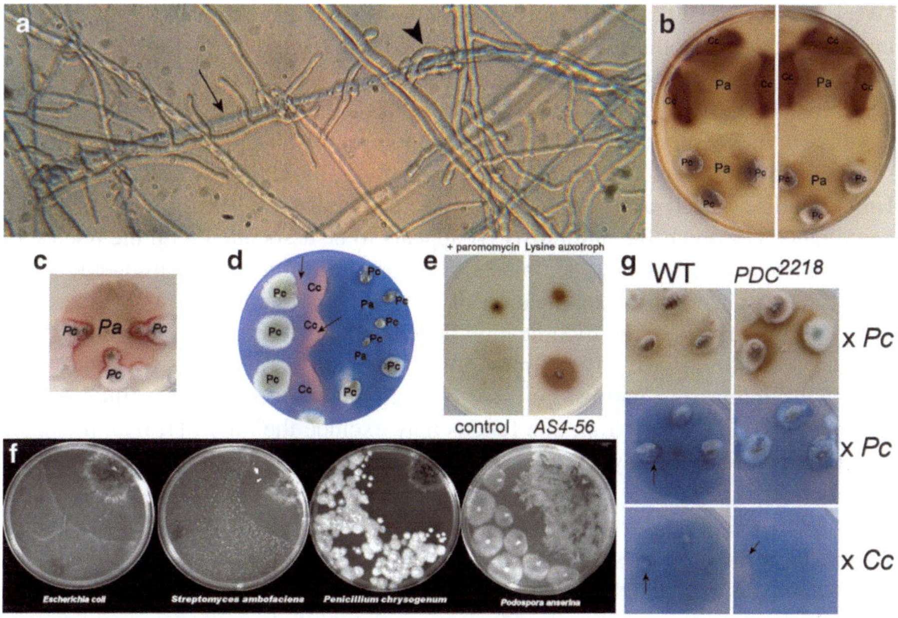

Fig. 1 (**a**) Hyphae of *P. anserina* have coiled (*arrowhead*) around a *C. globosum* hypha, which has been killed by HI (*arrow*). The dead *C. globosum* hypha is stained by Trypan blue, unlike the living *P. anserina* ones. (**b**) The DAB oxidation assay in the presence of horseradish peroxidase (*left*) and in the absence of peroxidase (*right*) for confrontations of *P. anserina* (*Pa*) with *C. cinerea* (*Cc*) and *P. anserina* with *P. chrysogenum* (*Pc*). The plates have been incubated in the same conditions except for the presence of peroxidase. (**c**) The tetrazolium assay reveals increased respiration in *P. chrysogenum* hyphae neighbouring those killed by HI. Plates were flooded with 0.1% 2,3,5-triphenyltetrazolium chloride and incubated for 3 h. Tetrazolium reduced to a pink formazan measure respiration activity (Ogur et al. 1957). (**d**) Trypan blue is degraded efficiently at the interface between *P. anserina* (*Pa*) and *C. cinerea* (*Cc*) and little or not at the interface between *P. anserina* and *P. chrysogenum* (*Pc*) or *C. cinerea* and *P. chrysogenum*. The *arrows* point at the halo zones indicative of the degradation of Trypan blue. (**e**) The DAB assay reveals an "oxidative burst" associated with slow growth in *P. anserina*. *Bottom left*, control with wild-type *P. anserina* growing on minimal media, no burst is detected. An "oxidative burst" is detected when *P. anserina* growth is impaired: *top left*, wild type grown on medium containing paromomycin; *top right*, a lysine auxotrophic mutant grown on medium containing low amount of lysine (no burst is detected for this mutant when grown on media containing optimal amount of lysine); *bottom right*, the *AS4-56* mutant that has a very impaired growth (Silar et al. 2000). Note that many additional factors that trigger slow growth result in "oxidative burst". The plates depicted in this figure have been incubated for less than 30 min, while the burst associated with HI is detected after 3–4 h. (**f**) Slow growth of the *AS4-56* mutant is associated with antibiosis towards a broad range of microorganisms, including *P. anserina* itself. *AS4-56* was inoculated on the *top right* allowed to grow for 2 weeks before inoculating the indicated challengers. Antibiosis is detected by the halo surrounding *AS4-56* in which no other microorganism is able to grow. (**g**) HI (i.e., the killing of *P. chrysogenum* hyphae) is abolished in the *PDC*^{2218} mutant, whilst DAB staining at the confrontation is greatly increased. *PDC*^{2218} appears also less sensitive to HI promoted by *C. cinerea*. *Arrows point* towards the killed hyphae visible at the confrontation between the challengers. In all experiments, *P. anserina* was inoculated at the center and challenged by three thalli

(Wicklow 1981, 1992). HI is only one of the different means by which mycelia interact with one another, which can be divided into competitions for nutrients – or exploitation competition – and combative interactions – or interference competition – (Wicklow 1981, 1992). Beside HI, Boddy (2000) defined three additional types of combative interactions between fungi, antagonism at distance through diffusible substances (also known as antibiosis), mycoparasitism and gross mycelial contact. Unlike antibiosis, HI appears to require the contact between the hyphae of both contestants. However, like antibiosis HI appears to be associated with the release of toxic substances. Moreover, antibiosis between grapevine phytopathogenic fungi has been associated with modification of ROS accumulation pattern (Freitas et al. 2009). In this instance, the modification depended upon the contestants, suggesting specific recognition. HI differs from mycoparasitism, in the lack of invasion of one of the interacting partner by the other, although some hyphal coiling maybe observed and, in the long run, one fungus may exclude the other. HI may thus be a less efficient form of mycelium invasion and predate mycoparasitism in fungal evolution.

Gross mycelial contact is associated with morphological changes of the thallus visible to the naked eye as well as redistribution of the mycelium. It may, like HI, be associated with hyphal death (mycelial interference; Dowson et al. 1988). This kind of interaction is associated with induction of phenoloxidases/laccases and related oxidase and peroxidase activities (Hiscox et al. 2010), a phenomenon previously observed during co-cultivation of wood decomposing fungi (Baldrian 2004; Iakovlev and Stenlid 2000; Score et al. 1997; Tsujiyama and Minami 2005; White and Boddy 1992). These enzymes are known to be involved in several redox processes including pigment production, xenobiotics biotransformation and lignocellulose degradation. They often produce and/or use reactive oxygen species (ROS) such as hydrogen peroxide. Recently, superoxide was detected by a nitroblue tetrazolium precipitation assay (NBT assay) at the interface between *Trametes versicolor* and three basidiomycetes during gross mycelial contact (Eyre et al. 2010). Intriguingly, the oxidation reaction of DAB detected at the confrontation with *P. anserina* and *C. cinerea* is detected in the absence of added peroxidase (Fig. 1b). Clearly, *P. anserina* and *C. cinerea* produced redox enzymes at the contact point with other fungal thalli as DAB is readily oxidized without adding exogenous enzymes. This questions what are the ROS detected by the DAB and NBT assays. It is possible that these assays actually measure a complex set of redox reactions involving not only peroxide and superoxide but other redox compounds and catalyzed by laccases and other redox enzymes (see below). For simplicity, I shall continue to use the term of "oxidative burst" to designate what is detected by these assays. Because, the last steps of fungal melanin biosynthesis are catalyzed by laccases, the redox reactions may participate both in the active killing of contestant hyphae during HI and in the production of protective pigments. For example, *P. anserina* often accumulates melanin at the zone of confrontation with fungal contestants, especially when facing *C. cinerea*. ROS could also participate in signalling since they are known to activate or repress many fungal signalling pathways (Scott and Eaton 2008).

Overall, there appears to be a large overlap between the different kinds of mycelium interactions in fungi. HI shares several properties with each of the three other defined interference competition mechanisms. It is therefore possible that all of them rely in part upon the same signalling pathways, possibly involving ROS, and resulting in the secretion of compounds permitting either to kill the contestant hyphae, to penetrate them or to engage in mycelium differentiation.

3 How Do Hyphae Die from HI?

3.1 The Morphological Manifestation of HI

The ultrastructure of hyphae dying by HI was studied in three species (Ikediugwu 1976a, b; Traquair and McKeen 1977). The morphological changes before death were very similar in the three cases. HI was associated with the disruption of the plasma and internal membrane systems, the cell wall was deteriorated, the cytoplasm was vacuolized and vesicles accumulated near the contact point. Especially, the plasma membrane invaginated extensively at the contact point as to create a large extra-plasmalemmal zone filled with vesicles. Coagulated cytoplasm may be observed near the contact point and is supposed to stop the spread of the damages of HI. The lack of penetration of the contestant hyphae was confirmed. Death was attributed to the disruption of the plasma membrane, resulting in leakage of the cytoplasm, thus confirming the loss of permeability of the dead hyphae initially observed (Ikediugwu and Webster 1970a).

Based on these ultrastructural studies, hyphal death appears to be an active process, i.e., some apoptose-like process. That death requires some metabolic activity from the dying partner can be easily visualized in the case of *P. chrysogenum* HI upon contacting *P. anserina* by using the tetrazolium assay (Ogur et al. 1957), which measure respiratory activity. As seen in Fig. 1c, a large increase in *P. chrysogenum* respiration is detected specifically at the contact point with *P. anserina*. Microscopic observation shows that the stained hyphae, i.e., the ones having the greatest respiration, are the *P. chrysogenum* hyphae just beside those that are dead.

3.2 What Is the Toxic HI Substance?

So far, only Ikediugwu and Webster (1970a) attempted without success to purify the substances responsible of HI. There are three types of candidates. Firstly, fungi are known to synthesize several secondary metabolites with antifungal activity. These include (+)-Isoepoxydon of *Poronia punctata* (Gloer and Truckenbrod 1988), the pyrine and pyridone compounds of *Physiosporinus sanguinolentus*

(Svensson et al. 2001), the sordarins produced by many ascomycetes species (Vicente et al. 2009) and cocktail of volatiles antimicrobials from *Muscodor albus* (Strobel et al. 2001). This list is incomplete and many secondary metabolites of fungal origin may exhibit antifungal and other activities (Leeder et al. 2011). Interestingly, production of volatiles organic compounds is known to be modified during mycelial interactions (Evans et al. 2008; Hynes et al. 2007). Secondly, fungi can produce and export polypeptides with antifungal activity (Selitrennikoff 2001). PAF, a small cysteine-rich secreted protein from *P. chrysogenum*, is the most studied one (Kaiserer et al. 2003; Marx et al. 1995, 2008). This protein penetrates hyphae (Oberparleiter et al. 2003) and interferes with several signalling pathways (Binder et al. 2010a, b). It promotes an apoptosis-like phenotype that looks very much like the hyphal death triggered by HI (Leiter et al. 2005). Peptides able to activate the vegetative incompatibility reaction, as the phcA protein from the bacterium *Pseudomonas syringae* are additional candidates (Wichmann et al. 2008). Last but not least, antifungal activity has been detected for some enzymes, namely glucose oxidases (Kim et al. 1990; Leiter et al. 2004). Glucose oxidase oxidizes glucose into D-glucono-δ-lactone and reduces dioxygen into peroxide. It was hypothesized that peroxide is the chemical toxic to the test fungi (Leiter et al. 2004). Yet, catalase was unable to protect fungi and vitamin C was protective only as low activity of glucose oxidase (Leiter et al. 2004), suggesting that toxicity could be effected by another mechanism. Whatever the mechanism of its toxicity towards fungi, glucose oxidase fits well with several properties of the HI substance: it is able to create a burst of peroxide and it is toxic. If peroxide is indeed the toxic molecule, its instability explains why it could not be purified. However, as will be discussed in Sect. 4, there exist *P. anserina* mutants with an increased "oxidative burst" upon meeting a contestant, but that are unable to kill. This is not expected if the enzyme (s) responsible for the "burst" are the one(s) responsible for hyphal death.

There are thus several candidates for the HI substance(s). As yet, none have been demonstrated to be indeed involved in hyphal death during HI. Moreover, it is possible that several substances act synergistically to trigger death and that the substance or cocktail of substances vary from one species to the other. Different substances could even be secreted in contact with different contestants. For example, contact between *C. cinerea* and *P. anserina* is associated with the release of enzyme(s) able to efficiently breakdown dyes such as Evans and Trypan blue. Contact of *P. chrysogenum* with *C. cinerea* is associated with a weak ability to breakdown the dye and contact with *P. anserina* is not (Fig. 1d).

3.3 How Do Fungi Recognize Each Other?

The DAB staining assay indicates that some filamentous fungi are able to recognize thalli from other species and differentiate them from those from the same species. Indeed, DAB oxidation by *P. anserina* and *C. cinerea* was detected only in non-self-confrontation, while self-confrontation, including those between vegetatively or sexually incompatible strains, did not promote the "oxidative burst". Note that those species that lack the "oxidative burst" may recognize non-self and engage a response not detected by the DAB assay. Because of the large spectrum of fungi able to elicit a response from *P. anserina* and *C. cinerea*, one wonders whether there are fungal specific receptors that recognize different contestants as those evidenced in plant and animals involved in recognizing pathogen associated molecular pattern (PAMP). Arguments for such specific receptors stem from the fact that *C. cinerea*, and not *P. anserina*, responds to soil bacteria (Silar 2005). In addition, highly specific response of fungi to contact with other living beings has already been documented. For example, production of secondary metabolites by *A. nidulans* is triggered by the intimate contact with a specific strain of *Streptomyces hygroscopicus* among the 58 tested soil-inhabiting actinomycetes (Schroeckh et al. 2009). So far the best candidates for such a task are the HNWD proteins involved in the vegetative incompatibility response that controls the fate of hyphae having undergone anastomosis. Indeed, they are structurally similar to animal and plant PAMP receptors and are also encoded by highly polymorphic genes present in multiple copies in fungal genomes (Paoletti and Saupe 2009). It has thus been proposed that they participate in some innate immunity of fungi (Paoletti and Saupe 2009). If this proves correct there would be a deep connection between the intraspecific and interspecific "self" versus "non-self" recognition processes of filamentous fungi. Other candidates are G-protein coupled receptors related to PTH11, a protein required for pathogenesis in *Magnaporthe grisea* (Kulkarni et al. 2005). These receptors are present as a multigenic family in genomes with a high number of copies (61 in *M. grisea*, 25 in *N. crassa* and we detected 57 in *P. anserina*) and can transduce signals from the outside of the cell. However, they appear specific to Pezizomycotina and could not account for "self" versus "non-self" recognition in basidiomycetes.

The fungal PAMP recognized by the putative HI receptors are also unknown. They depend upon living hyphae as mycelia inactivated by heat (Silar 2005) or killed by propylene oxide vapour (Ikediugwu and Webster 1970a) are not effective anymore in activating HI. Therefore, stable components of the cell wall, such as glucans or chitine, are likely not involved. The requirement for living hyphae to set up HI, at least to trigger the "oxidative burst", is somewhat puzzling and may call forth another hypothesis to explain HI. Indeed, alternatively to being a response being driven by specific receptors, HI could be due to the activation of some general stress response induced by a broad spectrum of fungi (and also soil bacteria in the case of *C. cinerea*). It is noteworthy that in *P. anserina* many stressful genetic and environmental conditions that promote slow growth are associated with a massive "oxidative burst" as detected by the DAB assay (P. Silar, unpublished data, Fig. 1e). When growth is very altered, as in the *AS4-56* mutant, this massive "oxidative burst" is associated with antibiosis towards a large spectrum of organisms, including *P. anserina* itself (Silar et al. 2000) (Fig. 1f). In a model based on these observations, recognition would not pass through specific receptors, but would rather be a general stress response triggered specifically by determinants present on or secreted by contestants. This would result in the secretion by *P. anserina* of a cocktail of enzymes, visualized by the DAB assay, enabling him to cope with the challenger. In this model, the HI substance(s) would be enzyme(s) rather than secondary metabolites or dedicated antifungal peptides. Stressful conditions would set on the HI response in the absence of challenger, leading to release of large amount of the toxic HI substance(s), thereby causing antibiosis.

4 How to Study HI?

Despite its prevalence and potential importance in shaping fungal communities (Wicklow 1981, 1992), there appear to be little study undertaken to understand the molecular basis of HI and related interference competition mechanisms. Molecular approaches have been initiated to identify genes whose expression is modified when mycelia encounter. The first study used *Heterobasidion annosum and P. sanguinolentus*. The latter fungus produces chemicals, which inhibit hyphal growth and spore germination of the former (Svensson et al. 2001); however, it is not known whether HI occurs when they contact. Differential display of mRNA from the non-interacting and interacting partners yielded 21 genes whose expression was altered during the interaction, 10 from *P. sanguinolentus* and 11 from *H. annosum* (Iakovlev et al. 2004). This study permitted to evidence that modification of gene transcription does occur during mycelium interaction. Yet, none of the genes coding for enzymes known to be activated when mycelia meet were recovered.

The second study monitored much more genes (nearly 2,000) and used *T. versicolor* in contact with *Stereum gausapatum, Bjerkandera adusta* and *Hypholoma fasciculare* (Eyre et al. 2010). These three species were chosen because their meetings with *T. versicolor* result in different outcomes: deadlock (i.e., the two mycelia stall when they encounter) with *B. adusta*, replacement by *H. fasciculare* and invasion of *S. gausapatum* mycelia. Unfortunately, presence of HI has not been evaluated during these interactions. Subtractive libraries identified genes regulated differently when *T. versicolor* grew alone as compared to when grown in the presence of *S. gaupatum*. These genes were then used to build microarrays, which were hybridized with probes constructed from *T. versicolor* mycelia in contact with each of the three challengers. When comparing the different contacts, over 50% of the genes were regulated solely in one of them and a small number of genes were found to be regulated in the same manner during the three gross mycelial contacts: 74 among the 1,103 genes differentially regulated in at least one of the contacts. This indicates that the modifications of the expression of the tested genes is dependent upon the challenger, suggesting that the signalling of these interactions is quite complex. Interestingly, fewer genes were up and down regulated during the contact with *H. fasciculare*, which is able to replace *T. versicolor*. Possibly, this fungus is not as well recognized as the others, preventing to build efficient responses to invasion by its mycelium. Among the genes differentially regulated those encoding ribosomal proteins and glycine-rich RNA binding proteins, involved in post-transcriptional gene regulation were most represented. This is in line with the morphological manifestation of gross mycelial contact, which necessitates synthesis of new proteins and regulation of gene expression. In these experiments, the genes coding the enzymes known to be up-regulated during the investigated mycelial contact (Hiscox et al. 2010) were not retrieved, indicating that despite the high number of genes investigating the picture of transcription regulation during contact is still incomplete. The mycelium used to extract for RNA in these experiments was not the one in direct contact with contestants, as to avoid contamination by RNA from the other species. With the advent of the new technologies for genome and transcriptome sequencing, we may expect rapidly a more comprehensive view of the gene expression modification in the region in direct contact with challengers, as well as those more remote from the interaction zone.

For a complete understanding of HI, the utilization of model organisms whose genes can be manipulated is mandatory. This will enable to clearly delineate the genes involved in HI through inactivation and over-expression. Because *P. anserina* and *C. cinerea* are long-established models to study fungal development and physiology, these two organisms should prove valuable to study HI. As yet, no study is conducted on HI with *C. cinerea*. On the contrary, several studies have already identified a few genes controlling HI in *P. anserina*. First, HI is dependent upon the PaNox1 NADPH oxidase, but not the PaNox2 and PaNox3 ones (Brun et al. 2009; Silar 2005). The mutants inactivated for PaNox1 display a much reduced "oxidative burst" at the contact zone with a contestant and are unable to kill efficiently *P. chrysogenum*. However, the mutants have greatly enhanced production of ROS (as detected by DAB oxidation) all over their thallus in the absence of contestant, indicating that the lack of burst may be due to the fact that they constitutively produce the involved enzymes. An additional interesting phenotype exhibited by the PaNox1 mutants is that it is not killed by *C. cinerea*, hence all aspects of HI in *P. anserina* are inhibited in the PaNox1 mutants. The involvement of Nox in HI is to be compared with the role of these enzymes in innate immunity in animals and hypersensitive response in plants (Bedard et al. 2007). It remains to be shown whether this is another instance of convergent evolution or the sign of an ancient use of Nox in "self" versus "non-self" recognition/innate immunity.

The PaNox1 mutants were recovered along with many others in screens aimed at understanding the Crippled Growth cell degeneration (Haedens et al. 2005). Some have phenotypes similar to those of PaNox1 (the "pink" mutants), whilst others exhibit different ones. Among the recovered pink mutants, some were affected in the PaASK1 MAP kinase kinase kinase (Kicka and Silar 2004). This protein is essential for HI, as the *PaASK1*-inactivated mutants have a phenotype very similar to those of the PaNox1 mutants (Silar 2005). Inactivation of the PaMKK1 MAP kinase kinase and PaMpk1 MAP kinase acting downstream of PaASK1 (Kicka et al. 2006) show that these kinases are also required for HI (P. Silar, unpublished data). Because PaNox1 is necessary for proper nuclear localization of PaMpk1 (Kicka et al. 2006), it is likely that PaNox1 and the MAP kinase cascade control HI through the same pathway. The two other MAP kinase pathways of *P. anserina* do not control HI (Lalucque et al. 2012). On the contrary, all the mutants identified as having the same phenotype as the PaNox1 and PaASK1 mutants are also defective

Table 1 *P. anserina* mutants affected in HI

	Pa × Pc		*Pa × Cc*
Mutants	Oxidative burst	Cell death	Cell death
IDC^{302}	Decreased	Decreased	Identical
IDC^{512}	Decreased	Slightly decreased	Identical
IDC^{520}	Increased all over the thallus	Decreased	Decreased
PDC^{2218}	Increased	Decreased	Slightly decreased

for HI, including the mutants of the *IDC1* gene that encodes a Pezizomycotina-specific protein of unknown function (Jamet-Vierny et al. 2007; P. Silar, unpublished data). They all exhibit the same phenotype, i.e., the "oxidative burst" is severely reduced, as are *C. chrysogenum* and *P. anserina* hyphal death, when confronted with *P. anserina* and *C. cinerea*, respectively. The affected genes most likely operate in the same signalling pathway as PaNox1 and PaMpk1. Interestingly, in *P. anserina*, this pathway controls the elaboration of fruiting bodies (Jamet-Vierny et al. 2007). There appears thus to be a connection between encountering a contestant and triggering of sexual reproduction, a fact well evidenced for many fungi, which often fruit when in contact with a challenger (Moreau 1954).

More interesting are the IDC^{302}, IDC^{512}, IDC^{520} and PDC^{2218} mutants recovered in the same screens as the PaNox1 mutants (Haedens et al. 2005). Indeed, in these mutants the "oxidative burst" is uncoupled from hyphal death (Fig. 1g and Table 1). In the IDC^{520} mutants, hyphal death is decreased while the DAB assay stains rapidly the whole thallus, even more rapidly than observed for the PaNox1 mutants. Many of the mutants isolated in the screens, including the IDC^{520} mutants, had an altered pattern of DAB staining on the thallus in the absence of contestant (Haedens et al. 2005). In the IDC^{302} and IDC^{512} mutants, the oxidative burst and *P. chrysogenum* hyphal death are decreased, while *P. anserina* hyphal death still occurs when confronted with *C. cinerea*. In other word, these mutants are not able to inflict HI anymore, while still being subjected to HI. In the PDC^{2218} mutant the "oxidative burst" is increased, while HI towards *P. chrysogenum* is abolished (Fig. 1g). In this mutant, the two aspects of HI that were always associated, i.e., the "oxidative burst" and HI towards *P. chrysogenum*, are separated. The recovery of such mutants should enable to better understand HI, especially when the affected genes will be identified. Presently, it strongly suggests that the enzyme(s) detected by the DAB assay may not be the one(s) having the killing activity towards *P. chrysogenum*. However, if the killing is due to a cocktail of enzymes with different activities, there is still the possibility that modifications in the proportion of the various enzymes may affect differently intensity of the burst and killing of *P. chrysogenum* hyphae.

A few additional genes were tested for their involvement in HI by reverse genetics, i.e., the *PaPls1* tetraspanin gene (Lambou et al. 2008), the *PaTrx1*, *PaTrx2* and *PaTrx3* genes encoding thioredoxins (Malagnac et al. 2007) and the *PaNat1* and *PaNat2* genes encoding arylamine N-acetyl transferases (Martins et al. 2009). None controls HI. PaPls1 acts along with the PaNox2 NADPH oxidase, which does not monitor HI. The thioredoxin are involved in redox regulation, and control sexual development. The arylamine N-acetyl transferases are involved in the detoxification of toxic arylamines, suggesting that the HI killing substance is unlikely to contain arylamines.

5 Potential Application of HI

At the present time, there is little application to use HI in the control of fungal population. However, Rotstop® is presently sold as a "biofungicide for biological stump treatment". It contains spores of *Phlebiopsis* (= *Peniophora*) *gigantea*. These can be applied on stumps of conifers on which they germinate. Treatment is recommended for Norway spruce and Scot pines just after trees have been felled. This prevents the invasion of stumps by *Heterobasidion annosum*, which has the ability to spread through roots to nearby healthy trees. These then die by butt or root rot. The protective effect was shown to be due to HI (Ikediugwu et al. 1970). This treatment, in application in Europe, is safe and offers an alternative to chemical application. Similar strategy could be implemented for controlling other phytopathogenic fungi (Cox and Scherm 2006; Nakasaki et al. 2007; Perello et al. 2002; White and Traquair 2006). However, usually more aggressive behaviours such as mycoparasitism are favoured to select for candidate. The example of Rotstop® clearly demonstrates that HI is sufficient for controlling a very destructive phytopathogen.

Another potential application of HI would result from the identification of the signalling pathways involved in HI. Indeed, HI is potentially associated with the production of secondary metabolites as well as enzymes involved in plant biomass and xenobiotics degradation. A better understanding of the signal involved in triggering HI could result in the enhanced production of these products, with potential applications in pharmaceuticals discovery and biosynthesis, biofuel production and bioremediation approaches. This is well evidenced in Fig. 1d, in which the toxic trypan blue used to detect cell death is most readily degraded at the interaction zone between *C. cinerea* and *P. anserina*.

Acknowledgement I would like to thank all members of my laboratory for helping me uncover various aspects of fungal biology, not only through experimental work, but also through imaginative discussions and moral support. Part of the work dealing with HI was supported by ANR grant n°ANR-05-Blan-0385-02.

References

Ainsworth AM, Rayner DM (1989) Hyphal and mycelial responses associated with genetic exchange within and between species of the basidiomycete genus *Stereum*. J Gen Microbiol 135:1643

Ainsworth AM, Rayner AD (1991) Ontogenetic stages from coenocyte to basidiome and their relation to phenoloxidase activity and colonization process in *Phanerochaete magnoliae*. Mycol Res 95:1414

Baldrian P (2004) Increase of laccase activity during interspecific interactions of white-rot fungi. FEMS Microbiol Ecol 50:245

Bedard K, Lardy B, Krause KH (2007) NOX family NADPH oxidases: not just in mammals. Biochimie 89:1107

Binder U, Chu M, Read ND, Marx F (2010a) The antifungal activity of the *Penicillium chrysogenum* protein PAF disrupts calcium homeostasis in *Neurospora crassa*. Eukaryot Cell 9:1374

Binder U, Oberparleiter C, Meyer V, Marx F (2010b) The antifungal protein PAF interferes with PKC/MPK and cAMP/PKA signalling of *Aspergillus nidulans*. Mol Microbiol 75:294

Boddy L (2000) Interspecific combative interactions between wood-decaying basidiomycetes. FEMS Microbiol Ecol 31:185

Brun S, Malagnac F, Bidard F, Lalucque H, Silar P (2009) Functions and regulation of the Nox family in the filamentous fungus *Podospora anserina*: a new role in cellulose degradation. Mol Microbiol 74:480

Cox KD, Scherm H (2006) Interaction dynamics between saprobic lignicolous fungi and *Armillaria* in controlled environments: exploring the potential for competitive exclusion of *Armillaria* on peach. Biol Control 37:291

DeVay JE (1956) Mutual relationships in fungi. Annu Rev Microbiol 10:115

Dowson CG, Rayner ADM, Boddy L (1988) The form and outcome of mycelial interactions involving cord-forming decomposer basidiomycetes in homogeneous and heterogeneous environments. New Phytol 109:423

Evans JA, Eyre CA, Rogers HJ, Boddy L, Müller CT (2008) Changes in volatile production during interspecific interactions between four wood rotting fungi growing in artificial media. Fungal Ecol 1(5):7

Eyre C, Muftah W, Hiscox J, Hunt J, Kille P, Boddy L, Rogers HJ (2010) Microarray analysis of differential gene expression elicited in *Trametes versicolor* during interspecific mycelial interactions. Fungal Biol 114:646

Freitas R, Rego C, Oliveira H, Ferreira RB (2009) Interactions among grapevine disease-causing fungi. The role of reactive oxygen species. Phytopathol Mediterr 48:117

Giovannetti M, Azzolini D, Citernesi AS (1999) Anastomosis formation and nuclear and protoplasmic exchange in arbuscular mycorrhizal fungi. Appl Environ Microbiol 65:5571

Gloer JB, Truckenbrod SM (1988) Interference competition among coprophilous fungi: production of (+)-Isoepoxydon by *Poronia punctata*. Appl Environ Microbiol 54:861

Haedens V, Malagnac F, Silar P (2005) Genetic control of an epigenetic cell degeneration syndrome in *Podospora anserina*. Fungal Genet Biol 42:564

Harper JE, Webster J (1964) An experimental analysis of the coprophilous fungus succession. Trans Br Mycol Soc 47:511

Hiscox J, Baldrian P, Rogers HJ, Boddy L (2010) Changes in oxidative enzyme activity during interspecific mycelial interactions involving the white-rot fungus *Trametes versicolor*. Fungal Genet Biol 47:562

Hynes J, Muller CT, Jones TH, Boddy L (2007) Changes in volatile production during the course of fungal mycelial interactions between *Hypholoma fasciculare* and *Resinicium bicolor*. J Chem Ecol 33:43

Iakovlev A, Stenlid J (2000) Spatiotemporal patterns of laccase activity in interacting mycelia of wood-decaying basidiomycete fungi. Microb Ecol 39:236

Iakovlev A, Olson A, Elfstrand M, Stenlid J (2004) Differential gene expression during interactions between *Heterobasidion annosum* and *Physisporinus sanguinolentus*. FEMS Microbiol Lett 241:79

Ikediugwu FEO (1976a) The interface in hyphal interference by *Peniophora gigantea* against *Heterobasidion annosum*. Trans Br Mycol Soc 66:291

Ikediugwu FEO (1976b) Ultrastructure of hyphal interference between *Coprinus heptemerus* and *Ascobolus crenulatus*. Trans Br Mycol Soc 66:281

Ikediugwu FEO, Webster J (1970a) Antagonism between *Coprinus heptemerus* and other coprophilous fungi. Trans Br Mycol Soc 54:181

Ikediugwu FEO, Webster J (1970b) Hyphal interference in a range of coprophilous fungi. Trans Br Mycol Soc 54:205
Ikediugwu FEO, Dennis C, Webster J (1970) Hyphal interference by *Peniophora gigantea* against *Heterobasidion annosum*. Trans Br Mycol Soc 54:307
Inoue K, Kanematsu S, Park P, Ikeda K (2011) Cytological analysis of mycelial incompatibility in *Rosellinia necatrix*. Fungal Biol 115:87
Jamet-Vierny C, Debuchy R, Prigent M, Silar P (2007) IDC1, a Pezizomycotina-specific gene that belongs to the PaMpk1 MAP kinase transduction cascade of the filamentous fungus *Podospora anserina*. Fungal Genet Biol 44:1219
Kaiserer L, Oberparleiter C, Weiler-Gorz R, Burgstaller W, Leiter E, Marx F (2003) Characterization of the *Penicillium chrysogenum* antifungal protein PAF. Arch Microbiol 180:204
Kicka S, Silar P (2004) PaASK1, a mitogen-activated protein kinase kinase kinase that controls cell degeneration and cell differentiation in *Podospora anserina*. Genetics 166:1241
Kicka S, Bonnet C, Sobering AK, Ganesan LP, Silar P (2006) A mitotically inheritable unit containing a MAP kinase module. Proc Natl Acad Sci USA 103:13445
Kim KK, Fravel DR, Papavizas GC (1990) Glucose oxidase as the antifungal principle of talaron from *Talaromyces flavus*. Can J Microbiol 36:760
Kulkarni RD, Thon MR, Pan H, Dean RA (2005) Novel G-protein-coupled receptor-like proteins in the plant pathogenic fungus *Magnaporthe grisea*. Genome Biol 6:R24
Lalucque H, Silar P (2003) NADPH oxidase: an enzyme for multicellularity? Trends Microbiol 11:9
Lalucque H, Malagnac F, Brun S, Kicka S, Silar P (2012) A non-mendelian MAPK-generated hereditary unit controlled by a second MAPK pathway in *Podospora anserina*. Genetics in the press.
Lambou K, Malagnac F, Barbisan C, Tharreau D, Lebrun MH, Silar P (2008) The crucial role during ascospore germination of the Pls1 tetraspanin in *Podospora anserina* provides an example of the convergent evolution of morphogenetic processes in fungal plant pathogens and saprobes. Eukaryot Cell 7:1809
Lara-Ortiz T, Riveros-Rosas H, Aguirre J (2003) Reactive oxygen species generated by microbial NADPH oxidase NoxA regulate sexual development in *Aspergillus nidulans*. Mol Microbiol 50:1241
Leeder AC, Palma-Guerrero J, Glass NL (2011) The social network: deciphering fungal language. Nat Rev Microbiol 9:440
Leiter E, Marx F, Pusztahelyi T, Haas H, Pocsi I (2004) *Penicillium chrysogenum* glucose oxidase – a study on its antifungal effects. J Appl Microbiol 97:1201
Leiter E, Szappanos H, Oberparleiter C, Kaiserer L, Csernoch L, Pusztahelyi T, Emri T, Pocsi I, Salvenmoser W, Marx F (2005) Antifungal protein PAF severely affects the integrity of the plasma membrane of *Aspergillus nidulans* and induces an apoptosis-like phenotype. Antimicrob Agents Chemother 49:2445
Malagnac F, Lalucque H, Lepere G, Silar P (2004) Two NADPH oxidase isoforms are required for sexual reproduction and ascospore germination in the filamentous fungus *Podospora anserina*. Fungal Genet Biol 41:982
Malagnac F, Klapholz B, Silar P (2007) PaTrx1 and PaTrx3, two cytosolic thioredoxins of the filamentous ascomycete *Podospora anserina* involved in sexual development and cell degeneration. Eukaryot Cell 6:2323
Martins M, Rodrigues-Lima F, Dairou J, Lamouri A, Malagnac F, Silar P, Dupret JM (2009) An acetyltransferase conferring tolerance to toxic aromatic amine chemicals: molecular and functional studies. J Biol Chem 284:18726
Marx F, Haas H, Reindl M, Stoffler G, Lottspeich F, Redl B (1995) Cloning, structural organization and regulation of expression of the *Penicillium chrysogenum* paf gene encoding an abundantly secreted protein with antifungal activity. Gene 167:167
Marx F, Binder U, Leiter E, Pocsi I (2008) The *Penicillium chrysogenum* antifungal protein PAF, a promising tool for the development of new antifungal therapies and fungal cell biology studies. Cell Mol Life Sci 65:445

Melzer MS, Ikeda SS, Boland GJ (2002) Interspecific transmission of double-stranded RNA and hypovirulence from *Sclerotinia sclerotiorum* to *S. minor*. Phytopathology 92:780

Moreau C (1954) Les genres SORDARIA et PLEURAGE, leurs affinités systématiques. Paul Lechevalier, Paris

Nakasaki K, Saito M, Suzuki N (2007) *Coprinellus curtus* (Hitoyo-take) prevents diseases of vegetables caused by pathogenic fungi. FEMS Microbiol Lett 275:286

Oberparleiter C, Kaiserer L, Haas H, Ladurner P, Andratsch M, Marx F (2003) Active internalization of the *Penicillium chrysogenum* antifungal protein PAF in sensitive *aspergilli*. Antimicrob Agents Chemother 47:3598

Ogur M, St John R, Nagai S (1957) Tetrazolium overlay technique for population studies of respiration deficiency in yeast. Science 125:928

Paoletti M, Saupe SJ (2009) Fungal incompatibility: evolutionary origin in pathogen defense? Bioessays 31:1201

Perello A, Simon MR, Arambarri AM (2002) Interactions between foliar pathogens and the saprophytic microflora of the wheat (*Triticum aestivum* L.) phylloplane. J Phytopathol 150:232

Roca MG, Davide LC, Davide LM, Mendes-Costa MC, Schwan RF, Wheals AE (2004) Conidial anastomosis fusion between *Colletotrichum* species. Mycol Res 108:1320

Schroeckh V, Scherlach K, Nutzmann HW, Shelest E, Schmidt-Heck W, Schuemann J, Martin K, Hertweck C, Brakhage AA (2009) Intimate bacterial-fungal interaction triggers biosynthesis of archetypal polyketides in *Aspergillus nidulans*. Proc Natl Acad Sci USA 106:14558

Score AJ, Palfreyman JW, White NA (1997) Extracellular phenoloxidase and peroxidase enzyme production during interspecific fungal interactions. Int Biodeter Biodegrad 39:225

Scott B, Eaton CJ (2008) Role of reactive oxygen species in fungal cellular differentiations. Curr Opin Microbiol 11:488

Selitrennikoff CP (2001) Antifungal proteins. Appl Environ Microbiol 67:2883

Silar P (2005) Peroxide accumulation and cell death in filamentous fungi induced by contact with a contestant. Mycol Res 109:137

Silar P, Rossignol M, Tahar R, Derhy Z, Mazabraud A (2000) Informational suppressor alleles of the eEF1A gene, fertility and cell degeneration in *Podospora anserina*. Mol Gen Genet 264:354

Strobel GA, Dirkse E, Sears J, Markworth C (2001) Volatile antimicrobials from *Muscodor albus*, a novel endophytic fungus. Microbiology 147:2943

Svensson M, Lundgren LN, Woods C, Fatehi J, Stenlid J (2001) Pyrone and pyridone compounds in the liquid culture of *Physisporinus sanguinolentus*. Phytochemistry 56:747

Traquair JA, McKeen W (1977) Hyphal interference by *Trametes hispida*. Can J Microbiol 23:1675

Tsujiyama S-I, Minami M (2005) Production of phenol-oxidizing enzymes in interaction between white-rot fungi. Mycoscience 46:268

Vicente F, Basilio A, Platas G, Collado J, Bills GF, Gonzalez del Val A, Martin J, Tormo JR, Harris GH, Zink DL, Justice M, Kahn JN, Pelaez F (2009) Distribution of the antifungal agents sordarins across filamentous fungi. Mycol Res 113:754

White NA, Boddy L (1992) Extracellular enzyme localization during interspecific fungal interactions. FEMS Microbiol Lett 98:75

White GJ, Traquair JA (2006) Necrotrophic mycoparasitism of *Botrytis cinerea* by cellulolytic and ligninocellulolytic Basidiomycetes. Can J Microbiol 52:508

Wichmann G, Sun J, Dementhon K, Glass NL, Lindow SE (2008) A novel gene, phcA from *Pseudomonas syringae* induces programmed cell death in the filamentous fungus *Neurospora crassa*. Mol Microbiol 68:672

Wicklow DT (1981) Interference competition and the organization of fungal communities. In: Wicklow DT, Carroll GC (eds) The fungal community. Its organization and role in the ecosystem, 1st edn. Marcel Dekker, New York, pp 351–375

Wicklow DT (1992) Interference competition. In: Wicklow DT, Carroll GC (eds) The fungal community. Its organization and role in the ecosystem, 2nd edn. Marcel Dekker, New York, pp 265–274

Sexual Pheromones in the Fungi

Silvia Polaino and Alexander Idnurm

Abstract The capability of sexual reproduction is distributed across the eukaryotes, including the fungi. A primary influence in the sexual interaction is the exchange of information mediated by diffusible molecules, called sexual pheromones. This chapter examines the biosynthesis of pheromones and the sexual responses induced by them in different branches of the fungal kingdom, with an emphasis on the early lineages. The best-studied species are members of the Dikarya and they use pheromones derived from peptide precursors. In contrast, members of the Mucoromycotina use apocarotenoids while the Blastocladiomycota use sesquiterpenes. Comparison between these pheromones establishes evolutionary trends among the fungal lineages.

1 Introduction

Most eukaryotic organisms can reproduce sexually, and in many species sexual and asexual reproduction coexist stably. Sex must confer benefits that allow its maintenance despite the speed and economy of asexual reproduction (Williams 1975; Maynard Smith 1978; Bell 1982). The most widely accepted advantage of sex is the production of genetic diversity, which facilitates adaptation to a changing environment, since sexual reproduction is the main source of recombinants.

In addition to a role in generating diversity, in the fungi sexual reproduction is also necessary for the production of specialized spore types that aid a species' ability to spread in the environment. The sexual process of the fungi is mediated, in part, by signal molecules called pheromones (Bölker and Kahmann 1993).

S. Polaino • A. Idnurm (✉)
Division of Cell Biology and Biophysics, School of Biological Sciences,
University of Missouri-Kansas City, 5100 Rockhill Road, Kansas City, MO 64110, USA
e-mail: polainoortss@umkc.edu; idnurma@umkc.edu

G. Witzany (ed.), *Biocommunication of Fungi*, DOI 10.1007/978-94-007-4264-2_11,

These messengers initiate sexual development of strains of the opposite sex. The etymology of *pheromone*, according to Karlson and Lüsher (1959), derives from the Greek words *pherein* (to transport) and *horman* (to stimulate). The first animal pheromone to be chemically characterized was *Bombykol*, released by the female silkworm to attract mates (Butenandt et al. 1961). Pheromones function across major lineages of life to mediate diverse aspects in cell-cell communication beyond just a role in mating. Given the abundance of fungi in terrestrial environments and their species diversity, pheromone communications are a vital part of fungal biology and evolution.

The topic of fungal pheromones has been reviewed a number of times (see for example Gooday and Adams 1993; Heitman et al. 2007; Jones and Bennett 2011; Kothe 2008). In this chapter we will briefly outline the pheromone signaling mechanisms in the different phyla of the eukaryote kingdom fungi – the Ascomycota, Basidiomycota and the basal lineages previously classified as the Zygomycota and the Chytridiomycota – and discuss the implications of these systems for communication between different species.

2 Pheromone Types

2.1 Pheromones in Ascomycetes

The yeast *Saccharomyces cerevisiae* has been extensively studied to understand the mechanisms of pheromone production and the responses to pheromones (Madhani 2007). Therefore, in this section and in the rest of this chapter we will refer to this species to represent the Ascomycetes. Within this phylum the pheromones from many other species have been characterized, such as *Aspergillus nidulans*, *Candida albicans*, *Magnaporthe oryzae*, *Neurospora crassa*, *Podospora anserina* and *Schizosaccharomyces pombe*.

S. cerevisiae sexual interaction occurs between two mating-type cells, α and **a**, controlled by *MAT*α and *MAT***a** alleles of the *MAT* locus (Herskowitz 1988; Coppin et al. 1997; Lee et al. 2010). Both alleles encode proteins that regulate the expression of cell-type specific genes through activation or repression. The products of the *MF*α*1/2* and *MFa1/2* genes are proteins that act as pre-pheromones or immature pheromones (Michaelis and Herskowitz 1988). They undergo a series of different steps for their maturation. The mature pheromones are of two classes, the α factor secreted by α cells and **a** factor by **a** cells.

The α factor is a peptide of 13 amino acids whose sequence is WHWLQLKPGQPMY (Fig. 1). MFα1/2 is not directly coded as such a small peptide, but as a large protein that is post-translationally modified to generate the functional α1 pheromones. The enzymes involved in their maturation process are the Kex2 protease, Kex1 carboxypeptidase and Ste13 diamino peptidase. Once released from the cell, α factor is recognized by the Ste2 receptor on the surface

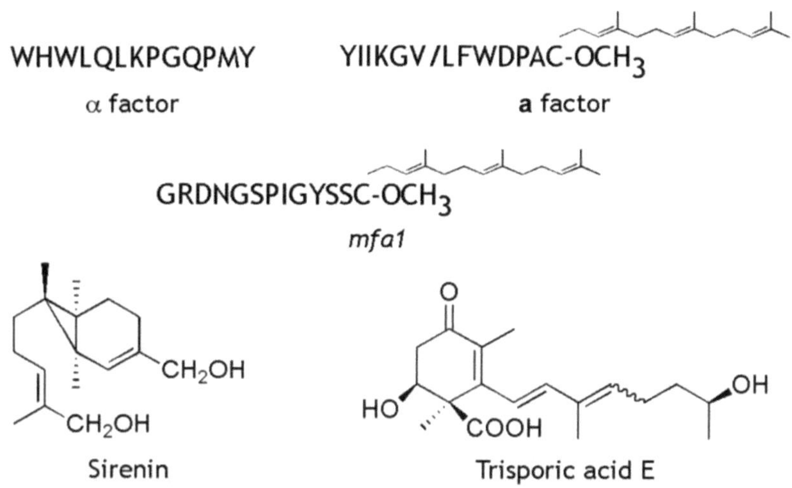

Fig. 1 Comparison of the pheromones of species belonging to different phyla of the kingdom fungi. Representative pheromones are the α factor and **a** factor of Ascomycete *S. cerevisiae*, *mfa1* of Basidiomycete *U. maydis*, Sirenin of chytrid *A. macrogynus*, and trisporic acid E of Mucoromycotina *P. blakesleeanus*. The Ascomycete and Basidiomycete pheromones are peptide-based, with the capitalized letters used representing the standard abbreviation for the amino acid residues

of **a** cells to initiate the mating signaling. The central amino acids, proline-8 and glycine-9 (underlined within the preceding sequence), are key to produce an optimal response of the receptor (Martin et al. 2011; Naider and Becker 2004).

The **a** factor is a peptide of 12 amino acids, and according to the pheromone precursor genes, *MFa1* and *MFa2*, the amino acid sequences vary by a single valine/leucine residue (Betz et al. 1987; Coppin et al. 2005). **a** factor differs in comparison to α factor in that it is associated with a lipid moiety (Fig. 1). The pre-pheromone has a conserved carboxy-terminal CAAX box, where C is cysteine, A is any aliphatic amino acid and X any residue. This cysteine residue is the site for transfer of a farnesyl group and appendage of a carboxymethyl group during the maturation process (Anderegg et al. 1988). A possible role of this lipid moiety, due to its hydrophobicity, may be to help the transport of pheromones. Like α factor, the pre-protein of is also cleaved to form the smaller mature **a** factor peptide.

While many Ascomycetes follow the pheromone system seen in *S. cerevisiae*, *Hypocrea jecorina* (teleomorph of *Trichoderma reesei*), some species of *Fusarium* and other members of the Hypocreales have pheromones that share characteristics of both the **a** and α factors of *S. cerevisiae*. These have been called hybrid pheromones as a result (Schmoll et al. 2010). Curiously, the pheromones of the Pucciniomycotina yeasts (Basidiomycota) also exhibit hybrid characteristics by being farnesylated and cleaved from precursor molecules (Akada et al. 1989). As a consequence, the evolutionary trajectory of the different classes of pheromones found in the Ascomycetes and Basidiomycetes is currently unclear.

2.2 Pheromones in Basidiomycetes

Pheromones have been identified and studied in detail in several species of Basidiomycetes, such as the mushroom-forming species *Schizophyllum commune* and *Coprinus cinereus*, the human pathogenic yeast *Cryptococcus neoformans*, the smuts *Ustilago hordei* and *Ustilago maydis* and also the red yeast *Rhodosporidium toruloides* (Raudaskoski and Kothe 2010). Indeed, the discovery of rhodotorucine A from *R. toruloides* and characterization during the 1980s, as well as the pheromones from another Basidiomycete *Tremella brasiliensis*, provided the first compelling evidence of a common evolutionary origin for the pheromones in Basidiomycetes and Ascomycetes, including the use of post-translational processing and farnesylation (Kamiya et al. 1978; Bölker and Kahmann 1993; Caldwell et al. 1995). Furthermore, cloning of the *R. toruloides* genes responsible illustrated that multiple copies of the pheromone precursor genes exist within the genome, an observation subsequently made in other Basidiomycetes (Akada et al. 1989). However, little more has been explored in this organism or its close relatives until the recent sequencing of Pucciniomycotina strains (Coelho et al. 2008, 2010). Details about the pheromones of the Basidiomycete *S. commune* can be found in the accompanying chapter by Kothe. Thus, we have chosen to use the corn smut fungus *U. maydis* as a representative species.

U. maydis has multiple mating types, unlike *S. cerevisiae* which has only two (bipolar). In *U. maydis*, sex is determined by two unlinked loci: *a* and *b*. The *a* locus has two alleles, *a1* and *a2*, while the *b* locus has many alleles (Vaillancourt et al. 1997; O'Shea et al. 1998). This genetic system is described as tetrapolar. Each allele of the *a* locus carries the genes that encode a sex-specific pheromone and the corresponding receptor (*mfa1* and *pra1*; *mfa2* and *pra2*).

U. maydis and Basidiomycetes in general produce only one class of pheromones: lipopeptides (Fig. 1). They are very similar in structure to the **a** factor of *S. cerevisiae* (Fowler et al. 1999). The maturation process of the pre-pheromones also involves the addition of a farnesyl and a carboxymethyl group to the cysteine residue of the CAAX box, to obtain lipopeptides of 9–13 amino acids (*mfa2* and *mfa1*, respectively).

2.3 Pheromones in Zygomycetes

The regulation and the molecular basis of sexual processes, including initial chemical signals, are shared by many species in the subphylum Mucoromycotina, particularly as established within the order Mucorales (Sutter 1987; Schimek et al. 2003). In this section and hereafter, we will focus on pheromones of the filamentous fungus *Phycomyces blakesleeanus*. Blakeslee (1904) characterized the sexuality of *Phycomyces* and other Mucorales, and defining the heterothallic species as having strains of either of two mating types, (+) and (−), morphologically indistinguishable and only identifiable by their mutual reaction. While *Phycomyces* is isogamous, i.e. gametes of both sexes contribute symmetrically to the sexual process (Spalla 1963), the two sexes produce different pheromones.

Burgeff (1924) showed that the sexual development of strains of the opposite sex starts with the exchange of chemical signals, called trisporic acids. These were the first pheromones discovered in fungi, and contrast in composition to those used in the Ascomycetes and Basidiomycetes (Fig. 1). These stimulate the development of a specialized hyphae, the zygophores, in single cultures (+) and (−). They also induce carotenogenesis that accompanies sexual activity. When mycelia are of different sex, sexual structures and mycelium in the sexual area intensify their color. In addition, these single cultures contain precursors that are converted to trisporic acids by cultures of the opposite sex (Sutter 1970; Sutter et al. 1973, 1974; Werkman and van den Ende 1973). Both trisporic acids and their precursors have been known for over 40 years and have been isolated from several organisms. From mated cultures of *Blakeslea trispora*, five classes of trisporic acids (A, B, C, D and E) have been isolated, the majority component being trisporic acid C (Caglioti et al. 1966). Other chemical signals have been described as trisporin B and C from single cultures of sex (−) of *B. trispora* and *Mucor mucedo*; trisporol B and C from single cultures of sex (−) of *B. trispora*; methyltrisporate B from single cultures of sex (+) of *B. trispora*; methyltrisporate C from single cultures of sex (+) of *B. trispora* and *M. mucedo*; dihydromethyltrisporate B and C from single cultures of sex (+) of *B. trispora* and *M. mucedo* (Bu'Lock et al. 1976; Sutter and Whitaker 1981); trisporic acid E (Fig. 1) from mated cultures of *M. mucedo* and *P. blakesleeanus* (Miller and Sutter 1984); apotrisporol C (Cainelli et al. 1967; Sutter and Whitaker 1981) and apotrisporin C from single and mated cultures of *B. trispora* (Sutter and Zawodny 1984) and apotrisporin E from mated cultures of *P. blakesleeanus* and *B. trispora* (Sutter 1986).

Sexual differentiation in Zygomycetes is a process of industrial interest, since it induces the production of carotenoids. Geranylgeranyl pyrophosphate is the direct precursor of all carotenoids, thus β-carotene is synthesized after seven enzymatic steps governed by two structural genes, *carRA* (Torres-Martínez et al. 1980; Arrach et al. 2001) and *carB* (Eslava and Cerdá-Olmedo 1974; Aragón et al. 1976). The trisporic acids and related compounds are similar to fragments of β-carotene and therefore are considered apocarotenoids. Strains lacking the β-carotene do not have sexual activity (Heisenberg and Cerdá-Olmedo 1968), further indicating that pheromones are apocarotenoids. During sexual interaction, β-carotene is split into three fragments, resulting in three families of apocarotenoids (Polaino et al. 2010). New growth conditions and techniques for separating apocarotenoids have been developed recently. This has allowed the identification of 13 compounds, of which 3 are new natural products (Polaino et al. 2012). Also recently, the *carS* gene has been identified from mutants that are sterile and are hyper-accumulators of β-carotene (Tagua et al. 2012). Its product is a β-carotene-cleaving oxygenase involved in the synthesis of apocarotenoids and is homologous to the *tsp3* gene characterized from *B. trispora* (Burmester et al. 2007; Von Lintig and Vogt 2000; Medina et al. 2011).

The diversity of trisporoids produced by the Mucorales is likely a consequence of several factors. The first may be the presence of multiple copies of carotenoid oxygenases in the genomes of the three Mucoromycotina species sequenced to date (Medina et al. 2011). Second, there is an intricate interplay between the two sexes in these fungi, in which the pheromones produced by one mating type may be chemically modified by the other (Schachtschabel et al. 2008; Schimek and Wöstemeyer 2009). There are two known enzymes in the postulated biosynthetic pathway (Sutter 1987) and that have had their genes identified: an NADP-dependent dehydrogenase, which catalyzes the oxidation of 4-dihydromethyltrisporate to methyltrisporate and probably also of 4-dihydrotrisporin to trisporin, and an esterase involved in the production of trisporic acids from methyltrisporate (Werkman 1976; Czempinski et al. 1996; Wetzel et al. 2009).

Regarding the biological effects of pheromones, the trisporic acids of *P. blakesleeanus* are the only metabolites known to cause the formation of zygophores in single cultures of either sex. Also there is unequivocal evidence of the relationship between sexuality and β-carotene synthesis (van den Ende 1968). When mycelia are of different sex, sexual structures and mycelium in the sexual area intensify their coloration. It was observed that acetate and other small carboxylic acids alter the metabolism and development during the sexual cycle, and are signals that inhibit sexual carotenogenesis and stimulate the formation of zygospores.

P. blakesleeanus is currently classified within the subphylum Mucoromycotina, a group newly defined with the latest taxonomic revision of the fungal kingdom (Hibbett et al. 2007). Three orders within the Mucoromycotina are the Mucorales (that includes *P. blakesleeanus*), Endogonales and Mortierellales. Pheromone signaling has been suggested to occur via trisporic acid mediated mechanisms in the Mortierellales (Schimek et al. 2003). However, BLAST analysis of the draft genome sequence of *Mortierella verticillata* at the Broad Institute and the genome of *M. alpina* (Wang et al. 2011) does not reveal the presence of homologs of the carotene biosynthesis genes required for this process in the Mucorales. Pheromone signaling thus remains to be further explored in the two other orders.

2.4 *Pheromones in Chytridiomycetes*

Pheromone signaling has been less well studied amongst the chytrids, a group historically defined as "lower fungi" that bear flagella at some stage of their lifecycles. We focus on the homothallic species *Allomyces macrogynus* (Blastocladiomycota).

A. macrogynus strains bear both male and female gametangia, and each of them results in production of their gametes. Both male gametangia and gametes have an orange color due to the production of β-carotene, which parallels what is seen with the Mucoromycotina species. However, it is not established that this is a pheromone precursor, as in the case of the Mucoromycotina. Instead, the female gametangia and gametes secrete a sesquiterpene pheromone into the environment, called sirenin (Fig. 1). It was the first fungal pheromone to have its structure determined (Machlis 1968). The male gametangia and gametes also secrete the pheromone, called parisin, but its structure or other chemical properties remain unknown (Pommerville and Olson 1987).

2.5 The Other "Lower" Fungi

In the sections above, the terms Ascomycete and Basidiomycete refer to monophyletic lineages, collectively joined as a monophyletic group known as the Dikarya. All Dikarya species are derived from a single progenitor, illustrated by shared pheromone characteristics. In contrast, "Zygomycete" and "Chytrid" are polyphyletic and refer to sets of lineages basal to the Dikarya. In this section we wish to draw attention to the phyla corresponding to the lower fungi, such as the Neocallimastigomycota, Blastocladiomycota, Microsporidia, Glomeromycota (Hibbett et al. 2007) and the Cryptomycota, the latter recently discovered (Jones et al. 2011). Additional subphyla are the Kickxellomycotina, Zoopagomycotina and Entomophthoromycotina. Little is known about mating in these fungi and nothing about the presence of pheromone signaling (Table 1). As such, these species represent key taxa in which to explore pheromone biosynthesis and signaling towards understanding the evolution and function of these molecules in the fungi.

3 Perception and Signal Transduction: Sensory Systems

3.1 Receptors and Signaling in Ascomycetes

Two types of receptors exist to perceive pheromones and transmit this signal, as seen for the *S. cerevisiae* receptor for the α factor (Ste2) and the receptor for the **a** factor (Ste3). Both are G protein-coupled receptors (GPCR). The better characterized is Ste2.

The Ste2 receptor has seven transmembrane domains, where the amino terminus is extracellular and the carboxyl terminus is intracellular. This latter region is directly involved in signal transduction. As described earlier, the central region of the α factor plays a key role in optimal response of its receptor, by placing its strategic ends so that the interaction and activation with the receptor occurs.

Once the pheromone interacts with the receptor and this activates the G protein, it triggers a mitogen-activated protein kinase (MAPK) signaling cascade (Lengeler et al. 2000; Bardwell 2005; Madhani 2007). The G protein is composed of subunits α, β and γ. The pheromone response is initiated through the βγ dimer, which interacts with three proteins (Ste20, Ste5 and Far1) important for signal transduction. The Ste20 protein initiates the process of phosphorylation through kinases Ste11, Ste7 and Fus3. This is mediated by the Ste5 protein, which binds to these

Table 1 Mating type and pheromone features of the main lineages of the fungi, and one specific representative species

Phylum	Mating-type cells	Mating system	Mating type locus alleles	Type of pheromone	Type of pheromone receptor	Gene(s) encoding pheromones/pheromone receptor
Ascomycota						
Saccharomyces cerevisiae	**a** α	Bipolar	*MAT***a**, *MAT*α	**a** factor (peptide) α factor (lipopeptide)	Ste3 (GPCR) Ste2 (GPCR)	*MFa1*, *MFa2*/*STE3* *MF*α*1*, *MF*α*2*/*STE2*
Basidiomycota						
Ustilago maydis	Multiple	Tetrapolar	*a* biallelic, *b* multiallelic	Lipopeptide	Pra1, Pra2 (GPCR)	*mfa1*, *mfa2*; *pra1*, *pra2*
Mucoromycotina						
Phycomyces blakesleeanus	+/−	Bipolar	*sexM*, *sexP*	Non-peptide	Unknown	*carRA*, *carB*, *carS*/*tsp3*
Chytridiomycota						
Allomyces macrogynus	Homothallic	N/A	Unknown	Non-peptide	Unknown	Unknown
Neocallimastigomycota	Unknown	Unknown	Unknown	Unknown	Unknown	Unknown
Blastocladiomycota	Unknown	Unknown	Unknown	Unknown	Unknown	Unknown
Microsporidia	Unknown	Unknown	*Sex*-related	Unknown	Unknown	Unknown
Glomeromycota	Unknown	Unknown	Unknown	Unknown	Unknown	Unknown
Cryptomycota	Unknown	Unknown	Unknown	Unknown	Unknown	Unknown

Note: For most fungal lineages, no information is available
N/A not applicable

proteins and controls their association with the plasma membrane. As a final step of the pathway, the phosphorylated Fus3 activates the proteins Ste12, necessary for the expression of mating genes, and Far1, involved in cell cycle arrest.

3.2 Receptors and Signaling in Basidiomycetes

The *pra1* and *pra2* genes encoding the pheromone receptors reside in the *a* mating type locus of *U. maydis*. These are also G protein-coupled receptors, as described above. A curious situation occurs in *Cryptococcus neoformans*, as it has a second type of G protein-coupled receptor that is activated without a pheromone. This type competes with the standard one to activate the signaling cascade (Xue et al. 2008).

The regulation of sexual development as well as virulence and filamentation in *U. maydis* occur through two coordinated pathways. One is the familiar pathway to the pheromone response and the other is the signal transduction pathway that is dependent on cyclic AMP (cAMP). The latter also uses a G protein-coupled receptor, cAMP and cAMP-dependent protein kinase (Lengeler et al. 2000). These same pathways have also been described in another pathogenic fungus, *C. neoformans*.

S. cerevisiae can initiate the pheromone signaling pathway using compatible sexual pheromones and receptors from the basidiomycete *S. commune* (Fowler et al. 1999). The substitution of exogenous pheromones and receptors is an ideal tool to elucidate aspects of the basis for ligand-receptor specificity and signaling, as well as highlighting the conservation seen in pheromone signaling in the Dikarya.

3.3 Receptors and Signaling in Zygomycetes and Chytridiomycetes

No receptors for apocarotenoids of Mucoromycotina species or receptors for sirenin or parisin of chytrids are known.

4 Evolutionary Considerations

4.1 Sexuality in Fungi: Pheromones as Determinants of Cell Type Identity

In fungi, sexual identity is controlled by a specialized region of the genome known as the *MAT* locus (Dyer 2008; Casselton 2008). This region shows conserved aspects as well as marked differences between Ascomycete and Basidiomycete fungi. Fungal species are often defined as either homothallic (self-fertile) or heterothallic (self-incompatible), although such definitions may be too strict because some species exhibit both modes of reproduction.

For instance, in discussing *S. cerevisiae*, this species has two mechanisms of sexual reproduction, the heterothallic version in which individuals are self-sterile, and homothallic version in which individuals are self-fertile (Pöggeler 2001; Lee et al. 2010; Madhani 2007). In the first mode, discussed above, sexual interaction occurs between two different mating types. In the second mode, homothallic natural strains change sex by copying to the *MAT* active locus the content of one of the silent genes HML and HMR, which contain the *MATα* and *MATa* alleles, respectively (Haber 1998). However, in Ascomycetes, the pheromones are not the genetic determinants of cell identity, but in some cases may function downstream of the *MAT* locus, if they are regulated under the control of the transcription factors encoded within *MAT*.

In contrast, the pheromones are part of cell type identity in some Basidiomycetes. Sex in the Basidiomycetes may be bipolar with two sexes (e.g. *Ustilago hordei*) or tetrapolar (e.g. *U. maydis*), in which two regions of different chromosomes, each with many alleles determine thousands of different sexes. The difference between bipolar and tetrapolar systems is whether or not the pheromones and their receptors are integral parts of cell type identity. There are several hypotheses on the origin of bipolar sexuality (Fraser et al. 2004). It has been demonstrated that the bipolar *MAT* locus of *U. hordei* formed through the fusion of the unlinked *a* and *b* loci observed in *U. maydis* (Bakkeren and Kronstad 1994; Bakkeren et al. 2006; Lee et al. 1999; Casselton and Olesnicky 1998). A similar event appears to have occurred in two bipolar *Cryptococcus* species in which the homeodomain proteins and pheromones and receptors are linked into a non-recombining *MAT* locus more than 100 kb in size (Fraser et al. 2004; Lengeler et al. 2002). Close relatives of these *Cryptococcus* species have tetrapolar characteristics (Metin et al. 2010; Rodriguez-Carres et al. 2010; Findley et al. 2012). In *U. maydis*, *C. neoformans* and *C. gattii*, pheromones may contribute in determining cell type, but are now physically linked to the transcription factor locus. The experimental substitution of different pheromone components in *C. neoformans* suggests that they still function in conferring cell type identity (Stanton et al. 2010).

In contrast to the examples from the genera *Ustilago* and *Cryptococcus*, the transition from the tetrapolar mushroom such as *Coprinus cinereus* to the bipolar mushroom species *Coprinus disseminatus* is due to a loss of function of the *B* locus (James et al. 2006; van Peer et al. 2011). The *A* locus carries homeodomain transcription factors and the *B* locus encodes for pheromones and pheromones receptors. Thus, the homeodomain transcription factors solely determine the mating type, without a role for pheromones or their receptors.

Recently, the sex-determining regions of *P. blakesleeanus, Mucor* spp. and *Rhizopus* spp. have been identified (Idnurm et al. 2008; Lee et al. 2008; Gryganskyi et al. 2010; Li et al. 2011; Wetzel et al. 2012). The *MAT* locus has two alleles, each with a single gene called *sexM* and *sexP,* respectively for the sex (−) and sex (+). The relationship between the apocarotenoids pheromones used in these species and the *sex* genes is currently unknown.

4.2 Co-evolution of Pheromones and Pheromone Receptors

For successful and specific signaling to occur, there must be an accurate interaction between a ligand and its receptor, in this case between fungal pheromones with their appropriate receptors. Pheromones and their receptors have coevolved in the different phyla of fungi. Within the basal fungal lineages, Zygomycetes utilize non-peptide pheromones derived from β-carotene (apocarotenoids). Chytridiomycetes also utilize non-peptide pheromones such as sesquiterpenes. In both groups, the pheromone receptors are unknown, so it is not possible to infer about the forces of evolution that would have shaped both components. The Ascomycetes have lipopeptides or peptides pheromones. Basidiomycetes also use lipopeptide pheromones. Presumably the lipopeptide was present in the progenitor of the Dikarya, although whether Basidiomycetes lost one class of the pheromones or the Ascomycetes gained a class is unresolved.

In the Dikarya, the pheromone receptors are known and are G protein-coupled receptors (GPCR). In humans, pheromones are both peptide and non-peptide in nature. Following this evolutionary trajectory, both basal fungi and metazoans share the use of non-peptide pheromones. This could suggest that these pheromones are part of a common ancestor of these two groups. Alternatively, the chemical properties of both may have been optimized for their environment, with some types more economical to make and distribute in aqueous versus terrestrial environments.

One point for consideration is how individuals recognize their correct species. There are as many as 5.1 million estimated species of fungi (Blackwell 2011): how do they maintain correct signaling with the right species? Clearly for this to occur it requires a level of specificity such that two different species are unable to undergo cell fusion or complete the sexual cycle. This first system may reflect the possible combination of pheromones available. Even with short pheromones of ~12 amino acids in length in the Ascomycetes and Basidiomycetes, this leaves 20^{12} combinations. While many of those are likely inherently non-functional or unable to interact with their receptor (Maynard Smith 1970), there is still enormous potential variation. The second is geographical aspects of speciation. In a single environment, thousands of different species will not be present. Thus, this largely eliminates the need for a unique system for each species (Fowler et al. 2001). A final consideration is that pheromones may also enable self-self recognition processes (Shen et al. 2002), adding another layer to the complexity of the co-evolution of pheromones and their receptors.

4.3 Pheromone Functions in Pathogenic Species

The preceding paragraph raises the possibility of different species interacting with one another through pheromone signaling. One of the most intriguing examples of this is the predicted role of pheromones required for pathogenicity in the Mucoromycotina. Three species, *Chaetocladium brefeldi*, *Dispira americana* and

Parasitella parasitica, parasitize other members of the Mucoromycotina, but only isolates that are of the opposite mating type are susceptible (Schultze et al. 2005; Wöstemeyer et al. 1995). As part of this process, the pathogens form mating-like structures with their hosts, and insert their DNA into the host species (Kellner et al. 1993). While it remains to be established formally that pheromones are essential for this process, the ability of different Mucoromycotina species to sense other species' pheromones is well-established, to the extent that it enables the assignment of the (+) and (−) mating types to new strains using tester strains of other species.

In the fungal kingdom, there are a large number of pathogenic species, particularly against plants, but also animals and other fungi. Most pathogenic fungi are capable of undergoing sexual reproduction. As noted above, sexual reproduction in fungi is governed by the mating type locus. Curiously, within the Ascomycetes, the most common human pathogens do not seem to undergo sexual reproduction or a regular basis, despite having all the sexual machinery available, as exemplified by the situation in *Candida albicans* and related *Candida* species, or *Aspergillus fumigatus* (Nielsen and Heitman 2007; Heitman 2010). By contrast, sexual reproduction prevails in plant pathogenesis by Basidiomycetes. There is a direct relationship between pathogenesis and dimorphism of some Basidiomycetes, like *U. maydis* (Morrow and Fraser 2009; Hsueh and Heitman 2008). Dimorphic Basidiomycetes, as their name suggests, have two alternate forms during their life cycle: a monokaryotic yeast form and a dikaryotic hyphal form, the latter usually involved in pathogenic processes. The transition from one form to another plays an essential role in the pathogenesis of these fungi. Establishing the pathogenic dikaryon state relies on pheromone communication. Furthermore, pheromones can also play a role after cell fusion, by aiding clamp cell formation that is necessary for extending the dikaryotic hyphal (Gola and Kothe 2003; Casselton and Olesnicky 1998).

A third link between pheromones and pathogenesis is in the Basidiomycete human pathogen *Cryptococcus neoformans*. In this species, mutation of the pheromone precursor genes or the Ste3 pheromone receptor in some strain backgrounds reduces virulence in animal models (Chang et al. 2003; Shen et al. 2002).

4.4 Conserved Themes in Fungal Pheromone Communication

Pheromones trigger a number of different physiological effects in fungi beyond the generalization that they facilitate mating. Within the plethora of effects caused by pheromones, there is one that occurs in many fungi. They arrest the cell cycle allowing the fusion of cells. For example, *P. blakesleeanus* does not form zygophores while there is vegetative growth (Drinkard et al. 1982). Ascomycete yeasts arrest their cell cycle in G_1 phase whereas Basidiomycete *U. maydis* cells undergo cell cycle arrest in G_2 phase. Arrest could be related to changes in cell morphology (García-Muse et al. 2003; Sprague and Thorner 1992), including the formation of cellular projections such as shmoos in *S. cerevisiae*, conjugation tubes, or enlarged cells.

A second common theme is that of processing pheromones from larger precursor molecules. This is observed with both types of peptide-based pheromones found in the Dikarya as well as the apocarotenoid pheromones found in the Mucoromycotina. This is an example of convergent evolution since the molecules and processing enzymes themselves are different between these two groups.

5 Concluding Comments

Fungal pheromones have been studied for nearly 90 years, during which time many advances have been made to demonstrate the importance of these molecules in the biology of fungi. A primary question that remains in understanding the evolution of mating systems in fungi is which component originated first: transcription factor dependent mating or pheromone/pheromone signaling. The pheromones and their receptors are known from some species. However, as highlighted by Table 1, this information is primarily from one single, albeit large, lineage of fungi. Other groups remain to be explored, promising to reveal new insights into this fundamental process of cell–cell communication in the fungi.

References

Akada R, Minomi K, Kai J, Yamashita I, Miyakawa T, Fukui S (1989) Multiple genes coding for precursors of rhodotorucine A, a farnesyl peptide mating pheromone of the basidiomycetous yeast *Rhodosporidium toruloides*. Mol Cell Biol 9:3491–3498

Anderegg RJ, Betz R, Carr SA, Crabb JW, Duntze W (1988) Structure of *Saccharomyces cerevisiae* mating hormone a-factor. Identification of S-farnesyl cysteine as a structural component. J Biol Chem 263:18236–18240

Aragón CMG, Murillo FJ, De La Guardia MD, Cerdá-Olmedo E (1976) An enzyme complex for the dehydrogenation of phytoene in *Phycomyces*. Eur J Biochem 63:71–75

Arrach N, Fernández-Martín R, Cerdá-Olmedo E, Avalos J (2001) A single gene for lycopene cyclase, phytoene synthase, and regulation of carotene biosynthesis in *Phycomyces*. Proc Natl Acad Sci USA 98:1687–1692

Bakkeren G, Kronstad JW (1994) Linkage of mating-type loci distinguishes bipolar from tetrapolar mating in basidiomycetous smut fungi. Proc Natl Acad Sci USA 91:7085–7089

Bakkeren G, Jiang G, Warren RL, Butterfield Y, Shin H, Chiu R, Linning R, Schein J, Lee N, Hu G, Kupfer DM, Tang Y, Roe BA, Jones S, Marra M, Kronstad JW (2006) Mating factor linkage and genome evolution in basidiomycetous pathogens of cereals. Fungal Genet Biol 43:655–666

Bardwell L (2005) A walk-through of the yeast mating pheromone response pathway. Peptides 26:339–350

Bell G (1982) The masterpiece of nature: the evolution and genetics of sexuality. Croom Helm, London

Betz R, Crabb JW, Meyer HE, Wittig R, Duntze W (1987) Amino acid sequences of *a*-factor mating peptides from *Saccharomyces cerevisiae*. J Biol Chem 262:546–548

Blackwell M (2011) The fungi: 1, 2, 3 … 5.1 million species? Am J Bot 98:426–438

Blakeslee AF (1904) Sexual reproduction in the Mucorineae. Proc Am Acad Arts Sci 40:205–319

Bölker M, Kahmann R (1993) Sexual pheromones and mating responses in fungi. Plant Cell 5:1461–1469

Bu'Lock JD, Jones BE, Winskill N (1976) The apocarotenoid system of sex hormones and prohormones in Mucorales. Pure Appl Chem 47:191–202

Burgeff H (1924) Untersuchungen über Sexualität und Parasitismus bei Mucorineen. Bot Abh 4:1–135

Burmester A, Richter A, Schultze K, Voelz K, Schachtschabel D, Boland W, Wöstemeyer J, Schimek C (2007) Cleavage of β-carotene as the first step in sexual hormone synthesis in zygomycetes is mediated by a trisporic acid regulated β-carotene oxygenase. Fungal Genet Biol 44:1096–1108

Butenandt A, Beckamnn R, Hecker E (1961) On the sex attractant of silk-moths. I. The biological test and the isolation of the pure sex-attractant bombykol. Hoppe Seylers Z Physiol Chem 324:71–83

Caglioti L, Cainelli G, Camerino B, Mondelli R, Prieto A, Quilico A, Salvatori T, Selva A (1966) The structure of trisporic-C acid. Tetrahedron 22(Suppl):175–187

Cainelli G, Grasselli P, Selva A (1967) Struttura dell'acido trisporico B. La Chimica e L'Industria 49:628–629

Caldwell GA, Naider F, Becker JM (1995) Fungal lipopeptide mating pheromones: a model system for protein prenylation. Microbiol Rev 59:406–422

Casselton LA (2008) Fungal sex genes-searching for the ancestors. Bioessays 30:711–714

Casselton LA, Olesnicky NS (1998) Molecular genetics of mating recognition in basidiomycete fungi. Microbiol Mol Biol Rev 62:55–70

Chang YC, Miller GF, Kwon-Chung KJ (2003) Importance of a developmentally regulated pheromone receptor of *Cryptococcus neoformans* for virulence. Infect Immun 71:4953–4960

Coelho MA, Rosa A, Rodrigues N, Fonseca A, Gonçalves P (2008) Identification of mating type genes in the bipolar basidiomycetous yeast *Rhodosporidium toruloides*: first insight into the *MAT* locus structure of the *Sporidiobolales*. Eukaryot Cell 7:1053–1061

Coelho MA, Sampaio JP, Gonçalves P (2010) A deviation from the bipolar-tetrapolar mating paradigm in an early diverged basidiomycete. PLoS Genet 6:e1001052

Coppin E, Debuchy R, Arnaise S, Picard M (1997) Mating types and sexual development in filamentous ascomycetes. Microbiol Mol Biol Rev 61:411–428

Coppin E, de Renty C, Debuchy R (2005) The function of the coding sequences for the putative pheromone precursors in *Podospora anserina* is restricted to fertilization. Eukaryot Cell 4:407–420

Czempinski K, Kruft V, Wöstemeyer J, Burmester A (1996) 4-dihydromethyltrisporate dehydrogenase from *Mucor mucedo*, an enzyme of the sexual hormone pathway: purification, and cloning of the corresponding gene. Microbiology 142:2647–2654

Drinkard LC, Nelson GE, Sutter RP (1982) Growth arrest: a prerequisite for sexual development in *Phycomyces blakesleeanus*. Exp Mycol 6:52–59

Dyer PS (2008) Evolutionary biology: genomic clues to original sex in fungi. Curr Biol 18:207–209

Eslava AP, Cerdá-Olmedo E (1974) Genetic control of phytoene dehydrogenation in *Phycomyces*. Plant Sci Lett 2:9–14

Findley K, Sun S, Fraser JA, Hsueh YP, Averette AF, Li W, Dietrich FS, Heitman J (2012) Discovery of a modified tetrapolar sexual cycle in *Cryptococcus amylolentus* and the evolution of *MAT* in the *Cryptococcus* species complex. PLoS Genet 8:e1002528

Fowler TJ, Mitton MF, Vaillancourt LJ, Raper CA (2001) Changes in mate recognition through alterations of pheromones and receptors in the multisexual mushroom fungus *Schizophyllum commune*. Genetics 158:1491–1503

Fowler TJ, DeSimone SM, Mitton MF, Kurjan J, Raper CA (1999) Multiple sex pheromones and receptors of a mushroom-producing fungus elicit mating in yeast. Mol Biol Cell 10:2559–2572

Fraser JA, Diezmann S, Subaran RL, Allen A, Lengeler KB, Dietrich FS, Heitman J (2004) Convergent evolution of chromosomal sex-determining regions in the animal and fungal kingdoms. PLoS Biol 2:e384

García-Muse T, Steinberg G, Pérez-Martín J (2003) Pheromone-induced G_2 arrest in the phytopathogenic fungus *Ustilago maydis*. Eukaryot Cell 2:494–500

Gola S, Kothe E (2003) The little difference: in vivo analysis of pheromone discrimination in *Schizophyllum commune*. Curr Genet 42:276–283

Gooday GW, Adams DJ (1993) Sex hormones and fungi. Adv Microb Physiol 34:69–145

Gryganskyi AP, Lee SC, Litvintseva AP, Smith ME, Bonito G, Porter TM, Anishchenko IM, Heitman J, Vilgalys R (2010) Structure, function, and phylogeny of the mating locus in the *Rhizopus oryzae* complex. PLoS One 5:e15273

Haber JE (1998) Mating-type gene switching in *Saccharomyces cerevisiae*. Annu Rev Genet 32:561–599

Heisenberg M, Cerdá-Olmedo E (1968) Segregation of heterokaryons in the asexual cycle of *Phycomyces*. Mol Gen Genet 102:187–195

Heitman J (2010) Evolution of eukaryotic microbial pathogens via covert sexual reproduction. Cell Host Microbe 8:86–99

Heitman J, Kronstad JW, Taylor JW, Casselton LA (eds) (2007) Sex in fungi: molecular determination and evolutionary implications. ASM Press, Washington, DC

Herskowitz I (1988) Life cycle of the budding yeast *Saccharomyces cerevisiae*. Microbiol Rev 52:536–553

Hibbett DS, Binder M, Bischoff JF, Blackwell M, Cannon PF, Eriksson OE, Huhndorf S, James T, Kirk PM, Lücking R, Thorsten Lumbsch H, Lutzoni F, Matheny PB, McLaughlin DJ, Powell MJ, Redhead S, Schoch CL, Spatafora JW, Stalpers JA, Vilgalys R, Aime MC, Aptroot A, Bauer R, Begerow D, Benny GL, Castlebury LA, Crous PW, Dai Y-C, Gams W, Geiser DM, Griffith GW, Gueidan C, Hawksworth DL, Hestmark G, Hosaka K, Humber RA, Hyde KD, Ironside JE, Kõljalg U, Kurtzman CP, Larsson K-H, Lichtwardt R, Longcore J, Miadlikowska J, Miller A, Moncalvo J-M, Mozley-Standridge S, Oberwinkler F, Parmasto E, Reeb V, Rogers JD, Roux C, Ryvarden L, Sampaio JP, Schüssler A, Sugiyama J, Thorn RG, Tibell L, Untereiner WA, Walker C, Wang Z, Weir A, Weiss M, White MM, Winka K, Yao Y-J, Zhang N (2007) A higher-level phylogenetic classification of the fungi. Mycol Res 111:509–547

Hsueh YP, Heitman J (2008) Orchestration of sexual reproduction and virulence by the fungal mating-type locus. Curr Opin Microbiol 11:517–524

Idnurm A, Walton FJ, Floyd A, Heitman J (2008) Identification of the *sex* genes in an early diverged fungus. Nature 451:193–197

James TY, Srivilai P, Kües U, Vilgalys R (2006) Evolution of the bipolar mating system of the mushroom *Coprinellus disseminatus* from its tetrapolar ancestors involves loss of mating-type-specific pheromone receptor function. Genetics 172:1877–1891

Jones SK Jr, Bennett RJ (2011) Fungal mating pheromones: choreographing the dating game. Fungal Genet Biol 48:668–676

Jones MDM, Forn I, Gadelha C, Egan MJ, Bass D, Massana R, Richards TA (2011) Discovery of novel intermediate forms redefines the fungal tree of life. Nature 474:200–203

Kamiya Y, Sakurai A, Tamura S, Takahashi N (1978) Structure of rhodotorucine A, a novel lipopeptide, inducing mating tube formation in *Rhodosporidium toruloides*. Biochem Biophys Res Comun 83:1077–1083

Karlson P, Lüsher M (1959) 'Pheromones': a new term for a class of biologically active substances. Nature 183:55–56

Kellner M, Burmester A, Wöstemeyer A, Wöstemeyer J (1993) Transfer of genetic information from the mycoparasite *Parasitella parasitica* to its host *Absidia glauca*. Curr Genet 23:334–337

Kothe E (2008) Sexual attraction: on the role of fungal pheromone/receptor systems. Acta Microbiol Immunol Hung 55:125–143

Lee N, Bakkeren G, Wong K, Sherwood JE, Kronstad JW (1999) The mating-type and pathogenicity locus of the fungus *Ustilago hordei* spans a 500 kb region. Proc Natl Acad Sci USA 96:15026–15031

Lee SC, Corradi N, Byrnes EJ, Torres-Martinez S, Dietrich FS, Keeling PJ, Heitman J (2008) Microsporidia evolved from ancestral sexual fungi. Curr Biol 18:1675–1679

Lee SC, Ni M, Li W, Shertz C, Heitman J (2010) The evolution of sex: a perspective from the fungal kingdom. Microbiol Mol Biol Rev 74:298–340

Lengeler KB, Davidson RC, D'Souza C, Harashima T, Shen W-C, Wang P, Pan X, Waugh M, Heitman J (2000) Signal transduction cascades regulating fungal development and virulence. Microbiol Mol Biol Rev 64:746–785

Lengeler KB, Fox DS, Fraser JA, Allen A, Forrester K, Dietrich FS, Heitman J (2002) Mating-type locus of *Cryptococcus neoformans*: a step in the evolution of sex chromosomes. Eukaryot Cell 1:704–718

Li CH, Cervantes M, Springer DJ, Boekhout T, Ruiz-Vazquez RM, Torres-Martinez SR, Heitman J, Lee SC (2011) Sporangiospore size dimorphism is linked to virulence of *Mucor circinelloides*. PLoS Pathog 7:e1002086

Machlis L (1968) The response of wild type male gametes of *Allomyces* to sirenin. Plant Physiol 43:1319–1320

Madhani H (2007) From a to α: yeast as a model for cellular differentiation. Cold Spring Harbor Laboratory Press, Cold Spring Harbor

Martin SH, Wingfield BD, Wingfield MJ, Steenkamp ET (2011) Causes and consequences of variability in peptide mating pheromones of ascomycete fungi. Mol Biol Evol 28:1987–2003

Maynard Smith J (1970) Natural selection and the concept of a protein space. Nature 225:563–564

Maynard Smith J (1978) The evolution of sex. Cambridge University Press, Cambridge

Medina HR, Cerdá-Olmedo E, Al-Babili S (2011) Cleavage oxygenases for the biosynthesis of trisporoids and other apocarotenoids in *Phycomyces*. Mol Microbiol 82:199–208

Metin B, Findley K, Heitman J (2010) The mating type locus (*MAT*) and sexual reproduction of *Cryptococcus heveanensis*: insights into the evolution of sex and sex-determining chromosomal regions in fungi. PLoS Genet 6:e1000961

Michaelis S, Herskowitz I (1988) The **a**-factor pheromone of *Saccharomyces cerevisiae* is essential for mating. Mol Cell Biol 8:1309–1318

Miller ML, Sutter RP (1984) Methyl trisporate E. A sex pheromone in *Phycomyces blakesleeanus*? J Biol Chem 259:6420–6422

Morrow CA, Fraser JA (2009) Sexual reproduction and dimorphism in the pathogenic basidiomycetes. FEMS Yeast Res 9:161–177

Naider F, Becker JM (2004) The α-factor mating pheromone of *Saccharomyces cerevisiae*: a model for studying the interaction of peptide hormones and G protein-coupled receptors. Peptides 25:1441–1463

Nielsen K, Heitman J (2007) Sex and virulence of human pathogenic fungi. Adv Genet 57:143–173

O'Shea SF, Chaure PT, Halsall JR, Olesnicky NS, Leibbrandt A, Connerton IF, Casselton LA (1998) A large pheromone and receptor gene complex determines multiple *B* mating type specificities in *Coprinus cinereus*. Genetics 148:1081–1090

Pöggeler S (2001) Mating-type genes for classical strain improvements of ascomycetes. Appl Microbiol Biotechnol 56:589–601

Polaino S, Gonzalez-Delgado JA, Arteaga P, Herrador MM, Barrero AF, Cerdá-Olmedo E (2012) Apocarotenoids in the sexual interaction of Phycomyces blakesleeanus. Org Biomol Chem 10:3002–3009

Polaino S, Herrador M, Cerdá-Olmedo E, Barrero AF (2010) Splitting of β-carotene in the sexual interaction of *Phycomyces*. Org Biomol Chem 8:4229–4231

Pommerville J, Olson LW (1987) Evidence for a male-produced pheromone in *Allomyces macrogynus*. Exp Mycol 11:245–248

Raudaskoski M, Kothe E (2010) Basidiomycete mating type genes and pheromone signaling. Eukaryot Cell 9:847–859

Rodriguez-Carres M, Findley K, Sun S, Dietrich FS, Heitman J (2010) Morphological and genomic characterization of *Filobasidiella depauperata*: a homothallic sibling species of the pathogenic *Cryptococcus* species complex. PLoS One 5:e9620

Schachtschabel D, David A, Menzel K-D, Schimek C, Wöstemeyer J, Boland W (2008) Cooperative biosynthesis of trisporoids by the (+) and (−) mating types of the zygomycete *Blakeslea trispora*. ChemBioChem 15:3004–3012

Schimek C, Wöstemeyer J (2009) Carotene derivatives in sexual communication of zygomycete fungi. Phytochemistry 70:1867–1875

Schimek C, Kleppe K, Saleem AR, Voigt K, Burmester A, Wöstemeyer J (2003) Sexual reactions in Mortierellales are mediated by the trisporic acid system. Mycol Res 107:736–747

Schmoll M, Seibel C, Tisch D, Dorrer M, Kubicek CP (2010) A novel class of peptide pheromone precursors in ascomycetous fungi. Mol Microbiol 77:1483–1501

Schultze K, Schimek C, Wöstemeyer J, Burmester A (2005) Sexuality and parasitism share common regulatory pathways in the fungus *Parasitella parasitica*. Gene 348:33–44

Shen WC, Davidson RC, Cox GM, Heitman J (2002) Pheromones stimulate mating and differentiation via paracrine and autocrine signaling in *Cryptococcus neoformans*. Eukaryot Cell 1:366–377

Spalla C (1963) Ricerche sulla riproduzione sessuale in Mucorales. Studio della riproduzione sessuale in *Phycomyces blakesleeanus* e in *Choanephora circinans* e comparazione con quella di *Cunninghamella blakesleeana*, *C. elegans*, *Mucor hiemalis* e *M. racemosus*. Riv Patol Veg 3:189–198

Sprague GF, Thorner JW (1992) Pheromone response and signal transduction during the mating process of *Saccharomyces cerevisiae*. In: Broach JR, Pringle JR, Jones EW (eds) The molecular and cellular biology of the yeast *Saccharomyces*: gene expression. Cold Spring Harbor Laboratory Press, Cold Spring Harbor, pp 657–744

Stanton BC, Giles SS, Staudt MW, Kruzel EK, Hull CM (2010) Allelic exchange of pheromones and their receptors reprograms sexual identity in *Cryptococcus neoformans*. PLoS Genet 6: e1000860

Sutter RP (1970) Trisporic acid synthesis in *Blakeslea trispora*. Science 168:1590–1592

Sutter RP (1986) Apotrisporin-E: a new sesquiterpenoid isolated from *Phycomyces blakesleeanus* and *Blakeslea trispora*. Exp Mycol 10:256–258

Sutter RP (1987) Sexual development. In: Cerdá-Olmedo E, Lipson ED (eds) Phycomyces. Cold Spring Harbor Laboratory, Cold Spring Harbor, pp 317–336

Sutter RP, Whitaker JP (1981) Zygophore-stimulating precursors (pheromones) of trisporic acids active in (−)-*Phycomyces blakesleeanus*. Acid-catalyzed anhydro derivates of methyl 4-dihydrotrisporate-C and 4-dihydrotrisporate-C. J Biol Chem 256:2334–2341

Sutter RP, Zawodny PD (1984) Apotrisporin: a major metabolite of *Blakeslea trispora*. Exp Mycol 8:89–92

Sutter RP, Capage DA, Harrison TL, Keen WA (1973) Trisporic acid biosynthesis in separate plus and minus cultures of *Blakeslea trispora*: Identification by *Mucor* assay of two mating-type-specific components. J Bacteriol 114:1074–1082

Sutter RP, Harrison TL, Galasko G (1974) Trisporic acid biosynthesis in *Blakeslea trispora* via mating type-specific precursors. J Biol Chem 249:2282–2284

Tagua V, Medina HR, Martín-Domínguez R, Eslava AP, Corrochano LM, Cerdá-Olmedo E, Idnurm A (2012) A gene for carotene cleavage required for pheromone biosynthesis and carotene regulation in the fungus *Phycomyces blakesleeanus*. Fungal Genet Biol 49. doi:10.1016/j.fgb.2012.03.002

Torres-Martínez S, Murillo FJ, Cerdá-Olmedo E (1980) Genetics of lycopene cyclization and substrate transfer in β-carotene biosynthesis in *Phycomyces*. Genet Res 36:299–309

Vaillancourt LJ, Raudaskoski M, Specht CA, Raper CA (1997) Multiple genes encoding pheromones and a pheromone receptor define the Bβ1 mating-type specificity in *Schizophyllum commune*. Genetics 146:541–551

van den Ende H (1968) Relationship between sexuality and carotene synthesis in *Blakeslea trispora*. J Bacteriol 96:1298–1303

van Peer AF, Park S-Y, Shin P-G, Jang K-Y, Yoo Y-B, Park Y P, Lee D-M, Sung G-H, James TY, Kong W-S (2011) Comparative genomics of the mating-type loci of the mushroom *Flammulina velutipes* reveals widespread synteny and recent inversions. PLoS One 6:e22249

Von Lintig J, Vogt K (2000) Filling the gap in vitamin A research. Molecular identification of an enzyme cleaving β-carotene to retinal. J Biol Chem 275:11915–11920

Wang L, Chen W, Feng Y, Ren Y, Gu Z, Chen H, Wang H, Thomas MJ, Zhang B, Berquin IM, Li Y, Wu J, Zhang H, Song Y, Liu X, Norris JS, Wang S, Du P, Shen J, Wang N, Yang Y, Wang W, Feng L, Ratledge C, Zhang H, Chen YQ (2011) Genome characterization of the oleaginous fungus *Mortierella alpina*. PLoS One 6:e28319

Werkman BA (1976) Localization and partial characterization of a sex-specific enzyme in homothallic and heterothallic Mucorales. Arch Microbiol 109:209–213

Werkman BA, van den Ende H (1973) Trisporic acid synthesis in *Blakeslea trispora*. Arch Microbiol 90:365–374

Wetzel J, Burmester A, Kolbe M, Wöstemeyer J (2012) The mating-related loci *sexM* and *sexP* of the zygomycetous fungus *Mucor mucedo* and their transcriptional regulation by trisporoid pheromones. Microbiology 158:1016–1023

Wetzel J, Scheibner O, Burmester A, Schimek C, Wöstemeyer J (2009) 4-dihydrotrisporin-dehydrogenase, an enzyme of the sex hormone pathway of *Mucor mucedo*: purification, cloning of the corresponding gene, and developmental expression. Eukaryot Cell 8:88–95

Williams GC (1975) Sex and evolution. Monogr Popul Biol 8:3–200

Wöstemeyer J, Wöstemeyer A, Burmester A, Czempinski K (1995) Relationships between sexual processes and parasitic interactions in the host-pathogen system *Absidia glauca-Parasitella parasitica*. Can J Bot 73(Suppl 1):S243–S250

Xue C, Hsueh Y-P, Heitman J (2008) Magnificent seven: roles of G protein-coupled receptors in extracellular sensing in fungi. FEMS Microbiol Rev 32:1010–1032

Thoughts on Quorum Sensing and Fungal Dimorphism

Kenneth W. Nickerson, Audrey L. Atkin, Jessica C. Hargarten, Ruvini Pathirana, and Sahar Hasim

Abstract Farnesol has been best studied for its role in regulating fungal dimorphism. However, farnesol is also a lipid and in this review we analyze data relevant to farnesol's function and synthesis from the perspective of farnesol and bacterial endotoxins acting as membrane active compounds. This analysis implicates the possible roles of: (1) endotoxins in the regulation of farnesol production by *C. albicans*; (2) farnesol in the interactions between *C. albicans* and the host during disseminated infections; and (3) ubiquinones in the mechanisms for unusually high resistance to farnesol by some *C. albicans* cell types. Finally we discuss the implications that the use of farnesol as both a signaling molecule and to antagonize competing microbials species has for the regulation of HMG-CoA reductase, the enzyme that is the usual rate limiting step in sterol/lipid synthesis.

1 Introduction

The role of farnesol in the dimorphism of *Candida albicans* was discovered by Hornby et al. (2001), reviewed by Nickerson et al. (2006), and updated by Langford et al. (2009) and Hogan and Muhlschlegel (2011). Up to now the theme of farnesol research has been farnesol as a signaling molecule and how it affects fungal polymorphism via signal transduction (Nickerson et al. 2006). However, in the process the role of farnesol as a lipid has been somewhat neglected. Thus, the present review will focus on how the lipid nature of farnesol contributes to its roles in virulence and pathogenicity, mating, and the interactions of *C. albicans* with both host macrophages and competing microbes.

For the fungi, we define quorum sensing as any cell density dependent phenomenon which is mediated by an extracellular molecule which is produced and excreted by the

K.W. Nickerson (✉) • A.L. Atkin • J.C. Hargarten • R. Pathirana • S. Hasim
School of Biological Sciences, University of Nebraska, Lincoln, NE 68588-0666, USA
e-mail: knickerson1@unl.edu

G. Witzany (ed.), *Biocommunication of Fungi*, DOI 10.1007/978-94-007-4264-2_12,

fungus in question. The name is, of course, borrowed from the classic review by Fuqua et al. (1994) which discussed bacterial homoserine lactones. We have extended the terminology slightly by introducing quorum sensing molecules or QSMs (Nickerson et al. 2006). Critically, our use of the term QSM does not presuppose anything about its mode of action. In particular, it could include a situation where the fungal role is restricted to modifying an exogenously provided molecule, e.g. linoleic acid to 3-hydroxy tetradecaenoic acid (Nigam et al. 2010). The key point is that fungi also have mechanisms to sense their own population densities.

2 It Must Be Something in the Water

This story is both a cautionary tale and a possible area for future study on bacterial-fungal interactions. The first published report on farnesol as a quorum sensing molecule (QSM) for *C. albicans* (Hornby et al. 2001) was delayed for roughly 2 years by issues of water quality. In 1996 the Nickerson lab moved from an old building, soon to be torn down (Lyman Hall), to a new state-of-the-art research center named for University of Nebraska graduate and Nobel Prize winner George W. Beadle. At the time we were using an activity directed purification scheme to identify the molecule in spent media which blocked germ tube formation in *C. albicans*. The assay worked perfectly in Lyman Hall but did not work in our new facilities. We eventually found that if we purchased bottled distilled water at the local supermarket and used that water to prepare our growth media then the assay worked perfectly again.

What was different about the distilled waters provided in the two buildings? And did the building distilled/deionized water in Beadle prevent the QSM (farnesol) from being formed or inactivate it after it had been formed? In this regard, we know that farnesol is markedly sensitive to air oxidation and, consequently, we always store our farnesol stock solutions under nitrogen. Shchepin et al. (2003) showed that the 10, 11 epoxide of farnesol has only 3% of farnesol's QSM activity, and at the time of our move the city of Lincoln had just switched from chlorine to ozone treatment as the penultimate step in their water treatment procedures. Could there be any residual ozone carried over? However, this possibility was eliminated by mixing the two spent media, one with QSM activity and one without, and observing that the resulting QSM activity was undiminished.

Thus, we were left with discovering what difference in the two waters regulated QSM/farnesol production by *C. albicans*. This question has not yet been fully resolved. A chemical analysis of the respective waters was indicated. There were two precedents. First, waters can differ dramatically in their mineral contents. Consider, for instance, Table 12.1 of Okafor's text on Industrial Microbiology (Okafor 2007) which compares the mineral content of water in eight cities noted for their breweries. The concentrations of Ca^{2+}, Mg^{2+}, SO_4^{2-}, NO_3^-, Cl^-, and HCO_3^- in the respective waters differed by 44-, 62-, 212-, 62-, 120-, and 31-fold, respectively. Second, Cu^{2+} and Zn^{2+} cause morphological shifts in most dimorphic fungi, i.e. *Ceratocystis ulmi*, *Histoplasma capsulatum*, *Mucor rouxii*, *Sporothrix schenkii*, *Ustilago sphaerogena*, and *C. albicans*. The Zn^{2+} induced shifts are all in the same direction (mycelia to yeasts) but the concentrations of Zn^{2+} found to be effective varied. *C. ulmi* was typical in that it required 4–5 mM Zn^{2+} whereas *C. albicans* required only 10–20 μM Zn^{2+} (Yamaguchi 1975; Soll et al. 1981).

Accordingly, we obtained eight types of distilled water locally available in Lincoln and in Central Minnesota. Five of the eight permitted QSM accumulation, three did not. These eight water samples were analyzed by atomic absorption for 11 elements including Se and most of the transition metals. At that time ICP-MS was not available to us for the detailed analysis of the elemental composition of the waters. These analyses produced some surprises, such as two of the waters having 50-fold more Co^{2+} than the others, but there was no discernible correlation between elemental composition and QSM formation! We were stumped. At this point we decided to publish our data showing that farnesol was the QSM for *C. albicans* without having resolved the water issue. Thus, the Hornby et al. (2001) paper specifies that the GPP growth medium was always prepared with Kandiyohi distilled water (Kandiyohi Bottled Water Co., Willmar, Minn.). The Minnesota connection reflected the participation of visiting scientist Prof. Ellen Jensen from the College of St. Benedict's and St. John's University in Minnesota.

Some years later we learned that our standard water purification system (running distilled water through deionizing columns and activated charcoal columns followed by 0.22 μm filtration) might not remove bacterial endotoxins. We had not considered this possibility previously. Accordingly, we purchased an endotoxin detection kit and sampled the distilled/deionized water from three locations in the Beadle Center. They each contained 16–32 endotoxin units per ml. The Kandiyohi distilled water had no endotoxins and Crystal Glen distilled water had 1 endotoxin unit per ml. The other commercial distilled waters were no longer available. Samples of the Beadle distilled water taken prior to filtration contained 10–100 CFU/ml when plated on LB agar. As expected, these bacteria were Gram negative, presumably *Pseudomonads*. Our current scenario is that: rain water contains ca. 8 mg/L organics, *Pseudomonads* are notorious for the wide variety of organics they can metabolize (Stanier et al. 1966), and it is expected that large volumes of standing water with long residence times will experience some bacterial presence (McCoy 1980). Lyman Hall did not experience these problems because the distilled water had been obtained from condensed steam and our current supplier Super-Saver has a very short residence time.

This cautionary tale also suggests a direction of research opportunity. There is increasing evidence, summarized by Langford et al. (2009), that bacteria and fungi wage a sophisticated, molecular level battle with one another. In particular, farnesol treatment kills or inhibits many bacteria, i.e. *Acinetobacter baumannii*, *Pseudomonas aeruginosa*, *Staphylococcus aureus*, *Staphylococcus epidermidis*, and *Streptococcus mutans*, while at least five bacteria, *A. baumannii*, *Burkholderia cenocepacia*, *P. aeruginosa*, *Salmonella enterica*, and *Xanthomonas campestris*, secrete molecules which inhibit filamentation by *C. albicans* (see Langford et al. 2009). The supposition that *Pseudomonas* endotoxins block farnesol production by *C. albicans* (see above) is consistent with these observations and casts them in a new light. The secreted, filament inhibiting bacterial molecules which have been identified include: dodecanol, 2-dodecenoic acid, 11-methyl-2-dodecenoic acid,

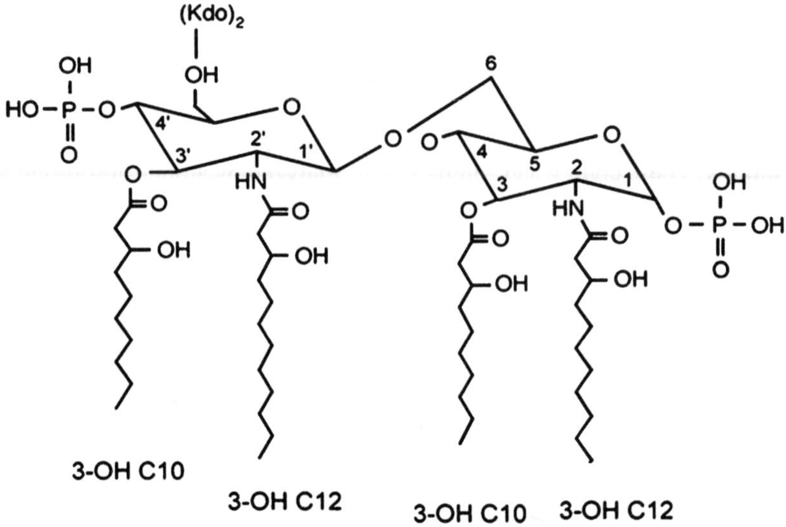

Fig. 1 Pseudomonas aeruginosa endotoxin structure (p245 of Wang and Quinn 2010)

and the C_{12}-acyl homoserine lactone. Each of these molecules makes a better structural analog for the C_{10}- and C_{12}-fatty acyl side chains of a *Pseudomonas* endotoxin (Fig. 1) than it would for farnesol.

With regard to location, *C. albicans* is most likely to encounter bacterial endotoxins in the anaerobic mammalian guts. Thus, it is entirely consistent that under anaerobic growth conditions *C. albicans* does not produce detectable farnesol or respond to added farnesol (Dumitru et al. 2004). It would be of interest to compare the effects of bacterial endotoxins on *C. albicans* growing under both aerobic and anaerobic conditions.

3 A Potential Role for Farnesol in *C. albicans* Host Interactions

3.1 Plasticity of Morphology: A Virulence Factor

Biologically speaking, the interactions of *C. albicans* and the human host are much like a never-ending game of cat-and-mouse: the human host laying down defense mechanisms to keep the fungus in check, and the fungus breaking these walls down for pure survival. Understanding this interplay is important as scientists seek to tip the balance away from the pathogenic *C. albicans*. Throughout the body, *C. albicans* has adapted mechanisms to gather, interpret, and respond to signals provided by the host and the diverse terrain the host environment poses, much of which is still unknown. One of the key responses is simply changing its morphology. *In vivo,* different morphologies of *C. albicans* have been associated with distinct degrees and locations of infection.

From the benign colonization of the skin typically by white or opaque phase cells to benign infections of the oral cavity by white phase cells, chronic vaginal infections by white phase yeast cells, the hyphal and pseudohyphal growth found in the gastrointestinal tract and in many disseminated systemic infections, this morphological plasticity appears to act as a tolerance mechanism to counteract the changes in pH, nutrient availability, microflora composition, and oxygen levels the fungus will encounter on and within its human host (Lachke et al. 2003; Sobel 1997; Sudbery 2011).

C. albicans easily colonizes many locations around an immune-competent host without doing much harm to the individual but, in an immunocompromised host, the capability to switch between morphologies acts as a virulence factor and is central to its pathogenicity. However all forms of *C. albicans* do not convey this level of virulence. Some forms are more susceptible to macrophages and the other defenses of the human immune system than others. This difference in morphological susceptibility could be an important opportunity for the development of antifungal drugs that block systemic Candidiasis.

3.2 Farnesol and Virulence

The secretion of farnesol also plays a role in host–pathogen interactions *in vivo*. Hornby et al. (2001) discovered that *in vitro* farnesol acted to block the yeast to mycelia transition. Thus, *in vitro* farnesol acted as a QSM. But what would its role be *in vivo*? At that time we presented two possibilities. Firstly, if farnesol acted *in vivo* as it did *in vitro*, then farnesol and its analogs should prove to be effective antifungals because the yeast to mycelia transition is essential for virulence. It was on this premise that 50 analogs of farnesol were synthesized and tested for their *in vitro* QSM potency (Shchepin et al. 2005). However, we also suggested that farnesol production by *C. albicans* might instead function as a virulence factor (Hornby et al. 2001) and for the mouse tail vein injection model that proved to be the case (Navarathna et al. 2007). *C. albicans* mutants which produced 85% less farnesol were five times less pathogenic to mice than their parent cells. Also, when farnesol was administered orally to the mice prior to infection, their mortality increased as did the colonization of kidneys (Navarathna et al. 2007).

These observations pose the dilemma of finding a mechanism whereby a molecule which blocks the yeast to mycelia transition can also act as a virulence factor. In this regard, we note that there is as yet no evidence for farnesol blocking the yeast to mycelia transition *in vivo* while there is evidence that farnesol behaves differently for surface infections, where it is protective (Hisajima et al. 2008), versus systemic infections where it is a virulence factor (Navarathna et al. 2007). A partial resolution of this dilemma comes from the realization that *in vitro*, in glass or plastic, excreted farnesol can accumulate whereas *in vivo* it would be soaked up by the mammalian cell membranes. Thus, different concentrations of farnesol should be present *in vitro* and *in vivo*.

3.3 A *C. albicans* Macrophage Chemoattractant: White Versus Opaque

Another partial resolution of this *in vitro* vs. *in vivo* dilemma concerns how farnesol affects the host innate immune system. The first level of defense the host has against candidal infection is through the innate immune system. Distinct morphologies elicit different responses by the host immune system. Both white and opaque cells are known to attract leukocytes to the site of infection, but only white cells produce and secrete a small molecular weight chemoattractant that draws the leukocyte directly towards the white cell (Geiger et al. 2004). Lohse and Johnson (2008) took this knowledge a step further by showing that not only were leukocytes more attracted to white cells than opaque cells, but because of the presence of a chemoattractant produced by the white cells, mouse macrophages engulfed white *C. albicans* cells much more efficiently than they did opaque cells (Lohse and Johnson 2008). Not only were the white cells engulfed at a higher rate but they were also less susceptible to killing by human macrophages and neutrophils than were opaque cells, possibly due to their increased capabilities of escape once phagocytosed or possibly due to another effect of the chemoattractant on the macrophages (Kolotila and Diamond 1990). The chemical identity of this chemoattractant is currently unknown, but the reason behind its continued secretion by the white form is intriguing. One likely candidate is farnesol (Langford et al. 2009). Macrophages are capable of detecting and responding to exogenous farnesol, specifically by stimulating secretion of pro-inflammatory and regulatory cytokines (IL-6, IL-1β, IL-10, and TNF-α; Ghosh et al. 2010). The production of these warning signals by macrophages is an important indicator of how the body ultimately hopes to clear the infection. Because of the cytotoxic effects farnesol has on macrophages (ROS and DNA fragmentation), farnesol suppresses the anti-Candida activity of macrophages (Abe et al. 2009), thus making it all the more difficult to eliminate the fungus early in infection.

This chain of events from attraction to engulfment to eventual killing of the macrophages is mediated at two points by different *C. albicans* morphologies. It is known that wild type, white cells of *C. albicans* can escape from mouse macrophages by switching to the hyphal morphology 6–8 h post-engulfment, and effectively puncturing the macrophage from within (Ghosh et al. 2009). Those strains with delayed or dysfunctional hyphal formation (through disruption of the arginine biosynthetic pathway for instance) were unable to survive within and escape from the macrophage (Ghosh et al. 2009). It remains a perplexing mystery why certain morphologies such as the opaque cells are better able to elude host immune defenses, or retaliate such as the hyphae, while other morphologies such as the white yeast cells seek to be found through the production of a potent chemoattractant. This aspect of farnesol production by *C. albicans* is in part counterintuitive of the way we think a fungus should behave but it does have precedent in the form of some pathogenic intracellular bacteria.

3.4 Bacterial Analogs for Host Evasion

Direct targeting of tissue phagocytes to the site of infection by pathogenic microorganisms, in the hopes of being phagocytosed, is not a novel concept in the realm of microbial infections. A comprehensive review of bacterial evasion strategies can be found in Flannagan et al. (2009). For some 'professional' intracellular bacteria, such as *Mycobacterium tuberculosis*, *Listeria monocytogenes*, and *Legionella pneumophila,* successful establishment of infection and dissemination throughout the host depends entirely on exploiting the natural responses of phagocytes. Following phagocytosis and entry into the phagosome, these bacteria have developed mechanisms to prevent further phagocyte killing and digestion, allowing for long term intraphagosomal survival within host cells, either through interfering with phagolysosome maturation, and the secretion of ROS and antibacterial proteases, or through counteracting the host cells' expression of MHC and loading of antigenic bacterial peptides, effectively eluding further host immune system detection.

The most researched of these pathogenic microbes that use macrophages to escape immune detection is *L. monocytogenes*. This model system could be used as a possible bacterial analog to understand what strategy *C. albicans* may utilize during infection. Phagocytosis of *L. monocytogenes* is mediated by a macrophage scavenger receptor that binds directly to the lipoteichoic acid on this Gram-positive bacterium (Dunne et al. 1994). Once within the phagosome of the macrophage, it uses an array of cholesterol-dependent cytolysins to prevent the further maturation of the phagosome by inhibiting its fusion with the lysosome. By sequestering inside the membrane vacuole, it is able to acquire the nutrients it requires to replicate directly from the host without host detection, prior to escape from the macrophage. Through this comparison, much can be learned about possible intracellular signaling interactions between the phagocyte and the intracellular fungi during the phagocytosis process. We note that during the 6 h between their engulfment and escape, *C. albicans* can spread through the body as the macrophages migrate.

4 Evolutionary Adaptations to Farnesol as a Signaling and Antagonistic Molecule

4.1 C. albicans Resistance to the Antifungal Effects of Farnesol

Regardless of their capacity to produce farnesol, many fungi respond to farnesol in that they are inhibited or killed by it, although the exact molecular mechanism of farnesol induced cell death is still under investigation. As examples, farnesol induced apoptosis in *Aspergillus nidulans* (Semighini et al. 2006), inhibited *Trichophyton rubrum* in co-culture with *C. albicans* (Jillson and Nickerson 1948), and antagonized many other fungi including *S. cerevisiae* (Machida et al. 1998). The exceptional resistance to farnesol shown by *C. albicans* is an interesting issue since it is the only known fungus which tolerates farnesol up to 300 μM while using it as a quorum sensing molecule. What makes *C. albicans* different from the rest of the fungi? We assume that *C. albicans* has a protective mechanism to safeguard itself from excessive farnesol, similar to the self defense mechanisms used by antibiotic-producing microorganisms to prevent them committing suicide by their own products. What's more, this resistance of *C. albicans* to farnesol is not just a constant suit of armor but a subtly variable protection. Anaerobically growing cells are resistant to farnesol right up to its solubility limit of 1–1.2 mM (Dumitru et al. 2004) while resistance is lost entirely in cells which have switched from the white phase to the opaque phase (Dumitru et al. 2007).

In terms of where to look for this variable resistance, the mitochondria are a likely target. As a result of aerobic respiration, all aerobic organisms produce Reactive Oxygen Species (ROS) which leads to oxidative destruction of cells. Consequently, all of these aerobes including yeasts developed efficient mechanisms to get rid of these unwelcome companions. In a study done to reveal the growth inhibitory effect of farnesol in *S. cerevisiae* (Machida et al. 1998), the level of farnesol induced ROS was found to increase in a dose dependent manner. Further, the inhibition of growth by farnesol could be prevented by the presence of antioxidants in the medium. Thus ROS generation leading to intracellular oxidative stress suggested involvement of the mitochondrial electron transport chain as the target of farnesol. This view on the primary means of farnesol mediated death in *S. cerevisiae* by generation of reactive oxygen species was further confirmed by Fairn et al. through a study on the genomic effects of the chemical compounds, farnesol and geraniol (chemogenomic profiling; Fairn et al. 2007).

4.2 *Ubiquinone and Its Variable Isoprenylation*

There is a growing interest in mitochondrial respiratory mechanisms, especially the participating components in its electron transport chain. Ubiquinone, also known as coenzyme Q (abbreviated as UQ) is an isoprenoid quinone which 'ubiquitously' exists in all living organisms, functioning as membrane bound, mobile electron carrier in the electron transport chain (Shinkarev 2006). The structure of UQ (Fig. 2) has a redox active benzoquinone ring and an all-*trans*-polyprenyl side chain of variable length. During biosynthesis, a host specific enzyme, polyprenyl diphosphate synthase successively adds isoprenoid chains to the benzoquinone ring, which ultimately determines the type of UQ present in that organism. UQ with 4–10 isoprenes are common in microbiology. The longer isoprenoid tails would have a more highly folded structure and thus be more firmly embedded in the mitochondrial membrane. In contrast, UQs with shorter isoprenoid length would have some difficulty in grabbing electrons as they are more exposed on the surface (Olgun et al. 2003). Thus, we suggest that the shorter UQs have a critical effect on cells because they can be either dislodged or inactivated by farnesol.

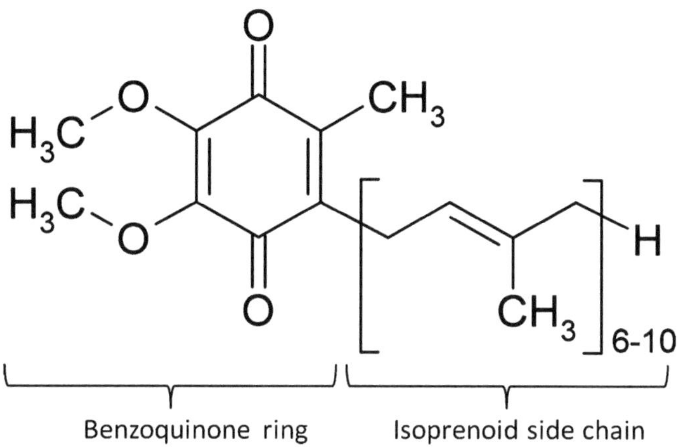

Fig. 2 The structure of ubiquinone. The number of isoprenoid side chains varies among different ubiquinones

For the organisms with which we are concerned, *S. cerevisiae* has UQ 6 and *C. albicans* has UQ 9 while almost all the rest of the *Candida* species produce UQ 7 (Suzuki and Nakase 2002). This variability in ubiquinone content depending on the length of isoprenoid side chain is thought to be taxonomically useful although there is as yet no specifically identified biological significance for these different chain lengths (Okada et al. 1998; Olgun et al. 2003). The farnesol sensitive *S. cerevisiae* has UQ 6 but research done with mutated strains by Okada et al. (1998) showed that the length of the UQ isoprenoid side chain did not act as a critical factor; no phenotypic variations were observed. However, the length of the isoprenoid tail may still have an unidentified function.

We hypothesize that *C. albicans*, as a part of evolving to being a farnesol excretor, shifted from using UQ 6 to using UQ 9 while all the other non-pathogenic Candidas stopped at UQ 7. If farnesol mediated cell death targets the ubiquinones, it is not surprising to observe the resistance of anaerobic cells of *C. albicans* to farnesol (Dumitru et al. 2004) since they lack mitochondrial respiration. Thus, a key question is whether there is a significant effect of UQ side chain length on farnesol sensitivity during aerobic respiration. This question could be answered by examining the farnesol sensitivity of *S. cerevisiae* which make UQ 9 instead of UQ 6 or *C. albicans* which made UQ 6/7 instead of UQ 9.

4.3 Possible Mechanisms for C. albicans White Cell Resistance to Farnesol

C. albicans can grow in both aerobic and anaerobic environments and it has two types of cells: white cells and opaque cells. These cell types respond to farnesol in different ways. Neither opaque cells nor anaerobic cells make farnesol (Dumitru et al. 2004, 2007). During anaerobic growth, *C. albicans* doesn't produce farnesol or respond to farnesol, even at concentration as high as 1.2 mM (Dumitru et al. 2004). Similarly, in the presence of farnesol, white cells are prevented from making germ tubes but they can tolerate farnesol at concentrations up to 300 μM, while opaque cells are lysed by farnesol at 50 μM (Dumitru et al. 2007). It is desirable to understand why the white cells of *C. albicans* are highly resistant to farnesol whereas opaque cells are very sensitive to it.

Different phases of white cell growth also differ in their tolerance to farnesol. Farnesol shows different activities towards *C. albicans* depending on the growth conditions and inoculum characteristics. In rich growth medium (YPD), *C. albicans* is very resistant to farnesol (Langford et al. 2010) while in defined media such as glucose-phosphate-proline (GPP) log-phase cells were significantly more sensitive to farnesol than were stationary phase cells. Consequently, when using an inoculum of stationary-phase cells, the growth curves are similar to those in YPD media (Langford et al. 2010) while using an inoculum of log phase cells resulted in significant delays due to farnesol induced cell lysis (Langford et al. 2010). Thus, the starting growth phase, media, and cell density of the inoculum are critical for the effect of farnesol on the cells. Finally, farnesol tolerance is an energy dependent process. Cells suspended in a farnesol buffer without an energy source lysed whereas those with an energy source did not (Langford et al. 2010). Together, the influence of cell type, growth conditions and inoculum characteristics suggest resistance to farnesol is an active and regulated process.

C. albicans may detoxify farnesol enzymatically. Using a spent medium assay for quorum sensing activity, i.e. the ability to block hypha formation, Hornby et al. (2001) found that the levels of farnesol present in the spent media roughly paralleled cell mass for 20 h after inoculation but then decreased rapidly after that. Farnesol contains three C=C double bonds and exists in four isomers of which only (E, E) farnesol has QSM activity. Farnesol is a very unstable molecule and air oxidation results in a hydroxyl or epoxide of farnesol which causes a dramatic decrease in QSM activity (Shchepin et al. 2003). Thus, the decrease in QSM activity observed by Hornby et al. (2001) could have been either enzymatic or spontaneous. Additionally, the morphological response to farnesol in *C. albicans* appears to be very sensitive to other minor changes in the structure of farnesol. For instance, farnesol with three isoprene units has three methyl branches. If the 2-methyl branch is either removed or enlarged to a 2-ethyl branch then the QSM activity of the resulting farnesol analog is 20-fold lower (Shchepin et al. 2005). Thus, the decreased activity of farnesol with time observed by Hornby et al. (2001) could result from the modification of farnesol to a new compound with a lower QSM activity and/or lower toxicity for the cells.

4.4 Membrane Differences in White and Opaque Cells

C. albicans strain WO-1 is capable of switching at high frequencies between two phenotypes, white and opaque cells. Among the many differences between white and opaque cells, it has been shown that these cells have variations in their lipid and sterol contents (Ghannoum et al. 1990). White cells have 4-fold and 7.7-fold higher sterol contents than do opaque cells in mid-exponential and stationary phase, respectively (Ghannoum et al. 1990). Additionally, white cells contain more free sterols and less of the steryl glycoside and steryl ester sterol derivatives (Ghannoum et al. 1990). Finally, the sterols present in mid-exponential and stationary phase cultures of white cells were qualitatively different. The sterols present in mid-exponential phase cultures of white cells were primarily lanosterol (48 wt%) and 24-methylene dihydrolanosterol (26.2 wt%), while for mid-exponential phase opaque cells they were ergosterol (49.3 wt%), lanosterol (33.2 wt%) and squalene (17.5 wt%). During stationary phase, ergosterol was the major sterol in both white and opaque cells (Ghannoum et al. 1990). These observations suggest the twin hypotheses that a high sterol content in the membrane protects cells from farnesol and that the percentages of total sterols and of ergosterol in particular will differ for mid-exponential phase cells grown with and without farnesol. It also leads to a focus on the regulation of HMG CoA reductase (HMGR), the rate limiting step in sterol biosynthesis.

Farnesol is an isoprenoid produced by dephosphorylation of farnesyl pyrophosphate, an intermediate in the isoprenoid pathway (Hornby et al. 2003). The isoprenoid pathway is responsible for synthesis of sterols from acetyl-CoA (Fig. 3). Acetyl-CoA is converted to mevalonate by reduction with NADPH by 3-hydroxy-3-methylglutaryl-CoA (HMG-CoA) reductase (HMGR). Mevalonate leads to the synthesis of farnesyl pyrophosphate, a branch point in the sterol biosynthetic pathway. Farnesol pyrophosphate serves as a precursor for sterol biosynthesis and nonsterol isoprenoids. Sterols are an important structural component of cellular membranes. Nonsterol isoprenoids include geranylgeranyl pyrophosphate and farnesol. Farnesyl pyrophosphate and geranylgeranyl pyrophosphate are involved in the prenylation of proteins. Isoprenoids are also precursors for the prenyl side chains of ubiquinone.

HMGR is often the rate-limiting enzyme for the sterol biosynthetic pathway. Human HMGR is regulated by transcription, protein degradation and phosphorylation (Table 1). Phosphorylation decreases the enzyme efficiency. The rapid degradation of mammalian HMGR is triggered by cellular sterols and farnesol, farnesyl pyrophosphate and geranylgeranyl pyrophosphate (reviewed in Joo and Jetten 2010; Burg and Espenshade 2011). These signals function both in vivo and in vitro as triggers for rapid degradation. In fungi, regulation of HMGR has been studied in the yeast *Saccharomyces cerevisiae* and the fission yeast *Schizosaccharomyces pombe. S. cerevisiae* has two HMGR isozymes, Hmg1 and Hmg2. Both isozymes are subject to feedback regulation by nonsterol products of the isoprenoid pathway (Table 1). Hmg1 is the primary source of HMGR in aerobically grown cells and in these conditions it is mainly regulated at the level of translation (Dimster-Denk et al. 1994). Hmg2, like its mammalian counterpart, is primarily regulated by protein turnover. In vivo, geranylgeranyl pyrophosphate is the signal regulating Hmg2 degradation (Theesfeld et al. 2011). However, in vitro farnesol also causes Hmg2p to undergo a change to a less folded structure (Shearer and Hampton 2005). This in vitro conformational change is the same as that observed

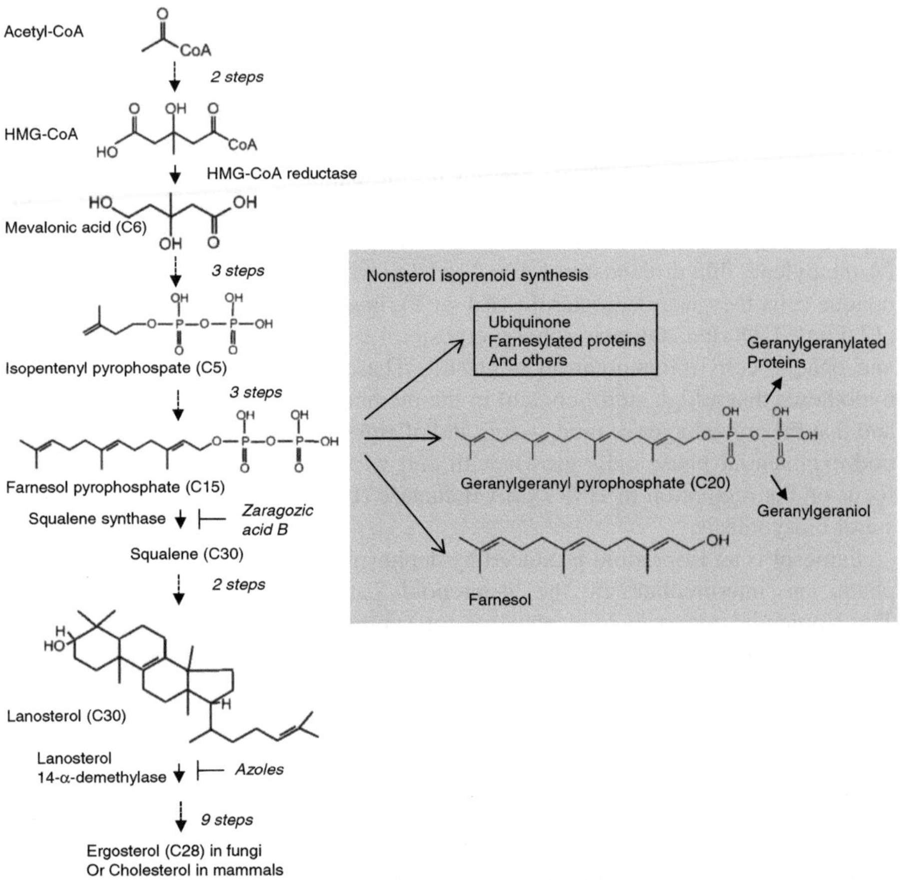

Fig. 3 The biosynthetic pathway for the synthesis of sterols and nonsterol isoprenoids. The molecules and enzymes discussed in the text are indicated. *Dashed arrows* represent multi-enzyme steps, with the number of reactions indicated in parentheses to the right of these *arrows*. The enzymes inhibited by zaragozic acid B, and the azoles (fluconazole, clotrimazole, ketoconazole or miconazole) are shown

in vivo which triggers the degradation of Hmg2p (reviewed in Hampton and Garza 2009). Different mechanisms are used to regulate HMGR by different fungi. For example, the fission yeast *Schizosaccharomyces pombe* has a single HMGR enzyme which is regulated by phosphorylation (Table 1; Burg et al. 2008).

Like *S. pombe, C. albicans* has a single gene for the HMGR enzyme. Currently regulation of the *C. albicans* HMGR has not been investigated, however an understanding of this regulation is important because the nature of the regulation could have profound effects on sterol and/or farnesol synthesis. *C. albicans* is unique amongst the fungi examined because some cell types make very high levels of farnesol while others do not make farnesol at all. This suggests the *C. albicans* HMGR is most likely resistant to feedback regulation by farnesol. Consistent with that prediction, farnesol production is elevated 8- to 40-fold in the presence

Table 1 Regulation of HMG-CoA reductase in different organisms

Organism	Enzyme	Type of regulation	Signal	Reference
Humans	HMGR	Transcription Phosphorylation Degradation	Sterols AMP/ATP ratio Farnesol, farnesyl pyrophosphate or geranylgeranyl pyrophosphate	Reviewed in Joo and Jetten (2010) and Burg and Espenshade (2011)
Saccharomyces cerevisiae	Hmg1	Translation	Mevalonate-derived molecule produced before the synthesis of squalene	Dimster-Denk et al. (1994)
	Hmg2	Degradation	Geranylgeranyl pyrophosphate (in vivo and in vitro) Also farnesol (in vitro)	Theesfeld et al. (2011), Garza et al. (2009), and Shearer and Hampton (2005)
Schizosaccharomyces pombe	Hmg1	Inactivated by phosphorylation	Nutrient stress Osmotic stress	Burg et al. (2008)

of sub-lethal levels of drugs that block sterol synthesis in fungi (Hornby et al. 2003; Hornby and Nickerson 2004). Treatment with zaragozic acid B, fluconazole, clotrimazole, ketoconazole, or miconazole all caused a dose dependent increase in farnesol levels. These drugs inhibit steps downstream of farnesyl pyrophosphate in the synthesis of ergosterol, the fungal sterol and after the branchpoint for non-sterol isoprenoids (Fig. 3). At the time (Nickerson et al. 2006), these elevated levels of farnesol were attributed to increased pool sizes for farnesyl pyrophosphate. Thus these results imply that HMGR is not feedback regulated by non-sterol isoprenoids.

5 Summary

Farnesol and bacterial endotoxins are lipids and in this review we analyzed data relevant to farnesol's function and synthesis from the perspective of farnesol and the endotoxins acting as membrane active compounds. This analysis implicated the possible roles of: (1) endotoxins in the regulation of farnesol production by *C. albicans*; (2) farnesol in the interactions between *C. albicans* and the host during disseminated infections; and (3) ubiquinones in the mechanisms for unusually high resistance to farnesol by some *C. albicans* cell types. Finally we discuss the implications that the use of farnesol as both a signaling molecule and to antagonize competing microbial species has for the regulation of HMG-CoA reductase, the enzyme that is the usual rate limiting step in sterol/lipid synthesis. In the future, to fully understand farnesol's function as a membrane active compound, three research directions should be pursued. (1) The role of lipid signaling molecules in interspecies communication between *C. albicans* and the typical microbial neighbors that it encounters in its normal habitats. (2) The role of farnesol in the interactions between *C. albicans* and the human innate immune system, and (3) The role of isoprenoids in regulation of HMG-CoA reductase of *C. albicans*. We expect that this new understanding will uncover the basic biological principles that underlie interspecies signaling by these lipid molecules.

References

Abe S, Tsunashima R, Iijima R, Yamada T, Maruyama N et al (2009) Suppression of anti-*Candida* activity of macrophages by a quorum-sensing molecule, farnesol, through induction of oxidative stress. Microbiol Immunol 53(6):323–330

Burg JS, Espenshade PJ (2011) Regulation of HMG-CoA reductase in mammals and yeast. Prog Lipid Res 50(4):403–410

Burg JS, Powell DW, Chai R, Hughes AL, Link AJ et al (2008) Insig regulates HMG-CoA reductase by controlling enzyme phosphorylation in fission yeast. Cell Metab 8(6):522–531

Dimster-Denk D, Thorsness MK, Rine J (1994) Feedback regulation of 3-hydroxy-3-methylglutaryl coenzyme A reductase in *Saccharomyces cerevisiae*. Mol Biol Cell 5 (6):655–665

Dumitru R, Hornby JM, Nickerson KW (2004) Defined anaerobic growth medium for studying *Candida albicans* basic biology and resistance to eight antifungal drugs. Antimicrob Agents Chemother 48(7):2350–2354

Dumitru R, Navarathna DH, Semighini CP, Elowsky CG, Dumitru RV et al (2007) In vivo and in vitro anaerobic mating in *Candida albicans*. Eukaryot Cell 6(3):465–472

Dunne DW, Resnick D, Greenberg J, Krieger M, Joiner KA (1994) The type I macrophage scavenger receptor binds to gram-positive bacteria and recognizes lipoteichoic acid. Proc Natl Acad Sci USA 91(5):1863–1867

Fairn GD, Macdonald K, McMaster CR (2007) A chemogenomic screen in *Saccharomyces cerevisiae* uncovers a primary role for the mitochondria in farnesol toxicity and its regulation by the Pkc1 pathway. J Biol Chem 282(7):4868–4874

Flannagan RS, Cosio G, Grinstein S (2009) Antimicrobial mechanisms of phagocytes and bacterial evasion strategies. Nat Rev Microbiol 7(5):355–366

Fuqua WC, Winans SC, Greenberg EP (1994) Quorum sensing in bacteria: the LuxR-LuxI family of cell density-responsive transcriptional regulators. J Bacteriol 176(2):269–275

Garza RM, Tran PN, Hampton RY (2009) Geranylgeranyl pyrophosphate is a potent regulator of HRD-dependent 3-hydroxy-3-methylglutrayl-CoA reductase degradtion in yeast. J Biol Chem 28(51):35368–35380

Geiger J, Wessels D, Lockhart SR, Soll DR (2004) Release of a potent polymorphonuclear leukocyte chemoattractant is regulated by white-opaque switching in *Candida albicans*. Infect Immun 72(2):667–677

Ghannoum MA, Swairjo I, Soll DR (1990) Variation in lipid and sterol contents in *Candida albicans* white and opaque phenotypes. J Med Vet Mycol 28(2):103–115

Ghosh S, Navarathna DH, Roberts DD, Cooper JT, Atkin AL et al (2009) Arginine-induced germ tube formation in *Candida albicans* is essential for escape from murine macrophage line RAW 264.7. Infect Immun 77(4):1596–1605

Ghosh S, Howe N, Volk K, Tati S, Nickerson KW et al (2010) *Candida albicans* cell wall components and farnesol stimulate the expression of both inflammatory and regulatory cytokines in the murine RAW264.7 macrophage cell line. FEMS Immunol Med Microbiol 60(1):63–73

Hampton RY, Garza RM (2009) Protein quality control as a strategy for cellular regulation: lessons from ubiquitin-mediated regulation of the sterol pathway. Chem Rev 109 (4):1561–1574

Hisajima T, Maruyama N, Tanabe Y, Ishibashi H, Yamada T et al (2008) Protective effects of farnesol against oral candidiasis in mice. Microbiol Immunol 52(7):327–333

Hogan DA, Muhlschlegel FA (2011) *Candida albicans* developmental regulation: adenylyl cyclase as a coincidence detector of parallel signals. Curr Opin Microbiol 14(6):682–686

Hornby JM, Nickerson KW (2004) Enhanced production of farnesol by *Candida albicans* treated with four azoles. Antimicrob Agents Chemother 48(6):2305–2307

Hornby JM, Jensen EC, Lisec AD, Tasto JJ, Jahnke B et al (2001) Quorum sensing in the dimorphic fungus *Candida albicans* is mediated by farnesol. Appl Environ Microbiol 67 (7):2982–2992

Hornby JM, Kebaara BW, Nickerson KW (2003) Farnesol biosynthesis in *Candida albicans* cellular response to sterol inhibition by zaragozic acid B. Antimicrob Agents Chemother 47:2366–2369

Jillson OF, Nickerson WJ (1948) Mutual antagonism between pathogenic fungi; inhibition of dimorphism in *Candida albicans*. Mycologia 40(3):369–385

Joo JH, Jetten AM (2010) Molecular mechanisms involved in farnesol-induced apoptosis. Cancer Lett 287(2):123–135

Kolotila MP, Diamond RD (1990) Effects of neutrophils and in vitro oxidants on survival and phenotypic switching of *Candida albicans* WO-1. Infect Immun 58(5):1174–1179

Lachke SA, Lockhart SR, Daniels KJ, Soll DR (2003) Skin facilitates *Candida albicans* mating. Infect Immun 71(9):4970–4976

Langford ML, Atkin AL, Nickerson KW (2009) Cellular interactions of farnesol, a quorum-sensing molecule produced by *Candida albicans*. Future Microbiol 4(10):1353–1362

Langford ML, Hasim S, Nickerson KW, Atkin AL (2010) Activity and toxicity of farnesol towards *Candida albicans* are dependent on growth conditions. Antimicrob Agents Chemother 54 (2):940–942

Lohse MB, Johnson AD (2008) Differential phagocytosis of white versus opaque *Candida albicans* by *Drosophila* and mouse phagocytes. PLoS One 3(1):e1473

Machida K, Tanaka T, Fujita K, Taniguchi M (1998) Farnesol-induced generation of reactive oxygen species via indirect inhibition of the mitochondrial electron transport chain in the yeast *Saccharomyces cerevisiae*. J Bacteriol 180(17):4460–4465

McCoy JW (1980) Microbiology of cooling water. Chemical Publishing Co., New York

Navarathna DH, Hornby JM, Krishnan N, Parkhurst A, Duhamel GE et al (2007) Effect of farnesol on a mouse model of systemic candidiasis, determined by use of a DPP3 knockout mutant of *Candida albicans*. Infect Immun 75(4):1609–1618

Nickerson KW, Atkin AL, Hornby JM (2006) Quorum sensing in dimorphic fungi: farnesol and beyond. Appl Environ Microbiol 72(6):3805–3813

Nigam S, Ciccoli R, Ivanov I, Sczepanski M, Deva R (2010) On mechanism of quorum sensing in *Candida albicans* by 3(R)-hydroxy-tetradecaenoic acid. Curr Microbiol 62(1):55–63

Okada K, Kainou T, Matsuda H, Kawamukai M (1998) Biological significance of the side chain length of ubiquinone in *Saccharomyces cerevisiae*. FEBS Lett 431(2):241–244

Okafor N (2007) Modern industrial microbiology and biotechnology. Science Publishers, Enfield, 530 p

Olgun A, Akman S, Tezcan S, Kutluay T (2003) The effect of isoprenoid side chain length of ubiquinone on life span. Med Hypotheses 60(3):325–327

Semighini CP, Hornby JM, Dumitru R, Nickerson KW, Harris SD (2006) Farnesol-induced apoptosis in *Aspergillus nidulans* reveals a possible mechanism for antagonistic interactions between fungi. Mol Microbiol 59(3):753–764

Shchepin R, Hornby JM, Burger E, Niessen T, Dussault P et al (2003) Quorum sensing in *Candida albicans*: probing farnesol's mode of action with 40 natural and synthetic farnesol analogs. Chem Biol 10(8):743–750

Shchepin R, Dumitru R, Nickerson KW, Lund M, Dussault PH (2005) Biologically active fluorescent farnesol analogs. Chem Biol 12(6):639–641

Shearer AG, Hampton RY (2005) Lipid-mediated, reversible misfolding of a sterol-sensing domain protein. EMBO J 24(1):149–159

Shinkarev VP (2006) Ubiquinone (coenzyme Q10) binding sites: low dielectric constant of the gate prevents the escape of the semiquinone. FEBS Lett 580(11):2534–2539

Sobel JD (1997) Vaginitis. N Engl J Med 337(26):1896–1903

Soll DR, Bedell GW, Brummel M (1981) Zinc and regulation of growth and phenotype in the infectious yeast *Candida albicans*. Infect Immun 32(3):1139–1147

Stanier RY, Palleroni NJ, Doudoroff M (1966) The aerobic *pseudomonads*: a taxonomic study. J Gen Microbiol 43(2):159–271

Sudbery PE (2011) Growth of *Candida albicans* hyphae. Nat Rev Microbiol 9(10):737–748

Suzuki M, Nakase T (2002) A phylogenetic study of ubiquinone-7 species of the genus *Candida* based on 18S ribosomal DNA sequence divergence. J Gen Appl Microbiol 48(1):55–65

Theesfeld CL, Pourmand D, Davis T, Garza RM, Hampton RY (2011) The sterol-sensing domain (SSD) directly mediates signal-regulated endoplasmic reticulum-associated degradation (ERAD) of 3-hydroxy-3-methylglutaryl (HMG)-CoA reductase isozyme Hmg2. J Biol Chem 286(30):26298–26307

Wang X, Quinn PJ (2010) Endotoxins: structure, function and recognition. Springer, New York

Yamaguchi H (1975) Control of dimorphism in *Candida albicans* by zinc: effect on cell morphology and composition. J Gen Microbiol 86(2):370–372

Part III
Transorganismic Communication

Metabolite-Mediated Interactions Between Bacteria and Fungi

Danielle M. Troppens and John P. Morrissey

Abstract In the complex environments in which they live, bacteria and fungi are in frequent contact. This leads to interactions that may be transient and occasional or constant and co-evolved. Although there are many forms interactions could take, the focus on this chapter is on those mediated by secreted metabolites. Antibiosis is one such interaction and some classical as well as novel interactions mediated by antibiotics are discussed. More recently, the concept of signal-mediated interactions between bacteria and fungi has gained currency. Some, or perhaps many, bacteria and fungi have developed the capacity to take advantage of, and in some cases, to subvert, signalling pathways in other organisms. This is addressed, in particular, the example of signalling between *Pseudomonas aeruginosa* and *Candida albicans*.

1 Introduction

The natural environments where one is likely to encounter fungi are typically complex communities of fungi, bacteria, protozoa and viruses and other organisms. Although fungi, like other microbes, may grow in pure isolated populations, this is not the norm. Similarly, while there are many well-studied examples of relatively simple one-to-one associations between a fungus and another organism; for example, a phytopathogenic fungus and the host plant, or a fungus and bacterium in a lichen; even in these cases, there are generally other microbes also interacting with the symbiosis in some way. When considered in this context, it is not unexpected that a fungus will interact with the other members of the niche in which it finds itself. These interactions can be benign or detrimental, mutualistic or competitive and are often of a dynamic nature – as environmental conditions change, so too can

D.M. Troppens • J.P. Morrissey (✉)
Microbiology Department, University College Cork, Food Science Building, Cork, Ireland
e-mail: j.morrissey@ucc.ie

G. Witzany (ed.), *Biocommunication of Fungi*, DOI 10.1007/978-94-007-4264-2_13,

the nature of the interaction. In many environments, fungi and bacteria share very similar niches – occupying the same spatial area, utilising similar resources, and encountering the same external pressures for survival. This coexistence over millions of years has resulted in the development of many bacterial–fungal interactions, some of which are the focus of specific chapters in this book. The scope of this chapter is restricted to two main types of bacteria–fungal interactions that involve known secreted metabolites: (1) antibiosis, where secreted secondary metabolites have negative effects on bacterial or fungal growth and (2) signalling interactions, where bacteria or fungi sense and react to the presence of other microbes in their environment.

2 Antagonistic Interactions Between Bacteria and Fungi

2.1 Antibiosis: Production of Antimicrobial Metabolites

Any consideration of antibiosis and its significance will focus on the soil environment as this is the natural habitat of many of the known antibiotic-producing microbes. Soil is generally a nutrient-limited environment and the capacity to antagonise competitors to gain advantage has driven the evolution of bioactive secondary metabolites. The diversity and importance of these secreted metabolites becomes very apparent in the rhizosphere of plants where polymicrobial communities of up to 10^9 bacterial cells and 10^6 fungal cells per gram of soil are possible. Some of the most common bacterial taxa that can be cultured from the rhizosphere are Pseudomonads, Bacilli and Actinomycetes, with many of these isolates being producers of bioactive secondary metabolites (Handelsman and Stabb 1996; Whipps 2001; Weller et al. 2002; Wheatley 2002; Sylvia et al. 2004). In the context of this chapter, we are especially interested in metabolites with anti-fungal potential and on fungal responses to these molecules (Fig. 1). *Pseudomonas* and *Bacillus* spp. have received particular attention because of the potential to exploit the antifungal properties of strains for biological control applications in agriculture (Weller 1988; van Elsas and Heijnen 1990; Handelsman and Stabb 1996; Keel et al. 1996; Bender et al. 1999; Whipps 2001). Several of the important *Pseudomonas* metabolites specifically target aspects of energy generation, notably the phenolic molecule 2,4-diacetylphloroglucinol (DAPG) (Cronin et al. 1997a, b; Gleeson et al. 2010), phenazines, which are analogues of flavin coenzymes, and the tryptophan derivative Pyrrolnitrin (Tripathi and Gottlieb 1969; Ligon et al. 2000). Because of the ubiquity of this function, some of these metabolites have broad spectrum of activities against many plant pathogenic fungi including *Fusarium*, *Gaeumannomyces* and *Botrytis* spp. Cyclic lipopeptides, which are also produced by *Pseudomonas* spp. have a different mode of action with viscosinamide, for example, showing activity against *Pythium ultimum* and *Rhizoctonia solani* due to strong biosurfactant properties (Nielsen et al. 1999, 2002;

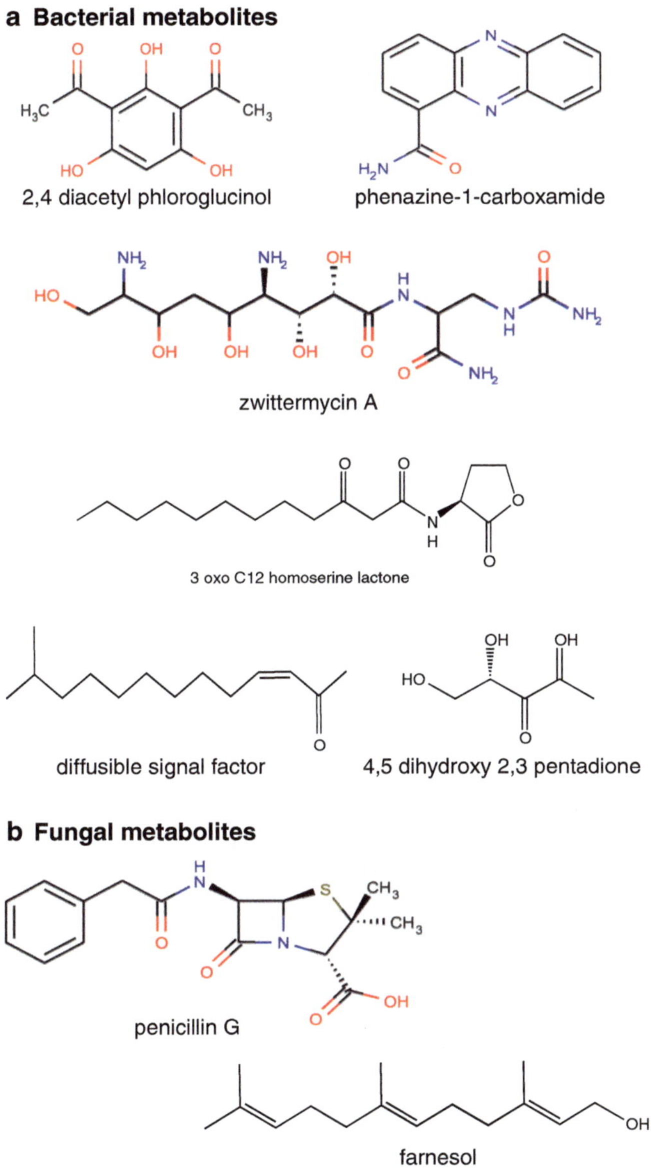

Fig. 1 Bioactive metabolites secreted by bacteria and fungi. The metabolites illustrated provide examples of some of the diversity of metabolites that play roles in mediating interactions between bacteria and fungi. Distinctions are not drawn between "antibiotics" and "signals" as it is clear that similar type of chemical backbones can be either of these depending on the context. Particular metabolites are mentioned in the main text

Raaijmakers et al. 2002). Similar targets area seen with iturin A and surfactin produced by *B. subtilis* (Asaka and Shoda 1996), and a group of compounds named xanthobaccins isolated from *Stenotrophomonas* spp. (Nakayama et al. 1999). Other metabolites produced by *Bacillus* spp. include zwittermycin A, which is active against *Phytophthora* and *Pythium* (Silo-Suh et al. 1994, 1998), and kanosamine (Milner et al. 1996). Species of *Streptomyces* are well-known for their production of antifungals, including amphotericin B and nystatin both belonging to the group of polyene antibiotics that target the fungal membrane. Fungi, of course, can also produce antibiotics, with the first clinical antibiotic, penicillin, identified and isolated from a fungus later identified as *Penicillium notatum* (Fleming 1929). Since then, scientists have identified a large number of penicillin-like compounds with varying activities. Cephalosporins, first isolated from the fungus *Cephalosporium acremonium*, form a subgroup of β-lactam antibiotics with the difference that they exhibit higher tolerance to penicillinases. There are a number of other antibacterial metabolites that have been isolated from fungi, e.g. naphthoquinones (Medentsev and Akimenko 1998) or the more recently identified pyrrocidines A and B, which were isolated from an unidentified filamentous fungus (He et al. 2002).

At one level, the production of antibiotics or similar bioactive metabolites as a competitive strategy appears very straightforward. There are, however, an increasing number of examples where the capacity to produce a bioactive metabolite has become an integral part of a much more complicated symbiotic interaction involving fungi. One fascinating example is the symbiosis of *Streptomyces* with fungus farming ants (*Acromyrmex* spp.). The context of this is that it has been known for a long time that leafcutter ants farm fungi (*Leucoagaricus* spp.) in underground chambers. From the late 1990s, however, it became clear that the existing concept of the seemingly bipartite symbiosis between the ants and their cultivar was too simple (Currie et al. 1999a, b, 2003a, b). Specifically, it was discovered that a parasitic fungus of the genus *Escovopsis* that could attack the farmed fungus was also present in the fungal gardens. The reason why this parasitic fungus did not destroy the fungal garden was that different species of Actinomycetes live in intimate mutualistic association with the ants and produce antimicrobials that protect the fungal garden (Currie et al. 1999a, b, 2003a, b; Cafaro and Currie 2005). These metabolites include the polyene candicidin (Haeder et al. 2009) and the cyclic depsipeptide, dentigerumycin (Oh et al. 2009). Other symbionts, such as black yeasts, are also present in the nest and it now seems that the antibiotics play an important role in community homeostasis and are far more important than simple antimicrobials (Little and Currie 2007, 2008). A second example is the case of endofungal bacteria. These are bacteria that are either facultative or obligate endosymbionts of fungi. Early work on these bacteria focused on some specific systems, for example, mycorrhizal fungi. In this case, bacteria associated both externally with the spores/hyphae and intracellularly within the hyphae, appear to play important roles in facilitating the infection of plants by AM fungi (Bonfante and Anca 2009). A further particularly intriguing bacterial–fungal endosymbiosis with implications for antagonistic interactions was first described in 2005

(Partida-Martinez and Hertweck 2005). In this instance, a bacterial endosymbiont of the Zygomycete fungus *Rhizopus microsporus* is responsible for producing a macrolide molecule, rhizoxin, long believed to be a fungal toxin (Sanders 2005). It became clear that this is not an isolated example when the same group demonstrated that a second *Rhizopus* "mycotoxin", rhizonin, is produced by a related endosymbiotic bacterium (Partida-Martinez et al. 2007a, b) and further examples were later demonstrated (Ibrahim et al. 2008; Lackner et al. 2009). The bacterial species in question have been named *Burkholderia rhizoxinica* and *Burkholderia endofungorum*, and share a close phylogenetic relationship (Partida-Martinez et al. 2007a, b). The recent publication and analysis of a complete genome sequence of *B. rhizoxinica* reveals a multitude of features that are likely to be important for interactions between the bacterium and the fungus within the fungal hyphae (Lackner et al. 2011a, b). There is less known about how the fungus adapts to its endosymbiont, though analysis of a large number of fungi suggests that mutations conferring toxin-resistance in an ancestral *Rhizopus* facilitated a parasitism-mutualism shift (Schmitt et al. 2008). Production of bioactive secondary metabolites by symbiotic associations of a bacterium and fungus is a fascinating twist on the antagonism/antibiotic paradigm (Fig. 1).

2.2 Antibiosis: Responses and Tolerance

In most studies on antagonism between bacteria and fungi, emphasis has been put on mechanisms of antibiotic production and its regulation in bacteria and much less attention on the fungal response. Fungi have developed efficient strategies to respond to and cope with a wide range of antagonistic traits by either enzymatic degradation or efflux mechanisms (for extensive review refer to Duffy et al. 2003). Efflux mechanisms are used by all studied organisms, from bacteria to humans, to cope with toxins and detrimental metabolites. In this context, ATP-binding cassette (ABC) transporters are well-known for their ability to use active efflux to protect fungal cells from toxic compounds. In phytopathogenic fungi, ABC transporters have been shown to be involved in defence mechanisms against plant toxins, bacterial metabolites and fungicides (de Waard 1997), and similar mechanisms are found in clinical fungi (Morschhauser 2010). The association of ABC transporters with increased tolerance in fungi to antibiosis caused by antagonistic bacteria has been shown in different fungal species in recent years. The plant pathogen *Botrytis cinerea* has a broad host range and infects fruits, flowers and other tissue of more than 200 plant species (Jarvis 1977). It was found that the *Botrytis cinerea* ABC transporter BcAtrB protects the fungus against the toxic effects of the phytoalexin resveratrol and the fungicide fenpiclonil (Schoonbeek et al. 2001) and that the bacterial antifungal DAPG and phenazines induce the expression of several ABC transporter including BcAtrB (Schoonbeek et al. 2002; Schouten et al. 2008). Another effective way to resist various toxic compounds is enzymatic degradation via a wide variety of enzymes such as acetyltransferases, hydrolases, hydratases, demethylases, and cytochrome P450-dependent monooxygenases. For example, modification of DAPG by secreted deacetylases in both *B. cinerea* and *Fusarium oxysporum* has been reported (Schouten et al. 2004, 2008) and some strains of *B. cinerea* secrete a laccase, BcLCC2, that can detoxify DAPG (Schouten et al. 2008). Laccase activity is induced in *Rhizoctonia solani* (black scurf in potato) in the presence of *P. fluorescens* (Crowe and Olsson 2001) and in the wood-rot fungus *Hypholoma fasculare* in the presence of *B. subtilis* (Griffith and Rayner 1994) suggesting that this mechanism of coping with a toxin may be widely distributed in fungi. Tolerance to the well-studied volatile compound hydrogen cyanide (HCN) that is involved in biocontrol is known to occur in a number of fungi (Keel et al. 1996). In some fungi, HCN tolerance is conferred by cyanide-resistant respiration for example in *Microcyelus ulei*, whereas others, such as *Fusarium lateritium*, *Stemphylium loti* and *Gloeocercospora sorghi*, are capable of converting HCN to formamide using the cyanide-inducible enzyme cyanide hydratase (Osbourn 1996). Cyanide hydratases have also been found in other fungal species suggesting that this is a common defence mechanism among pathogenic fungi. Other ways of fungi to defend themselves against antagonistic activities include detoxification of active oxygen species in *M. graminicola* caused by 1-hydroxyphenazine (Levy et al. 1992) and production of melanin and its incorporation into the fungal cell wall against the effect of hydrolytic enzymes. In *G. graminis* melanin content in hyphae directly links to the susceptibility of the fungus to lysis caused by enzymes of the antagonistic *Streptomyces lavendulae* (Tschudi and Kern 1979). In addition, other cell wall components such as hydropolysaccharides are thought to be involved in tolerance to cell wall degrading enzymes as was shown in chlamydospores of *Fusarium solani* (van Eck 1978).

3 Signalling Interactions Between Bacteria and Fungi

Since the discovery of bacterial quorum sensing in the 1990s, there has been a growing realization that microbes do not grow in isolation but interact extensively within their community to co-ordinate growth and gene expression. This type of cell–cell signaling has been extensively studied in bacteria and has been the subject of many reviews (Williams 2007). The best-understood systems are N-acyl-homoserine lactone (N-AHL) mediated signaling in Gram negative bacteria, with *Pseudomonas aeruginosa* being the model for much of what is known about regulation and function (Williams and Camara 2009). Other metabolites such as small peptides, fatty acids and alkyl quinolones can also be used as bacterial cell–cell signals in species-specific contexts (Holcombe et al. 2011). Although the principles of inter-cellular communication via mating pheromones has been long accepted in the fungal world (reference relevant chapters in this book), it is only in more recent years that the idea that a fungal populations may signal within itself has gained traction. The majority of data has come from studies of "quorum sensing" in *Candida albicans* (reference Nickerson's Chap. 12 and molecular aspects of cell–cell signaling in *C. albicans* have been the subject of recent reviews

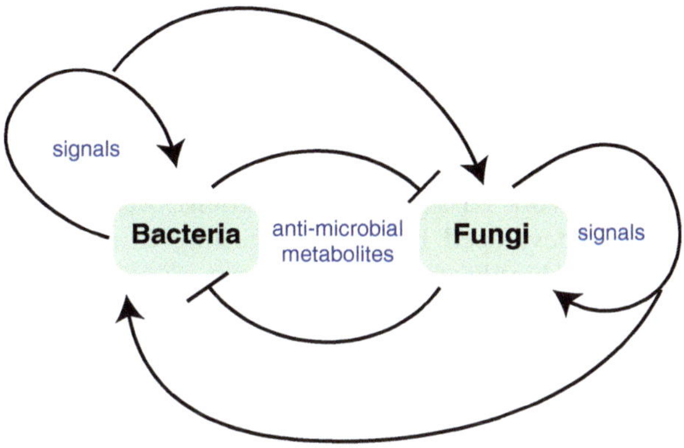

Fig. 2 Interactions between bacteria and fungi. The major classes of interaction discussed in the text are illustrated. Antagonistic interactions are mediated by antimicrobial metabolites whereas signals are predominantly within a species but can also be between bacteria and fungi. In all cases, responses include changes in gene expression, physiology and phenotype and often these programmed responses involve co-opted pathways as described in the text

(de Sordi and Muhlschlegel 2009; Hogan and Sundstrom 2009; Hall and Muhlschlegel 2010; Peleg et al. 2010; Holcombe et al. 2011)). The system that is best understood uses the molecule farnesol as a morphogenic regulator. Farnesol, which is synthesized during planktonic growth, inhibits the yeast to hyphal switch in *C. albicans* by disrupting the Ras1p-cAMP-PKA pathway (de Sordi and Muhlschlegel 2009; Hall and Muhlschlegel 2010). That environmentally-responsive pathway integrates multiple external signals and stimuli and modulates phenotypes such as morphogenesis, biofilm formation and stress response (Hogan and Sundstrom 2009; Deveau and Hogan 2011) (Fig. 2).

It had long been known that bacteria and fungi could be isolated from the same environments and the general assumption was that interactions between these organisms were either benign or antagonistic. There was a sea change in thinking in 2004 when it was reported that bacterial quorum-sensing molecules could modulate physiology in *C. albicans* (Hogan et al. 2004). More specifically, long chain homoserine lactones of the type produced by *P. aeruginosa* impaired the switch from yeast to hyphal growth in *C. albicans*. Subsequently, it was found that this could be reproduced with supernatants from clinical strains of *P. aeruginosa* – but only in cases where the strain was producing N-AHLs (McAlester et al. 2008). Thus, in a mixed culture, dimorphism of *C. albicans* is modulated by bacterial signals. Elucidation of the mechanism of interaction proceeded in parallel with new insights on farnesol signaling in *C. albicans* and it is now accepted that long-chain homoserine lactones such as 3oC12 HSL and farnesol are structurally similar and can function as analogues in the other species. This explains why addition of exogenous farnesol can complement a *P. aeruginosa lasI* mutant that fails to make 3oC12 HSL (Cugini et al. 2010), and why long chain HSLs disrupt the Ras1p-cAMP-PKA pathway by mimicking the action of farnesol (Davis-Hanna et al. 2008). It has since emerged, however, that *P. aeruginosa* produces additional signals that exert different phenotypic effects on *C. albicans*. The most significant of these is a homoserine lactone-independent inhibition of biofilm formation that has been observed (Holcombe et al. 2010). Full details of the specific signals and mechanism need to be determined but some viable candidate molecules such as bacterial fatty acid derivatives (Davies and Marques 2009) and alkyl quinolones (Reen et al. 2011) have been identified. Work on signaling interactions between bacteria and other fungi is not as well developed but it is interesting to note that inhibition by homoserine lactones of biofilm formation in *Aspergillus fumigatus* has also been observed (Mowat et al. 2010). This suggests that this type of single-mediated interaction may be widespread.

By designating an interaction as being of a "signaling" nature, there is an implicit implication that the interaction modulates phenotypes and behaviour. In particular in the biomedical setting, a number of studies have now confirmed that this is the case. Within the oral cavity, metabolites and molecules secreted by different Actinomycetes and *Streptococci* can variously stimulate or inhibit *Candida* growth, morphogenesis and biofilm formation (Holcombe et al. 2011). Inhibition of the yeast to hyphal switch by a similar mechanism described for homoserine lactones is likely to occur, in this case mediated by a class of fatty acid derivatives known as "diffusible signal factor" or DSF. These metabolites were first identified in *Xanthamonas* but seem to be quite widespread in bacteria, including oral streptococci (Vilchez et al. 2010; Ryan and Dow 2011). Another chemical family of molecules, collectively known as "autoinducer 2" (AI-2), seems to play a role in promoting hyphal development and biofilm formation. AI-2 molecules are derivatives of 4,5 dihydroxy 2,3 pentadione (DPD), itself an intermediate of amino acid metabolism synthesised by the *LuxS* gene (Federle 2009). Although, the precise status of AI-2 as secreted microbial signal remains controversial, there are now strong genetic and physiological data that support a role for DPD-derivatives in communication between bacteria and fungi. The cystic fibrosis (CF) lung is another human niche where there are data supporting the view that interactions between *Pseudomonas* and *Candida* affect phenotypes. This is best illustrated by a retrospective study that established that CF patients co-infected with both *P. aeruginosa* and *C. albicans* has poorer disease outcomes than patients infected with either alone (Chotirmall et al. 2010). A picture has now emerged where an environment like the CF lung is a complex milieu of secreted signals such as farnesol, homoserine lactones, fatty acids, alkyl quinolones and antibiotics, with multiple competing effects on phenotypic behaviour of bacteria and fungi.

4 Conclusions

Bacteria and fungi have co-evolved in many different ecological niches, both marine and terrestrial, and have developed strategies to control one-another when competing for the same nutrients. This is the basis of antibiosis. Since the discovery of antibiotics, there has been a lot attention paid to the production of secondary metabolites by individual bacteria and fungi. More recent work is starting to highlight how interaction between antagonists can trigger the synthesis of antibiotics, and the mechanisms that underpin this are an area of ongoing research. For example, the concepts of cryptic loci for secondary metabolites and the activation of expression in response to stimuli from competitors are addressed elsewhere in this book (Axel Brackhage's chapter). In a different, but related, concept it is now emerging that some antimicrobials are produced only in complex symbiotic scenarios such as the highly evolved endosymbiotic *Burkholderia* of some fungi or the ant-associated *Pseudonocardia* Actinomycetes. Thus, for such examples, some type of communication or signalling must be taking place. Many of these details remain elusive but there are some cases of inter-domain signalling between bacteria and fungi that are revealing possible mechanisms. It is intriguing to see, for example, that *Pseudomonas* uses its own signals to mimic fungal signals thereby modifying gene expression and phenotypes in *Candida*. It can be expected that many more such examples will emerge in the coming years. There is no doubt that our understanding of signalling and communication across the microbial world is still in its infancy.

Acknowledgements Work in John Morrissey's laboratory is supported by Science Foundation Ireland grant number 08/RFP/GEN1319.

References

Asaka O, Shoda M (1996) Biocontrol of *Rhizoctonia solani* damping-off of tomato with *Bacillus subtilis* RB14. Appl Environ Microbiol 62:4081–4085

Bender C, Rangaswamy V, Loper J (1999) Polyketide production by plant-associated pseudomonads. Annu Rev Phytopathol 37:175–196

Bonfante P, Anca IA (2009) Plants, mycorrhizal fungi, and bacteria: a network of interactions. Annu Rev Microbiol 63:363–383

Cafaro MJ, Currie CR (2005) Phylogenetic analysis of mutualistic filamentous bacteria associated with fungus-growing ants. Can J Microbiol 51:441–446

Chotirmall SH, Greene CM, McElvaney NG (2010) *Candida* species in cystic fibrosis: a road less travelled. Med Mycol 48(Suppl 1):S114–S124

Cronin D, Moenne-Loccoz Y, Fenton A, Dunne C, Dowling DN, O'Gara F (1997a) Role of 2,4-diacetylphloroglucinol in the interactions of the biocontrol pseudomonad strain F113 with the potato cyst nematode *Globodera rostochiensis*. Appl Environ Microbiol 63:1357–1361

Cronin D, Moënne-Loccoz Y, Fenton A, Dunne C, Dowling DN, O'Gara F (1997b) Ecological interaction of a biocontrol *Pseudomonas fluorescens* strain producing 2,4-diacetylphloroglucinol with the soft rot potato pathogen *Erwinia carotovora* subsp. *atroseptica*. FEMS Microbiol Ecol 23:95–106

Crowe JD, Olsson S (2001) Induction of laccase activity in *Rhizoctonia solani* by antagonistic *Pseudomonas fluorescens* strains and a range of chemical treatments. Appl Environ Microbiol 67:2088–2094

Cugini C, Morales DK, Hogan DA (2010) Candida albicans-produced farnesol stimulates *Pseudomonas quinolone* signal production in LasR-defective *Pseudomonas aeruginosa* strains. Microbiology 156:3096–3107

Currie CR, Mueller UG, Malloch D (1999a) The agricultural pathology of ant fungus gardens. Proc Natl Acad Sci USA 96:7998–8002

Currie CR, Scott JA, Summerbell RC, Malloch D (1999b) Fungus-growing ants use antibiotic-producing bacteria to control garden parasites. Nature 398:701–704

Currie CR, Wong B, Stuart AE, Schultz TR, Rehner SA, Mueller UG, Sung GH, Spatafora JW, Straus NA (2003a) Ancient tripartite coevolution in the attine ant-microbe symbiosis. Science 299:386–388

Currie CR, Bot ANM, Boomsma JJ (2003b) Experimental evidence of a tripartite mutualism: bacteria protect ant fungus gardens from specialized parasites. Oikos 101:91–102

Davies DG, Marques CN (2009) A fatty acid messenger is responsible for inducing dispersion in microbial biofilms. J Bacteriol 191:1393–1403

Davis-Hanna A, Piispanen AE, Stateva LI, Hogan DA (2008) Farnesol and dodecanol effects on the *Candida albicans* Ras1-cAMP signalling pathway and the regulation of morphogenesis. Mol Microbiol 67:47–62

de Sordi L, Muhlschlegel FA (2009) Quorum sensing and fungal-bacterial interactions in *Candida albicans*: a communicative network regulating microbial coexistence and virulence. FEMS Yeast Res 9:990–999

Deveau A, Hogan DA (2011) Linking quorum sensing regulation and biofilm formation by *Candida albicans*. Methods Mol Biol 692:219–233

de Waard M (1997) Significance of ABC transporters in fungicide sensitivity and resistance. Pestic Sci 51:271–275

Duffy B, Schouten A, Raaijmakers JM (2003) Pathogen self-defense: mechanisms to counteract microbial antagonism. Annu Rev Phytopathol 41:501–538

Federle MJ (2009) Autoinducer-2-based chemical communication in bacteria: complexities of interspecies signaling. Contrib Microbiol 16:18–32

Fleming A (1929) On the antibacterial action of cultures of a *Penicillium*, with special reference to their use in the isolation of *B. influenzae*. Br J Exp Pathol 10:226–236

Gleeson O, O'Gara F, Morrissey JP (2010) The *Pseudomonas fluorescens* secondary metabolite 2,4 diacetylphloroglucinol impairs mitochondrial function in *Saccharomyces cerevisiae*. Antonie Van Leeuwenhoek 97:261–273

Griffith G, Rayner A (1994) Interspecific interactions and mycelial morphogenesis of *Hypholoma fasciculare* (Agaricaceae). Nova Hedwigia 59:47–75

Haeder S, Wirth R, Herz H, Spiteller D (2009) Candicidin-producing Streptomyces support leaf-cutting ants to protect their fungus garden against the pathogenic fungus Escovopsis. Proc Natl Acad Sci USA 106:4742–4746

Hall RA, Muhlschlegel FA (2010) A multi-protein complex controls cAMP signalling and filamentation in the fungal pathogen *Candida albicans*. Mol Microbiol 75:534–537

Handelsman J, Stabb EV (1996) Biocontrol of soilborne plant pathogens. Plant Cell 8:1855–1869

He H, Yang H, Bigelis R, Solum E, Greenstein M, Carter G (2002) Pyrrocidines A and B, new antibiotics produced by a filamentous fungus. Tetrahedron Lett 43:1633–1636

Hogan DA, Sundstrom P (2009) The Ras/cAMP/PKA signaling pathway and virulence in *Candida albicans*. Future Microbiol 4:1263–1270

Hogan DA, Vik A, Kolter R (2004) A *Pseudomonas aeruginosa* quorum-sensing molecule influences *Candida albicans* morphology. Mol Microbiol 54:1212–1223

Holcombe LJ, McAlester G, Munro CA, Enjalbert B, Brown AJ, Gow NA, Ding C, Butler G, O'Gara F, Morrissey JP (2010) *Pseudomonas aeruginosa* secreted factors impair biofilm development in *Candida albicans*. Microbiology 156:1476–1486

Holcombe LJ, O'Gara F, Morrissey JP (2011) Implications of interspecies signaling for virulence of bacterial and fungal pathogens. Future Microbiol 6:799–817

Ibrahim AS, Gebremariam T, Liu M, Chamilos G, Kontoyiannis D, Mink R, Kwon-Chung KJ, Fu Y, Skory CD, Edwards JE Jr, Spellberg B (2008) Bacterial endosymbiosis is widely present among zygomycetes but does not contribute to the pathogenesis of mucormycosis. J Infect Dis 198:1083–1090

Jarvis W (1977) *Botryotinia* and *Botrytis* species: taxonomy, physiology, and pathogenicity: a guide to the literature. Canada Department of Agriculture, Harrow

Keel C, Weller DM, Natsch A, Defago G, Cook RJ, Thomashow LS (1996) Conservation of the 2,4-diacetylphloroglucinol biosynthesis locus among fluorescent Pseudomonas strains from diverse geographic locations. Appl Environ Microbiol 62:552–563

Lackner G, Mobius N, Scherlach K, Partida-Martinez LP, Winkler R, Schmitt I, Hertweck C (2009) Global distribution and evolution of a toxinogenic Burkholderia-Rhizopus symbiosis. Appl Environ Microbiol 75:2982–2986

Lackner G, Moebius N, Partida-Martinez L, Hertweck C (2011a) Complete genome sequence of *Burkholderia rhizoxinica*, an endosymbiont of *Rhizopus microsporus*. J Bacteriol 193:783–784

Lackner G, Moebius N, Partida-Martinez LP, Boland S, Hertweck C (2011b) Evolution of an endofungal lifestyle: deductions from the *Burkholderia rhizoxinica* genome. BMC Genomics 12:210

Levy E, Eyal Z, Chet I, Hochman A (1992) Resistance mechanisms of *Septoria tritici* to antifungal products of *Pseudomonas*. Physiol Mol Plant Pathol 40:163–171

Ligon J, Hill D, Hammer P, Torkewitz N, Hofmann D, Kempf H, van Pee K (2000) Natural products with antifungal activity from Pseudomonas biocontrol bacteria. Pest Manage Sci 56:688–695

Little AE, Currie CR (2007) Symbiotic complexity: discovery of a fifth symbiont in the attine ant-microbe symbiosis. Biol Lett 3:501–504

Little AE, Currie CR (2008) Black yeast symbionts compromise the efficiency of antibiotic defenses in fungus-growing ants. Ecology 89:1216–1222

McAlester G, O'Gara F, Morrissey JP (2008) Signal-mediated interactions between *Pseudomonas aeruginosa* and *Candida albicans*. J Med Microbiol 57:563–569

Medentsev AG, Akimenko VK (1998) Naphthoquinone metabolites of the fungi. Phytochemistry 47:935–959

Milner JL, Silo-Suh L, Lee JC, He H, Clardy J, Handelsman J (1996) Production of kanosamine by *Bacillus cereus* UW85. Appl Environ Microbiol 62:3061–3065

Morschhauser J (2010) Regulation of multidrug resistance in pathogenic fungi. Fungal Genet Biol 47:94–106

Mowat E, Rajendran R, Williams C, McCulloch E, Jones B, Lang S, Ramage G (2010) *Pseudomonas aeruginosa* and their small diffusible extracellular molecules inhibit *Aspergillus fumigatus* biofilm formation. FEMS Microbiol Lett 313:96–102

Nakayama T, Homma Y, Hashidoko Y, Mizutani J, Tahara S (1999) Possible role of xanthobaccins produced by *Stenotrophomonas* sp. strain SB-K88 in suppression of sugar beet damping-off disease. Appl Environ Microbiol 65:4334–4339

Nielsen TH, Christophersen C, Anthoni U, Sorensen J (1999) Viscosinamide, a new cyclic depsipeptide with surfactant and antifungal properties produced by *Pseudomonas fluorescens* DR54. J Appl Microbiol 87:80–90

Nielsen TH, Sorensen D, Tobiasen C, Andersen JB, Christophersen C, Givskov M, Sorensen J (2002) Antibiotic and biosurfactant properties of cyclic lipopeptides produced by fluorescent *Pseudomonas* spp. from the sugar beet rhizosphere. Appl Environ Microbiol 68:3416–3423

Oh DC, Poulsen M, Currie CR, Clardy J (2009) Dentigerumycin: a bacterial mediator of an ant-fungus symbiosis. Nat Chem Biol 5:391–393

Osbourn AE (1996) Preformed antimicrobial compounds and plant defense against fungal attack. Plant Cell 8:1821–1831

Partida-Martinez LP, Hertweck C (2005) Pathogenic fungus harbours endosymbiotic bacteria for toxin production. Nature 437:884–888

Partida-Martinez LP, de Looss CF, Ishida K, Ishida M, Roth M, Buder K, Hertweck C (2007a) Rhizonin, the first mycotoxin isolated from the zygomycota, is not a fungal metabolite but is produced by bacterial endosymbionts. Appl Environ Microbiol 73:793–797

Partida-Martinez LP, Groth I, Schmitt I, Richter W, Roth M, Hertweck C (2007b) *Burkholderia rhizoxinica* sp. nov. and *Burkholderia endofungorum* sp. nov., bacterial endosymbionts of the plant-pathogenic fungus *Rhizopus microsporus*. Int J Syst Evol Microbiol 57:2583–2590

Peleg AY, Hogan DA, Mylonakis E (2010) Medically important bacterial-fungal interactions. Nat Rev Microbiol 8:340–349

Raaijmakers JM, Vlami M, de Souza JT (2002) Antibiotic production by bacterial biocontrol agents. Antonie Van Leeuwenhoek 81:537–547

Reen FJ, Mooij MJ, Holcombe LJ, McSweeney CM, McGlacken GP, Morrissey JP, O'Gara F (2011) The pseudomonas quinolone signal (PQS), and its precursor HHQ, modulate interspecies and interkingdom behaviour. FEMS Microbiol Ecol 77(2):413–428

Ryan RP, Dow JM (2011) Communication with a growing family: diffusible signal factor (DSF) signaling in bacteria. Trends Microbiol 19:145–152

Sanders IR (2005) Microbiology: conspirators in blight. Nature 437:823–824

Schmitt I, Partida-Martinez LP, Winkler R, Voigt K, Einax E, Dolz F, Telle S, Wostemeyer J, Hertweck C (2008) Evolution of host resistance in a toxin-producing bacterial-fungal alliance. ISME J 2:632–641

Schoonbeek H, Del Sorbo G, de Waard MA (2001) The ABC transporter BcatrB affects the sensitivity of Botrytis cinerea to the phytoalexin resveratrol and the fungicide fenpiclonil. Mol Plant Microbe Interact 14:562–571

Schoonbeek HJ, Raaijmakers JM, de Waard MA (2002) Fungal ABC transporters and microbial interactions in natural environments. Mol Plant Microbe Interact 15:1165–1172

Schouten A, van den Berg G, Edel-Hermann V, Steinberg C, Gautheron N, Alabouvette C, de Vos CH, Lemanceau P, Raaijmakers JM (2004) Defense responses of *Fusarium oxysporum* to 2,4-diacetylphloroglucinol, a broad-spectrum antibiotic produced by *Pseudomonas fluorescens*. Mol Plant Microbe Interact 17:1201–1211

Schouten A, Maksimova O, Cuesta-Arenas Y, van den Berg G, Raaijmakers JM (2008) Involvement of the ABC transporter BcAtrB and the laccase BcLCC2 in defence of *Botrytis cinerea* against the broad-spectrum antibiotic 2,4-diacetylphloroglucinol. Environ Microbiol 10:1145–1157

Silo-Suh LA, Lethbridge BJ, Raffel SJ, He H, Clardy J, Handelsman J (1994) Biological activities of two fungistatic antibiotics produced by *Bacillus cereus* UW85. Appl Environ Microbiol 60:2023–2030

Silo-Suh LA, Stabb EV, Raffel SJ, Handelsman J (1998) Target range of zwittermicin A, an aminopolyol antibiotic from *Bacillus cereus*. Curr Microbiol 37:6–11

Sylvia D, Fuhrmann J, Hartel P, Zuberer D (2004) Principles and applications of soil microbiology. Prentice Hall, Upper Saddle River

Tripathi RK, Gottlieb D (1969) Mechanism of action of the antifungal antibiotic pyrrolnitrin. J Bacteriol 100:310–318

Tschudi S, Kern H (1979) Specific lysis of the mycelium of *Gaeumannomyces graminis* by enzymes of *Streptomyces lavendulae*. In: Schippers B, Gams W (eds) Soilborne plant pathogens. Academics, New York, pp 611–615

van Eck W (1978) Chemistry of cell walls of *Fusarium solani* and the resistance of spores to microbial lysis. Soil Biol Biochem 10:155–157

van Elsas JD, Heijnen CE (1990) Methods for the introduction of bacteria into soil: a review. Biol Fertil Soils 10:127–133

Vilchez R, Lemme A, Ballhausen B, Thiel V, Schulz S, Jansen R, Sztajer H, Wagner-Dobler I (2010) *Streptococcus mutans* inhibits *Candida albicans* hyphal formation by the fatty acid signaling molecule trans-2-decenoic acid (SDSF). ChemBioChem 11:1552–1562

Weller DM (1988) Biological control of soilborne plant pathogens in the rhizosphere with bacteria. Annu Rev Phytopathol 26:379–407

Weller DM, Raaijmakers JM, Gardener B, Thomashow LS (2002) Microbial populations responsible for specific soil suppressiveness to plant pathogens. Annu Rev Phytopathol 40:309–348

Wheatley RE (2002) The consequences of volatile organic compound mediated bacterial and fungal interactions. Antonie Van Leeuwenhoek 81:357–364

Whipps JM (2001) Microbial interactions and biocontrol in the rhizosphere. J Exp Bot 52:487–511

Williams P (2007) Quorum sensing, communication and cross-kingdom signalling in the bacterial world. Microbiology 153:3923–3938

Williams P, Camara M (2009) Quorum sensing and environmental adaptation in *Pseudomonas aeruginosa*: a tale of regulatory networks and multifunctional signal molecules. Curr Opin Microbiol 12:182–191

Viruses, Fungi and Plants: Cross-Kingdom Communication and Mutualism

Rusty J. Rodriguez and Marilyn Roossinck

Abstract Plants in natural ecosystems are symbiotic with fungal endophytes that have profound effects on host ecophysiology. In addition, symbiotic fungi may contain viruses that influence fungal physiology and alter the outcome of plant–fungal symbioses. For example, *Dichanthelium lanuginosum* is a plant that thrives in geothermal soils, tolerating root zone temperatures up to 60°C. Thermotolerance is dependent on the fungal endophyte *Curvularia protuberata*. Remarkably, the ability of *C. protuberata* to confer heat tolerance requires a double stranded RNA virus harbored by the fungus. When *C. protuberata* (containing the virus) and *D. lanuginosum* are grown independent of each other, they only tolerate temperatures up to 38°C. This is a clear example of how a symbiosis achieves something that the individual components cannot. In this chapter, we describe how this three-way symbiosis allows three organisms to survive in an environment they cannot tolerate on their own and explain what is known about the inter-organismal communication responsible for this mutualism.

1 Introduction

Biocommunication may occur between different organisms, genomes or biochemical pathways and result from electronic, physical or chemical interactions. More than 3.5 billion years ago life on earth began to develop as a result of

R.J. Rodriguez (✉)
Western Fisheries Research Center, International Symbiosis Society,
U.S. Geological Survey, 6505 NE 65th, Seattle, WA 98115, USA

Department of Biology, US Geological Survey, Symbiogenics, Seattle, WA 98125, USA
e-mail: rjrodriguez@symbiogenics.org

M. Roossinck
Departments of Plant Pathology and Biology, Pennsylvania State University,
University Park, PA 16802, USA
e-mail: mjr25@psu.edu

G. Witzany (ed.), *Biocommunication of Fungi*, DOI 10.1007/978-94-007-4264-2_14,

biocommunication, and adapted to a changing global environment with increased levels of organismal complexity. As single-celled life evolved it paved the way for cell-to-cell and cell-to-virus biocommunication. The evolution of multicellular life brought about more complex interactions. Between 3.5 billion and 800 million years ago bacterial photosynthesis is thought to have increased oxygen in the atmosphere to a level that allowed for remarkable biological developments: aerobic respiration and photosynthesis (Knauth and Kennedy 2001; Finlay and Esteban 2009; Payne et al. 2011). These developments were followed by an explosion of biological diversity and complexity. Collectively, these evolutionary events put a form of biocommunication in play that would become a fundamental aspect of multicellular life on earth: symbiosis.

One of the more remarkable aspects of symbiosis is the biocommunication between genetically distant organisms that allows them to achieve levels of fitness they cannot achieve alone. The outcome of symbioses span a continuum from mutualism (positive for both partners) to parasitism (beneficial to one and detrimental to the other); the only difference appears to based on communication (Hirsch 2004). Although the communication between pathogens and hosts has been studied extensively, comparatively little effort has been invested in understanding mutualisms.

Since the initial description of symbiosis by de Bary in 1879 scientists viewed it as a phenomenon involving two partners, one macroscopic and one microscopic (de Bary 1879). Since then, it has become apparent that symbiotic associations may involve multiple organisms from different kingdoms or domains of life. For example, certain insects are known to associate with both fungi and bacteria (Pinto-Tomas et al. 2009). In some cases the significance of different associated organisms is well defined and in others it is not.

Viruses are probably the best studied entities in biology and were responsible for great advancements in molecular biology and genomic research. However, the vast majority of research has focused on viruses as pathogens that inflict severe diseases in plants and animals. Recent studies indicate that there is more to viruses than simply causing diseases (Xu et al. 2008). Although the concept of viruses as mutualists is relatively new, we hypothesize that positive interactions between viruses and other organisms is common (Roossinck 2011).

The interactions between fungi and plants have been studied for at least 150 years, with most research from the perspective that these organisms are individuals. However, it is common for fungi and plants to harbor asymptomatic viruses (Ghabrial and Suzuki 2008; Morsy et al. 2010; Roossinck 2010) and few ecological functions have been attributed to them.

In this chapter we discuss recent findings that describe how communication among a virus, fungus and plant result in a mutualism. The mutualism allows all three entities to survive in geothermal soils that impose temperature, moisture and chemical stress. Three levels of symbiotic communication are described: plant: fungal, fungal:virus, and plant:fungal:virus. We have generated strains of one fungal endophyte that is virus free for comparative studies to determine which symbiont(s) is involved in the communication responsible for specific mutualistic benefits. Some of the mutualistic benefits and the role of virus and endophyte in biocommunication are described below.

2 Communication Between Class 2 Fungal Endophytes and Plants

Although the communication between plants and fungi has been studied for some time, the communication between plants and Class 2 endophytes (Rodriguez et al. 2009a) has only recently become a research focus. This is partly because compared to mycorrhizae, endophytes reside entirely within plant tissues and emerge only during plant senescence, making them more difficult to study. For example, mycorrhizal fungi both penetrate into roots and grow out into the rhizosphere making mycelia readily available for analysis. Nonetheless, it is possible to assess plant endophyte communication by assessing physiological processes and genetic expression in host plants. There are three aspects of plant physiology that are involved symbiotic communication: growth and development, metabolic efficiency, and stress tolerance. Endophytes can have profound impacts on all these physiological processes and it is likely that all plants in natural ecosystems are symbiotic with endophytes (Petrini 1986). Moreover, the fossil record indicates that plant-fungal associations have been in existence for more than 400 MY and it has been proposed that fungi played a significant role in the movement of plants onto land (Pirozynski and Malloch 1975; Remy et al. 1994; Redecker et al. 2000; Krings et al. 2007).

2.1 *Growth and Development*

Many microorganisms such as bacteria, mycorrhizae and fungal endophytes are known to significantly enhance the growth of plants. For example, rice seedlings colonized with the Class 2 endophyte *Fusarium culmorum* (isolate FcRed1) grows 5–10 times faster than nonsymbiotic seedlings (Rodriguez et al. 2009b; Redman et al. 2011). The growth enhancement occurs in both shoot and root systems with symbiotic plants achieving greater biomass than nonsymbiotic plants (Redman et al. 2002, 2011; Rodriguez et al. 2009b). The mechanism(s) of endophyte enhanced growth responses are not known and although endophytes are able to produce plant hormones on culture medium (Redman et al. 2011), it is not known if they produce hormones *in planta* or if hormones alone are responsible for growth responses. It appears that viruses are not involved in symbiotically induced plant growth promotion as endophytes free of fungal viruses elicit the same growth responses (Márquez et al. 2007; Rodriguez et al. 2008).

In addition to plant growth promotion, endophytes may alter carbon allocation and seedling development (Rodriguez et al. 2009b). Symbiotic seedlings preferentially allocate carbon to root systems until root hairs form before they allocate significant amounts of carbon to shoot growth (Fig. 1). Depending on day length, nonsymbiotic plants either allocate carbon equally to root and shoot growth or predominantly to shoot growth. Clearly, the best strategy for allocation of resources

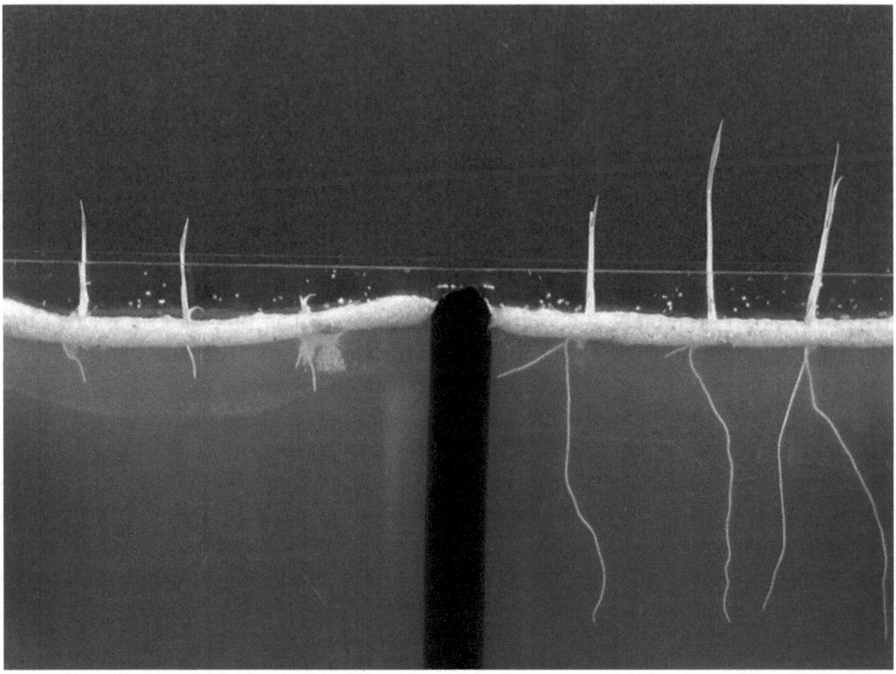

Fig. 1 Differential development of three nonsymbiotic (*left*) and symbiotic (*right*) rice plants after 4 days of growth in a agar medium with 16 h light/day. The *white layer* at the top of the medium is sand and the medium was poured between two glass plates separated by 0.5 in. with tygon tubing

in seedlings is to get roots established to the point of nutrient acquisition before expanding shoots. The influence of fungal endophytes on plant development occurs in the presence or absence of viruses so the communication is presumed to be fungal-specific.

Fungal endophytes may also influence sexual reproduction in plants (Miglia et al. 2007; Rodriguez et al. 2009b). For example, in Salt Creek Canyon, Utah, two subspecies of big sagebrush (*Artemesia tridentata*) occur at elevations of <1,830 m (basin, *A. t.* ssp. *tridentata*) and >1,880 m (mountain, *A. t.* ssp. *vaseyana*) with a hybrid zone occurring between the parental populations (Fig. 2). There is no documentation of parentals or hybrids invading each other's habitat and a reciprocal transplant study revealed a relationship between fungal endophytes, host genomes and indigenous soils that regulates plant reproduction. Plants had to have their indigenous endophytes and be in their native habitat to reproduce sufficiently; hybrids have greatly reduced fitness and reproduction in either parental soil as do parental plants in either the hybrid or the heterologous parental soil. In these studies, the viral component of the fungal endophytes was not assessed so it is not possible to assign symbiotic communication to a specific symbiont or combination of symbionts.

Fig. 2 A field site in Central Utah where two big sagebrush (*Artemesia tridentata*) subspecies grow in close proximity and form a stable hybrid zone. The *circles* define the zones of the basin subspecies (*B*; *A. t.* ssp. *tridentata)* occurring at elevations of $<$1,830 m, the mountain subspecies (*M*; *A. t.* ssp. *vaseyana*) occurring at elevations of $>$1,880 m, and a hybrid zone (*H*) occurring between the parental populations

2.2 Metabolic Efficiency

One mutualistic benefit of fungal endophytes is that they alter water relations and final biomass of both monocot and eudicot plants. All of the Class 2 endophytes tested so far reduce water consumption in plants by as much as 50% compared to nonsymbiotic plants. Although the mechanism of reduced water consumption is not known, gas exchange measurements with a LiCor 6400 indicate that it does not appear to stem from differential control of stomatal activity (Woodward et al. 2012).

In addition to water consumption changes, the endophytes may also increase plant biomass by as much as 50% (Rodriguez et al. 2008; Redman et al. 2011). Decreased water consumption and increased biomass can be interpreted as the symbiotic plants having more photosynthate or greater metabolic efficiency than nonsymbiotic plants. Photosynthetic rates measured by gas exchange revealed no significant difference between symbiotic and nonsymbiotic plants suggesting no differences in photosynthate levels (Woodward et al. 2012). Metabolic activity measured by differential scanning calorimetry (DSC) also revealed no significant differences between symbiotic and nonsymbiotic plants (Woodward et al. 2012). We surmise from these data that physiological differences between symbiotic and nonsymbiotic plants are due to other factors such as photo system efficiency.

The symbiotic communication responsible for the physiological effects by fungal endophytes appears to be fungal-specific. Comparative studies demonstrated that a virus free strain has the same affects as virus containing strains (Márquez et al. 2007; Rodriguez et al. 2008).

2.3 Abiotic Stress Tolerance

One remarkable aspect of Class 2 fungal endophytes is their ability to confer abiotic stress tolerance to plants (Rodriguez et al. 2009a). This occurs in a habitat-specific manner so that endophytes from geothermal plants confer heat but not salt tolerance, and endophytes from coastal plants confer salt but not heat tolerance (Rodriguez et al. 2008). We define this phenomenon as Habitat-Adapted Symbiosis (HAS) and postulate that it is a ubiquitous epigenetic aspect of plant ecology, adaptation and evolution. The most extensive work in HAS involves the geothermal plant *Dichanthelium lanuginosum* (tropical panic grass) and the fungal endophyte *Curvularia protuberata* (Redman et al. 2002, 2011; Márquez et al. 2007; Rodriguez et al. 2008). Field and laboratory studies revealed that *D. lanuginosum* is not able to adapt itself to the heat stress of geothermal habitats. Instead, *D. lanuginosum* forms a symbiosis with a Class 2 fungal endophyte that adapts the plant to thermal stress. An interesting aspect of this system is that both the plant and endophyte tolerate up to 38°C when grown independently of each other. However, as a symbiosis the partners can tolerate root zone temperatures up to 70°C.

Fungi often contain double stranded RNA (dsRNA) viruses but, few functions have been attributed to the viruses (see Sect. 3). *C. protuberata* also contains a dsRNA virus that has profound impacts on the *D. lanuginosum/C. protuberata* symbiosis (Márquez et al. 2007). Comparative studies between virus containing and virus free strains of *C. protuberata* demonstrated that the virus was required for heat tolerance. Therefore, a cross-kingdom three-way symbiosis was necessary for plants adapting to geothermal soils.

3 Communication Between Viruses and Fungi

Viruses by definition are obligate symbionts, although they are not always defined this way, and their symbiotic relationships have been studied almost exclusively in the context of pathology. However, a number of viruses also establish mutualistic relationships with their hosts (Roossinck 2011). Although most fungal viruses do not cause disease in their hosts, the most well-studied fungal–virus biocommunications are between fungi and pathogenic viruses. For example, *Cryphonectria parasitica*, the causative agent of chestnut blight, has lower fecundity and slower growth rates when infected by *Cryphonectria hypovirus*-1 (CHV-1) (Milgroom and Cortesi 2004). Virus-infected fungi are also hypovirulent on chestnut trees.

A number of studies have compared transcriptomes of virus-infected and virus-free *C. parasitica* and/or the chestnut hosts (Allen et al. 2003; Deng et al. 2007; Shang et al. 2008). Other studies of fungal–virus communications are summarized in two recent reviews that discuss the relationships among plant–pathogenic fungi and their viruses (Ghabrial and Suzuki 2009; Pearson et al. 2009). However, very little work has been done on mutualistic viruses in general, and on mutualistic fungal viruses in particular. Single-celled fungi (yeasts) can harbor mutualistic viruses that help them invade new territory. The virus produces a toxin that the host fungus is immune to, but that kills off sensitive yeast, opening the way for invasion (Schmitt and Breinig 2006).

Curvularia protuberata requires its virus, Curvularia thermal tolerance virus (CThTV), in order to confer thermal tolerance to plants (Márquez et al. 2007). Studies to understand the communication between the virus and the fungus in the context of the holobiont have proven very difficult. However, comparisons between cultured *C. protuberata* with and without virus have led to some insights about the virus–fungus communications that may be responsible for heat tolerance. Although the fungus cannot grow in culture at the high temperatures found in geothermal soils, it can grow at 38°C. Under these conditions, the virus-free isolate produced fewer viable spores. In addition, at 30°C the virus-free strain grew slower. Subtractive cDNA libraries provided an analysis of transcripts induced by virus-infection under both control temperatures (25°C) and under heat stress (38°C). Under heat stress virus-induced increased gene synthesis occurred in the trehalose and melanin biosynthesis pathways (a sugar and a pigment, respectively, that have been implicated as osmoprotectants), reactive oxygen detoxification enzymes, the taurine catabolism pathway (a key player in carbon use regulation), and some heat-shock proteins (Morsy et al. 2010). Any or all of these may be involved in the communications that occur between the fungus and the virus during heat stress within the plants. It is also possible that the virus releases factors that are transferred directly to the plant.

4 Summary

The *D. lanuginosum*-*C. protuberata*-CThTV symbiosis is a great example of how symbiosis can achieve functionalities that the individual entities are unable to express. For *D. lanuginosum* to survive in geothermal soils, it must be colonized with *C. protuberata* and for *C. protuberata* to confer heat tolerance, it must be colonized by CThTV. Since CThTV cannot survive without a host, this mutualism is based on an inextricable link between symbionts representing different kingdoms. One of the more enlightening aspects of this system is the mutualistic role of CThTV. Viruses are commonly viewed as negative parasites responsible for many diseases. Yet, the role of most viruses in nature is unknown. We have found that many plants require fungal endophytes for abiotic stress tolerance and hypoth esize that three-way symbioses between plants, fungi and viruses are ubiquitous in nature and are critical to plant ecology, adaptation and evolution. Yet to be elucidated are the biochemical and molecular bases of the biocommunication responsible for the symbiosis.

Acknowledgements Funding was provided by USGS, NSF (0414463&0950447), US/IS BARD (3260-01C) and ARO (54120-LS).

References

Allen TD, Dawe AL, Nuss DL (2003) Use of cDNA microarrays to monitor transcriptional responses of the chestnut blight fungs *Cryphonectria parasitica* to infection by virulence-attenuating hypoviruses. Eukaryot Cell 2:1253–1265

de Bary A (1879) Die Erschenung Symbiose. In: Trubner KJ (ed) Vortrag auf der Versammlung der Naturforscher und Artze zu Cassel, pp 1–30

Deng F, Allen TD, Hillman BI, Nuss DL (2007) Comparative analysis of alterations in host phenotype and transcript accumulation following hypovirus and mycoreovirus infections of the chestnut blight fungus *Cryphonectria parasitica*. Eukaryot Cell 6:1286–1298

Finlay BJ, Esteban GF (2009) Can biological complexity be rationalized? Bioscience 59:333–340

Ghabrial SA, Suzuki N (2008) Fungal viruses. In: Granoff A, Webster R (eds) Encyclopedia of virology. Elsevier, Amsterdam, pp 284–291

Ghabrial SA, Suzuki N (2009) Viruses of plant pathogenic fungi. Annu Rev Phytopathol 47:353–384

Hirsch AM (2004) Plant-microbe symbioses: a continuum from commensalism to parasitism. Symbiosis 37:345–363

Knauth LP, Kennedy MJ (2001) The late Precambrian greening of the Earth. Nature 460:728–732

Krings M, Taylor TN, Hass H, Kerp H, Dotzler N, Hermsen EJ (2007) Fungal endophytes in a 400-million-yr-old land plant: infection pathways, spatial distribution, and host responses. New Phytol 174:648–657

Márquez LM, Redman RS, Rodriguez RJ, Roossinck MJ (2007) A virus in a fungus in a plant – three way symbiosis required for thermal tolerance. Science 315:513–515

Miglia KJ, McArthur ED, Redman RS, Rodriguez RJ, Zak JC, Freeman DC (2007) Genotype, soil type, and locale effects on reciprocal transplant vigor, endophyte growth, and microbial functional diversity of a narrow sagebrush hybrid zone in Salt Creek, Canyon, Utah. Am J Bot 94:425–436

Milgroom MG, Cortesi P (2004) Biological control of chestnut blight with hypovirulence: a critical analysis. Annu Rev Phytopathol 42:311–338

Morsy MR, Oswald J, He J, Tang Y, Roossinck MJ (2010) Teasing apart a three-way symbiosis: transcriptome analyses of *Curvularia protuberata* in response to viral infection and heat stress. Biochem Biophys Res Commun 401:225–230

Payne JL, McClain CR, Boyer AG et al (2011) The evolutionary consequences of oxygenic photosynthesis: a body size perspective. Photosynth Res 107:37–57

Pearson MN, Beever RE, Boine B, Arthur K (2009) Mycoviruses of filamentous fungi and their relevance to plant pathology. Mol Plant Pathol 10:115–128

Petrini O (1986) Taxonomy of endophytic fungi of aerial plant tissues. In: Fokkema NJ, van den Heuvel J (eds) Microbiology of the phyllosphere. Cambridge University Press, Cambridge, pp 175–187

Pinto-Tomas AA, Anderson MA, Suen G et al (2009) Symbiotic nitrogen fixation in the fungus gardens of leaf-cutter ants. Science 326:1120–1123

Pirozynski KA, Malloch DW (1975) The origin of land plants a matter of mycotrophism. Biosystems 6:153–164

Redecker D, Kodner R, Graham LE (2000) Glomalean fungi from the Ordovician. Science 289:1920–1921

Redman RS, Sheehan KB, Stout RG, Rodriguez RJ, Henson JM (2002) Thermotolerance conferred to plant host and fungal endophyte during mutualistic symbiosis. Science 298:1581

Redman RS, Kim YO, Woodward CJDA, Greer C, Espino L, Doty SL, Rodriguez RJ (2011) Increased fitness and adaptation of rice plants to cold, drought and salt stress via habitat adapted symbiosis: a strategy for mitigating impacts of climate change. PLoS One 6(7): e14823. doi:10.1371/journal.pone.0014823

Remy W, Taylor TN, Hass H, Kerp H (1994) Four hundred-million-year-old vesicular arbuscular mycorrhizae. Proc Natl Acad Sci USA 91:11841–11843

Rodriguez RJ, Henson J, van Volkenburgh E et al (2008) Stress tolerance in plants via habitat-adapted symbiosis. Int Soc Microb Ecol 2:404–416

Rodriguez RJ, White JFJ, Arnold AE, Redman RS (2009a) Fungal endophytes: diversity and functional roles. New Phytol 182:314–330

Rodriguez RJ, Freeman DC, McArthur ED, Kim YO, Redman RS (2009b) Symbiotic regulation of plant growth, development and reproduction. Commun Integr Biol 2:1–3

Roossinck MJ (2010) Lifestyles of plant viruses. Philos Trans R Soc B 365:1899–1905

Roossinck MJ (2011) The good viruses: viral mutualistic symbioses. Nat Rev Microbiol 9:99–108

Schmitt MJ, Breinig F (2006) Yeast viral killer toxins: lethality and self-protection. Nat Rev Microbiol 4:212–221

Shang J, Wu X, Lan X et al (2008) Large-scale expressed sequence tag analysis for the chestnut blight fungus *Cryphonectria parasitica*. Fungal Genet Biol 45:319–327

Woodward C, Hansen L, Beckwith F, Redman RS, Rodriguez RJ (2012) Symbiogenics: An epigenetic approach to mitigating impacts of climate change on plants. HortScience, In Press.

Xu P, Chen F, Mannas JP, Feldman T, Sumner LW, Roossinck MJ (2008) Virus infection improves drought tolerance. New Phytol 180:911–921

Communication Between Plant, Ectomycorrhizal Fungi and Helper Bacteria

Aurélie Deveau, Jonathan M. Plett, Valérie Legué, Pascale Frey-Klett, and Francis Martin

Abstract Development of mutualistic symbioses between ectomycorrhizal fungi and their host trees involves multiple gene networks that are involved in a complex series of interdependent, sequential developmental steps. Through secreted signals and nutrient interactions, rhizospheric bacteria play a major role in the development of mycorrhizal symbioses. Current research into symbiosis development and functioning is aimed at understanding these plant–microbe interactions in the framework of environmental, developmental and physiological processes that underlie colonization and morphogenesis. After a brief introduction to the ectomycorrhizal symbiosis, the present chapter aims (1) to highlight recent work on the early signal exchange taking place between symbionts and their associated bacteria, and (2) to sketch out the way that functional genomics is altering our thinking about how soil microbes alter host functioning during ectomycorrhizal root development.

Keywords Auxins • Effectors • Secretome • Gene expression • Helper bacteria • Mycorrhizosphere • Root hairs • Transduction pathways

1 Introduction

In the rhizospheric soil of forests and woodlands, a diverse community of prokaryotic and eukaryotic micro-organisms proliferate, compete and interact with tree roots and each other for nutrients such as nitrogen, phosphorus and carbon. Ectomycorrhizal (ECM) fungi, which form part of the basidiomycetes and ascomycetes, are one of the major classes of microbial species within the rhizosphere that competes with other soil-borne micro-organisms for carbon compounds

A. Deveau (✉) • J.M. Plett • V. Legué • P. Frey-Klett • F. Martin
Ecogenomics of Interactions Lab, UMR 1136 INRA Nancy University,
Interactions Arbres/Micro-Organismes, INRA-Nancy, 54280 Champenoux, France
e-mail: deveau@nancy.inra.fr; fmartin@nancy.inra.fr

G. Witzany (ed.), *Biocommunication of Fungi*, DOI 10.1007/978-94-007-4264-2_15,

found within, or that are secreted from, plant roots. ECM fungal hyphae colonize lateral root tips of host trees to form a novel composite organ, the so-called ectomycorrhizal root tip, which is the site of mutualistic nutrient and carbon transfers between the two symbionts. The increased nutrient supply obtained from this fungal interaction allows tree species in boreal and temperate forests to grow efficiently in suboptimal environments (Smith and Read 2008). Ectomycorrhizal root tips are structurally characterized by (i) the presence of an extensive extramatrical mycelial network whose function is to prospect the soil for mineral nutrients such nitrogen and phosphate; (ii) a mantle of fungal hyphae that forms a sheath around the root and acts mainly as a storage compartment; and (iii) a network of hyphae growing in the apoplastic space of the rhizodermis (in angiosperms) and cortex (in conifers) to form a structure called the Hartig net (Fig. 1). The fungus gains access to sugars from the plant across the large surface area of the Hartig net while nutrient and water uptake of the plant is mediated via the extraradical hyphal web of the mycobiont. Formation of the symbiotic root tips involves a sequence of complex and overlapping ontogenic processes between the fungal hyphae and host roots including increased rhizogenesis, enhanced hyphal branching, aggregation of the proliferating hyphae onto the root surface, arrest of meristematic activity in roots surrounded by the fungal mantle, and radial elongation of epidermal cells. These dramatic morphological changes are accompanied by the onset of novel transcripts and protein patterns (Veneault-Fourrey and Martin 2011) and coordinated metabolic re-organizations (Martin 2007) in fungal and plant cells leading to the functioning symbiosis.

Despite morphological differences between roots colonized by different ECM species (e.g. color, shape etc.), overall there are pronounced developmental similarities between roots colonized by a range of ECM fungi. This suggests that key or conserved developmental genetic programs are triggered in both symbiotic partners during the colonization process. No matter the species, the mycobiont must have the ability to recognize and to become associated with its host, escape the host defense surveillance, and establish bi-directional nutrient transfers. Similarly, the host must be able to differentiate mutualistic colonizers versus pathogenic organisms, to be able to vary root organogenesis and to closely monitor the exchange of nutrients with the fungal symbiont, to name a few. In contrast to arbuscular mycorrhizal (AM) symbiosis (Kosuta et al. 2003; Requena et al. 2007; Heupel et al. 2010; Maillet et al. 2011), the nature of the signaling molecules and the molecular basis of signal perception and transduction in ectomycorrhizal root tips are unknown or not well defined. Identifying the processes that regulate the information flow between ectomycorrhizal fungi and host roots is, however, an active field of research. AM fungi are the most ancient form of fungal mutualists known to exist while the ECM fungi are a group of mutualists that evolved a number of independent times from saprotrophic ancestors starting approximately 100–110 million years ago (Hibbett and Matheny 2009). Host plants release into the rhizosphere critical metabolites that are able to trigger spore germination (Fries et al. 1987), growth of hyphae towards the root (Horan and Chilvers 1990) and the early developmental steps of mycorrhiza formation (Jambois et al. 2005).

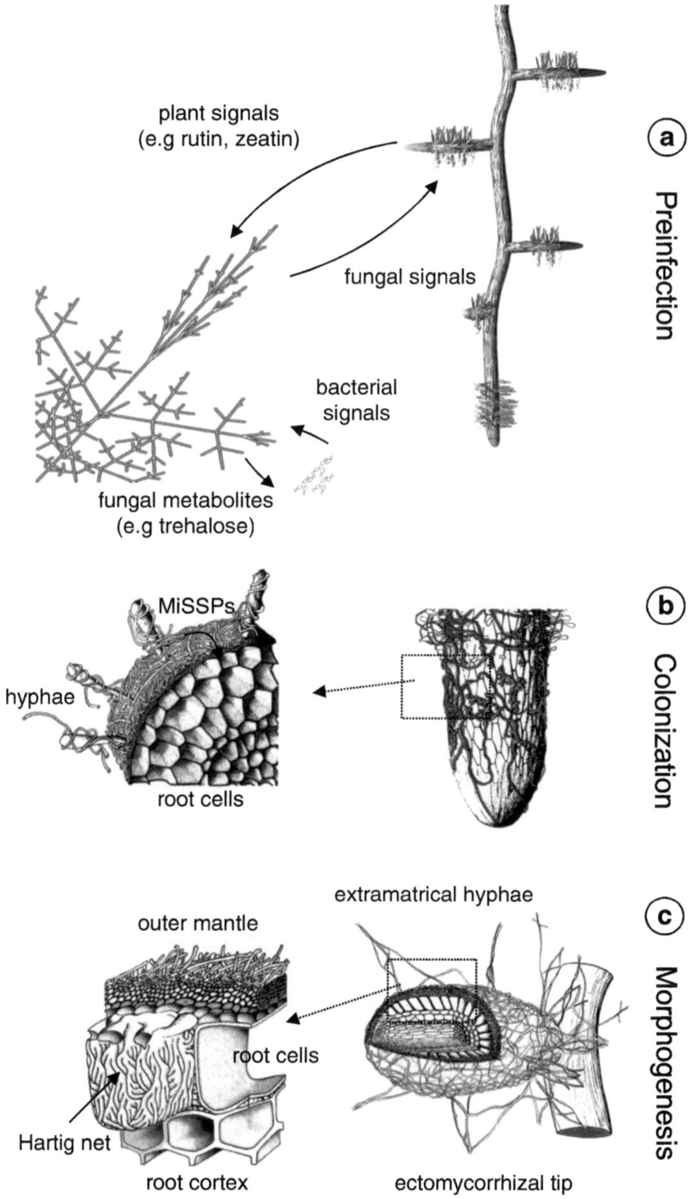

Fig. 1 General overview of the signals which can occur between the three actors: plant, ECM, helper bacteria. (**a**) The preinfection stage: host roots and helper bacteria release signals in the rhizosphere which stimulate fungal growth and hyphal branching; conversely, the hyphae release chemicals which induce changes in root morphology by regulating auxin signaling in the root tips. The fungus also produces metabolites which attract helper bacteria to its vicinity. (**b**) The colonization stage. Running hyphae attach to root cells and start secreting small molecules (MiSSPs) which enter plant cells and modulate plant defenses. (**c**) Morphogenesis: the massive and rapid aggregation of hyphae around the root lead to the formation of a pseudoparenchyma, the mantle is followed by the penetration of hyphae between epidermal cells and cortical cells and formation of the Hartig net

Molecules that pass between the host and fungus during the early stages of an interaction may control key steps in the symbiosis development including tropism of hyphae for host tissues through rhizospheric signals; attachment and invasion of host tissues by hyphae; induction of organogenetic programs in both fungal and root cells (hormones, effector-like proteins and secondary signals); facilitating survival of the mycobiont despite plant defense responses (effector-like proteins, proteases); and coordinating strategies for exchanging carbon and other metabolites for *in planta* colonization and for balancing growth of the soil fungal web with its role in gathering minerals from the soil. To date, published studies have provided an initial molecular understanding of how plant and fungal signals combine to determine the ectomycorrhiza development.

Ectomycorrhizal root development, thus, involves multiple signaling pathways and gene networks that are implicated in a complex series of sequential, interdependent, steps. Our current understanding of the complex signals involved in these processes are discussed in the present chapter.

2 Dominating the Conversation: The Role of ECM Secreted Proteins in Controlling the Plant Host

Successful adaptation to an ecosystem, or of a particular niche within an ecosystem, requires skill on the part of an organism in modeling the local environment to suit its growth and reproduction. To greater or lesser extents, all organisms have the ability to impact their surroundings to suit their life cycle. AM and ECM fungi are a fascinating examples of two classes of microbes that have refined the ability of modulating host function in order to create an environment in which they may safely grow and obtain the nutrients necessary for further growth and reproduction. How these fungi modulate the function of a host organism in order to colonize these roots, however, is an area of intense speculation and research. As outlined in the introduction, colonization of root tissues by mutualistic fungi is a very invasive process, and yet very few plant defenses are raised against the invading hyphae. In this section, we will consider the role of secreted signals from ECM and AM fungi in the structuring of their environment during the colonization of plant tissues. We will use as a basis of comparison the role of secreted proteins in pathogenic interactions in performing similar tasks.

In the pre-genomic age for mycorrhizal fungi, there was lively debate in academic circles concerning how mutualistic fungi were able to colonize host plants such that they avoided plant defenses and established a bi-directional flux of nutrients between the organisms. One theory postulated that the biology of the fungus was sufficiently different from pathogens that the plant could differentiate between the two classes of organisms (Redman et al. 2001). Similarly, another hypothesis postulated that there must be a balanced molecular cross-talk between the partners in order to shift the balance from a parasitic interaction to a mutualistic one (Paszkowski 2006). Both of these theories assumed that mutalistic fungi were very different from pathogenic organisms and that, during their evolution from saprotrophic ancestors, this class of fungi developed a novel means of communicating their beneficial presence to the plant host. It was unknown at the time what form this new means of communication took.

Currently, the most developed theories and understanding of communication between fungi and plants concern pathogenic interactions. Specifically, a large amount of research has focused on understanding the role of the small secreted proteins (SSPs) used by pathogenic fungi as signals to communicate with, or control, their plant hosts. These pathogenic signals, called effectors, are able to restructure the functioning of plant cells to further help pathogen growth on or within the plant tissue. Depending on the lifestyle of the pathogen, the role of these effectors varies greatly. Effectors of biotrophic pathogens, for example, will attempt to restructure the plant defensive system such that fungal tissues go undetected and the fungus may feed off of the living tissues. Necrotrophic pathogens, conversely, will actively seek to induce a hypersensitive response and, through this, enhance cell death. In either case, a subset of pathogenic effectors exists that are able to enter the plant cell where they will typically localize to the nucleus and alter the transcriptome of the host cell to suit their purposes (Bonas et al. 1989; Zhu et al. 1998; Yang et al. 2000; Deslandes et al. 2003; Schornack et al. 2004).

Prior to the release of the first ECM fungal genomes in 2008 (*Laccaria bicolor*; Martin et al. 2008) and 2010 (*Tuber melanosporum*; Martin et al. 2010), it was unknown if SSPs would form the basis of signals used by mutualistic fungi in communication with the host, or, if present, what role they would serve. Interestingly, on this point, the two genomes sequenced were diametrically opposed. While both genomes encoded several hundred putative SSPs, only the transcriptome of *L. bicolor* exhibited regulation of a number of these SSPs during the establishment of symbiosis while *T. melanosporum* did not have any significantly regulated SSPs in mature mycorrhizal root tips (Martin et al. 2008, 2010). Additionally, of the SSPs regulated in *L. bicolor* (named MiSSPs for MYCORRHIZA iNDUCED SMALL SECRETED PROTEINS), a number contained some of the hallmarks of pathogenic effector proteins (e.g. secretion signals, cell entry motifs, known DNA binding domains). These findings began to change the thinking of how mutualistic fungi colonized plant tissues, and raised the possibility that mutualists might use methods of communicating with (or controlling) their host that were remarkably close to those signals used by pathogens. Unfortunately, the lack of homology between the majority of the MiSSPs of *L. bicolor* and known or characterized pathogenic effectors has severely hampered our ability to judge their roles during plant colonization. Only recently has the characterization of the first two MiSSPs, MiSSP7 from the ECM fungus *L. bicolor* (Plett et al. 2011) and SP7 from the AM fungus *Glomus intraradices* (Kloppholz et al. 2011), been published allowing for new insight into who might have the upper hand during the establishment of symbiosis.

Of the genes predicted to encode MiSSPs in *L. bicolor*, MiSSP7 was found to be the most highly upregulated in mature mycorrhizal root tips (Martin et al. 2008).

MiSSP7 is a 7 kDa protein with a very high pI, a secretion signal and a sequence that bears no homology to any known protein. Its production is induced by root, and continues from early colonization stages through to the establishment of a mature mycorrhizal root tip (Plett et al. 2011). Upon secretion of MiSSP7, it is imported into plant cells via PI-3-P mediated endocytosis. This latter form of endocytosis is an entry mechanism also used by pathogenic effectors to gain entry to cells via binding of an RXLR or RXLR-like region of the protein to membrane phospholipids (Dou et al. 2008; Kale et al. 2010; Van West et al. 2010). Once in the plant cell, MiSSP7 localizes to the plant nucleus where it affects the transcription of a number of genes associated with cell wall and root architecture modifications, oxidative stress and defense (Plett et al. 2011). This evidence would suggest that ECM fungi are using one or more of their secreted proteins to gain control of plant cell signaling networks to colonize plant tissues. SP7, a protein also induced during mycorrhization between *G. intraradicies* and its host plant, shares some characteristics with MiSSP7. SP7 is also produced upon receipt of plant based signals. SP7 also appears to enter the plant cell through interactions with membrane phospholipids, although this is thought to be mediated by a series of imperfect, hydrophilic repeats on the amino end of the protein rather than by an RXLR-like motif as found in pathogens and MiSSP7. In the host plant cell, SP7 localizes to the plant nucleus where it interacts with Ethylene Response Factor 19 (ERF19) to repress plant defense signaling (Kloppholz et al. 2011). When SP7 was heterologously expressed in the hemibiotrophic fungus *Magnaporthe oryzae*, it acted to prolong the biotrophic phase and suppress the HR response within plant cells. Therefore the two proteins, MiSSP7 and SP7, appear to closely mirror pathogenic effectors in their role and localization within plant tissues.

But how do these results inform our understanding of mutualism and the role of communication during the establishment of symbiosis? These papers, from the first ECM genomes to the recent characterizations of MiSSP7 and SP7, have finally given us the break-through needed to begin to understand, at least theoretically, the development of mutualism and the tools used by ECM fungi to interact with plants. First, these results demonstrate that ECM fungi (and AM fungi), utilize the same basic mode of communication to interact with their plant partners as pathogenic fungi – the SSPs (Fig. 2). Second, rather than communicating to the plant that they are "friendly" or "good", these organisms are taking over or subverting plant defense signaling pathways to avoid detection and to prolong the biotrophy phase of their existence in plant tissues. This is a much stronger, and proactive, position for the fungus than originally thought and would suggest that the fungus has the upper hand over the plant. Thirdly, it is very interesting to find that both AM and ECM fungi, which do not share a recent common ancestor, have developed, or share, the same symbiotic toolbox to enter into a symbiotic relationship with plants. This means that despite the large sequence divergence between the effector SSPs of mutualists and pathogens, they are operating on the same basic signaling pathways to achieve the same basic goal – dominance over the plant. As it is very unlikely that the plant is a silent partner in this exchange it will be interesting to see in the coming

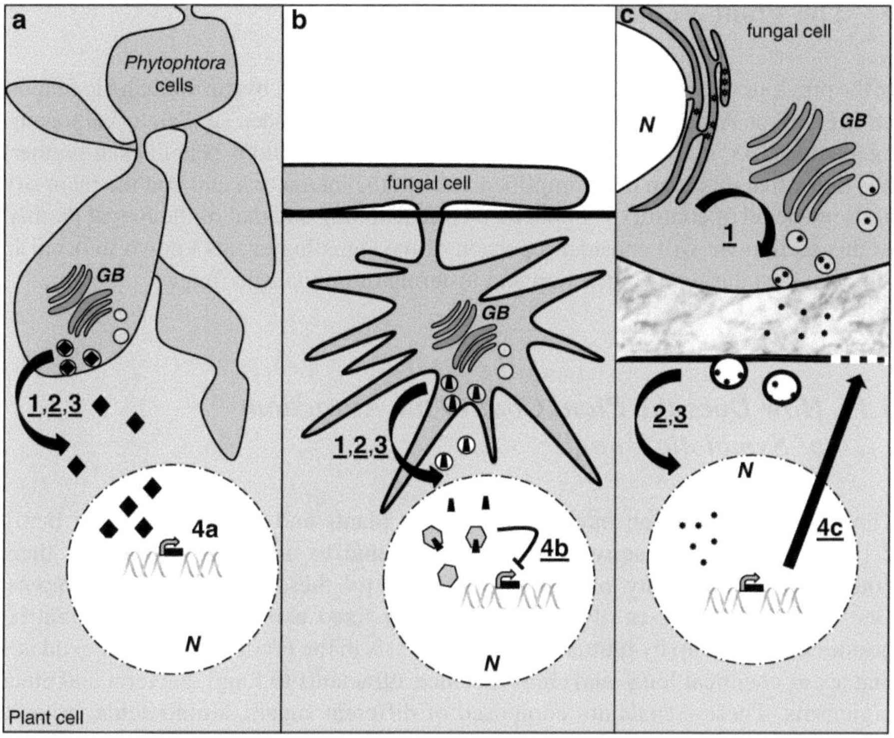

Fig. 2 Effectors from mutualistic and pathogenic show similar modes of action. Schematic representation of three different fungi that are interacting with plant cells. (**a**) The Nuk6 and Nuk7 effectors (*black diamonds*) are produced and secreted from the haustoria of the pathogen *Phytophthora infestans* (step 1). Once the proteins are imported into the plant cell they localize to the plant nucleus (steps 2, 3; Kanneganti et al. 2007) where their role during pathogenesis is unclear (Step 4a). (**b**) The SP7 effector (*rectangular shape*) from the mutualistic arbuscular mycorrhizal fungus *Glomus intraradices* is secreted from arbuscules within plant cells and taken up into the plant cytoplasm (steps 1 and 2). Once in plant cells SP7 localizes to the plant nucleus (step 3) where it associates with ERF19 (*octogonal shape*) which causes the suppression of defense gene production (Step 4b; Kloppholz et al. 2011). (**c**) The MiSSP7 effector protein (*star shape*) from the mutualistic ectomycorrhizal fungus *L. bicolor* is secreted into the apoplastic space of the root (step 1) where it is taken up into the plant via endocytosis (step 2). Once in the plant cell, the protein localizes to the nucleus (step 3) where it alters the expression of genes associated with cell wall maintenance (Step 4b; Plett et al. 2011). *GB* Golgi body, *N* plant nucleus

years what role the other MiSSPs encoded by ECM and AM fungi play during the establishment of symbiosis and how/if plants counter the attempt of mutualistic fungi to overtake plant cell function. These results will further inform our theoretical model whereby mutualistic fungi use modified pathogen-like effectors to communicate with their plant host. It will also be important to ascertain how ECM fungi that do not encode MiSSPs, like *T. melanosporum*, communicate with their hosts.

3 The Plant Voice

In the previous section we considered the role of signals sent by the mutualistic fungus, be it ECM or AM, in stifling the plant 'voice' in a manner similar to pathogenic interactions. As, by its very definition, mutualistic interactions benefit both partners it is likely that this is an over simplified view of the interaction and that the plant still has some level of control over both its own functioning and that of the fungal partner. In this section we will consider a portion of the signaling events known to occur in the plant host during the attraction and fostering of mutualistic fungi.

3.1 How Does the Plant Control the Attraction of Symbiotic Fungi?

The capacity to form an interaction between plants and rhizosphere fungi firstly depends upon the capacity of root to sense and to attract microbes and then, potentially, the capacity of host plant to control these microbes as symbionts. One of the major losses of photosynthetically fixed carbon into the soil matrix, besides that taken up by mutualistic organisms, is in the production of root exudates that act as chemical long- and short-distance attractants to fungi, bacteria and other organisms. These signals are composed of different sugars, amino acids, organic acids, fatty acids and growth factors, some of which control the functioning of fungal hyphae (Wen et al. 2007). Plant based compounds such as rutin (a flavonoid), zeatin and strigolactones have all been identified as signals that alter the branching and growth orientation of ECM and AM fungal hyphae such that, it is hypothesized, the fungus will have an increased opportunity to encounter a host root (Gogala 1991; Lagrange et al. 2001; Martin et al. 2001; Gomez-Roldan et al. 2008; Akiyama et al. 2005). As an attractant, rutin has received attention for its role in acting as a signaling agent to both mycorrhizal fungi and phytopathogenic fungi (Lagrange et al. 2001; Kalinova and Radova 2009). Other flavonoids have also been shown to induce the expression of protein signals necessary for both symbiotic and pathogenic organism colonization of plant tissues. These include the induction of MiSSP7 and Nod factors in mutualistic organisms (Peters et al. 1986; Plett and Martin 2012) and of *PDA1* in the soil pathogen *Nectria haematococca* MPVI (Khan and Straney 1999). Strigolactones are also a very interesting case study as they have a multiplicity of effects on fungal biology. While these compounds are originally thought to have evolved a role in the control of plant functioning (e.g. inhibition of shoot branching and lateral root formation; Gomez-Roldan et al. 2008; Kapulnik et al. 2011), they also modify the energetics of AM fungal hyphae by inducing mitochondrial and nuclear division and increases in energy metabolism (Besserer et al. 2006, 2008). A role of signaling molecules in the control of ECM fungi has yet to be demonstrated but has not been excluded as a possibility. Therefore the plant appears to be secreting a wide variety of signals, some which affect both beneficial and pathogenic organisms (i.e. rutin) and others which are only known to affect mutualistic organisms (i.e. strigolactones). Together these data would suggest that, while the plant can attract and control mutualistic fungi, it produces non-specific signals that can also be acted upon by pathogenic organisms. Therefore, as healthy plants are not generally overwhelmed by pathogens, the plant must also have a certain capacity to sense and to control attacks from pathogens. Recent publications have pointed to the role of physical or chemical barriers in protection of the plant against pathogens.

Notably, the production by exfoliating cells of a root cap slime containing a complex mix of extracellular DNA, exoenzymes and nutrients would play a key role in the innate defenses of the roots (Hawes et al. 2000, 2011; Wen et al. 2009).

3.2 Re-structuring of Root System Development in Response to ECM Fungi

From the first contact between fungal hyphae and the root system of a receptive host plant, either physically or upon the first exchange of signals, one of the first physiological changes observed on the plant side is the formation of new lateral roots followed by a reduction of the root cap and of the size of the apical meristem zone (Massicote et al. 1987; Felten et al. 2011). These lateral roots will be then colonized by the ECM fungus. Changes in root development are regulated by a balance of plant hormones including auxin, cytokinin, and ethylene (review, Depuydt and Hardtke 2011). In particular, support for the role of auxin and auxin homeostasis has been given during the formation of microbe induced lateral roots (Charvet-Candela et al. 2002; Niemi et al. 2002; Reddy et al. 2006; Felten et al. 2009). Auxin homeostasis can be achieved through synthesis, conjugation and degradation (Vanneste and Friml 2009). Research would suggest that all three levels of auxin control are involved in the induction of lateral rooting during mycorrhizal colonization. Indeed, modulation of the expression of genes involved in auxin synthesis (*PtASA, PtTAA* genes; J. Felten, personal communication), transport (*PtPIN* genes, Felten et al. 2009) and conjugation (*PpGH3-16*, *PtGH3* genes; Reddy et al. 2006; Charvet-Candela et al. 2002; Felten et al. 2009) have been described during the colonization of *Pinus pinaster* and *Populus spp.* by several ECM fungi. Interestingly, these studies demonstrate that the accumulation of fungal auxin inside plant tissues triggers the rapidity and the final outcome of colonization. Hence, from these data, it may be possible that the plant actively adjusts auxin homeostasis during the colonization of fungi. But is this reaction under the control of the plant alone or is it influenced by signals from the fungus? The majority of studies considering the role of auxin in lateral root formation during fungal colonization only looked at roots in direct contact. Therefore it was difficult to determine if what was occurring in the plant was controlled by the plant or if it was influenced by direct contact with the fungus. It was not until two recent studies

(Splivallo et al. 2009; Felten et al. 2009) that studied auxin homeostasis and lateral root formation in indirect contact with ECM fungi, that it was discovered that lateral root formation and auxin gradients changed in a similar manner to root systems in direct contact with the fungus. The current working model in ECM host plants is that a/some diffusible effector(s) from ECM fungi, be they auxin or another effector-like compound, activate PIN auxin transporters such that basipetal auxin transport is enhanced. This, in turn, primes the creation of an increased number of lateral root initials in the expansion zone that will later give rise to mature lateral roots (Felten et al. 2009). Therefore, it is likely that the induction of lateral roots, in this case, is in response to fungal signals and is thus not a phenomenon purely regulated by the plant itself. The identity of the fungal signals that control auxin flux within plants have yet to be identified. Interestingly, as with effector and effector-like proteins, phytopathogenic fungi and bacteria also play on the auxin signaling network of their plant hosts. The obligate biotroph *Plasmodiophora brassicae,* for instance, causes auxin and amide conjugates of auxin to accumulate at higher concentrations in the infected plant tissues when compared to non-infected roots (Ludwig-Müller et al. 1993, 1996). The plant pathogenic bacteria *Pseudomonas syringae* produces an effector protein that promotes host auxin biosynthesis (Chen et al. 2007) that concurrently increases the pathogenicity of *P. syringae* (Chen et al. 2007), an effect that is likely due to the antagonistic role of the auxin signaling pathway in repressing the salicylic acid defense system (Wang et al. 2007).

Therefore, as observed with the MiSSP effector proteins, there is a distinct overlap between the colonization ‘toolbox’ used by pathogenic and mutualistic fungi. In this section we considered the role of the plant in altering the behaviour of mutualistic fungi. However, in both the attraction and the hosting of ECM fungi, the plant appears to play a subservient role to the fungus as the fungus appears to be able to produce signals that manipulate host function. Thus, increasingly, while there is a balanced nutrient exchange in this relationship, the idea of the equality of the partners in controlling the dialogue during the interaction is very much in question, although this may be an over-simplification. For example, while no SSPs in ECM host species have been identified to date that act to raise a defense against fungal control signals, in non-ectomycorrhizal species such as *Arabidopsis thaliana* and *Glycine max* a number of peptides have recently been characterized that auto-induce the plant defense system, or amplify its response, in response to pathogen attack (Huffaker et al. 2006; Pearce et al. 2010; Yamaguchi et al. 2010, 2011). Therefore, the defense system of the plant is still able to respond to challenges by foreign organisms. Further, *Medicago truncatula* is able to differentiate between AM fungi that provide a large quantity of nutrients from other species that do not, and ‘reward’ the more productive AM species with more sugar (Kiers et al. 2011). This ability to differentiate between mycorrhizal fungi that try to cheat and those that do not suggests that the plant has a sophisticated perception mechanism to detect and control nutrient transfer to symbionts hosted within its tissues. A similar ability to discriminate between ECM symbionts has not been demonstrated to date. Together, these studies demonstrate that the plant is not entirely manipulated by the fungal symbiont, but rather that it has a voice in the interaction.

4 Spying, Dialoging or Ignoring: When Bacteria Interact with ECM Fungi

In natural ecosystems, ectomycorrhizal roots host not only fungal mycelia but also intricate bacterial communities. In this mycorrhizal complex, fungi, plant and bacteria interact physically and functionally to impact plant nutrition and gross production (Frey-Klett et al. 2007). Bacteria colonize every layer of the mycorrhizal root tip, from the external mantle to the Hartig net and even the inside of hyphae. The composition of bacterial communities associated with ectomycorrhizal root tips diverge both functionally and taxonomically from communities of the bulk soil, suggesting that the mycorrhizal root tip create an ecological niche for some bacteria (Frey-Klett et al. 2005). In this niche, intense molecular interactions between bacteria and fungi and between the micro-organisms and the plant occur. To date, the molecular mechanisms of these interactions remain poorly understood and our knowledge is almost exclusively restricted to the interactions between mycorrhiza helper bacteria (MHB) and mycorrhizal fungi. These bacteria improve the performance of ECM fungi both in pre-symbiotic stages as well as in the promotion of mycorrhizal root tip formation and the functioning of mycorrhiza. MHB aid the growth and survival of free-living mycelium in a number of ways (Brulé et al. 2001; Founoune et al. 2002; Schrey et al. 2005). For example, *L. bicolor* S238N, which cannot product thiamine (an essential enzymatic co-factor), needs to obtain it from an external source. While thiamine may be obtained from a plant partner, in pre-symbiotic growth, thiamine produced by MHB can complement this fungal deficiency (Deveau et al. 2010). MHB then aid in the colonization of root tissues by the induction of lateral roots (Aspray et al. 2006), suppression of plant defenses (Lehr et al. 2007) or augmentation of fungal resistance to stress (Vivas et al. 2003). Finally, the bacterium can potentially improve the ability of the mycorrhiza to acquire nutrients by stimulating exoenzymatic activities (Diedhiou et al. unpublished). In return, the mycorrhizal fungus also influence the activity (Riedlinger et al. 2006) or the survival of the helper bacteria (Deveau et al. 2010). Here, we detail some of these mechanisms at the molecular level and discuss how much these interactions can be considered as a dialogue between ectomycorrhizal fungi and their bacterial associates (Fig. 3).

4.1 *How MHB Find and Control Their Fungal and Plant Partners*

Current research suggests that some MHB and other bacteria colonizing mycorrhiza are attracted by fungal signals (Frey et al. 1997; Uroz et al. 2007; Deveau et al. 2010). The helper strain *P. fluorescens* BBc6R8 is chemoattracted by a specific disaccharide accumulated in the mycelium of the ECM *L. bicolor* S238N, the trehalose (Deveau et al. 2010). Because the MHB strain can use this disaccharide

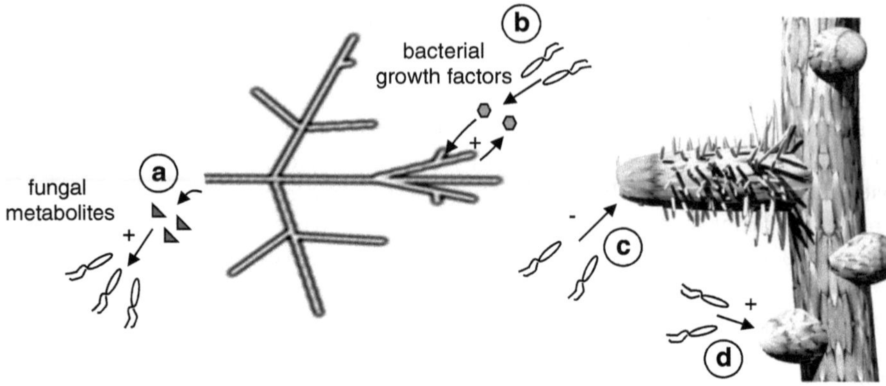

Fig. 3 Model of the mechanisms of the interactions between ectomycorrhizal fungi and helper bacteria. (**a**) Fungal metabolites (e.g. trehalose) attract helper bacteria and enhance bacterial growth and survival in the hyphosphere. (**b**) Helper bacteria produce growth factors (e.g. thiamine, auxofuran) which stimulate the growth and the survival of the fungus thus increasing the probability of encounter between the mycelium and the roots. In addition, bacterial signals stimulate the branching of hyphae. This could improve mycorhiza formation since branching of the hyphae is involved in the early steps of the colonization of the roots. Helper bacteria can also interfere with root physiology by suppressing plant defenses (**c**) or inducing short root formation (**d**)

as a sole source of carbon, in contrast to many other bacteria found within the bulk soil (Frey et al. 1997), it has been proposed that the hyphosphere (the region around fungal hyphae) of the fungus could provide a specialized niche for the bacterium where it could exert its effect. In addition, by transporting photoassimilates from the plant into the mycorrhizosphere and changing patterns of amino acids secretion, the fungus would stimulate the development of specific bacterial communities (Johansson et al. 2004). In each case, the bacteria are attracted by nutritional signals which may not be intentionally released by the fungus. But the enrichment of the mycorrhizosphere with mutualistic bacteria may suggest that there are more specific signals still to be characterized that specifically target MHB.

Once the bacteria reach the hyphosphere, MHB exert their effect through the production of soluble or volatile signaling compounds that can have very different effects on the development of ECM fungi. For example, the excretion of auxofuran, a secondary metabolite produced by *Streptomyces* strain AcH505, promotes the growth of the ECM fungus *Amanita muscaria*. Although the growth promotion effect is not specific to *A. muscaria*, and the compound is constitutively produced by the bacterium, an interesting feedback loop occurs in which the fungus stimulates the production of auxofuran by the bacterium thereby further stimulating its own growth (Riedlinger et al. 2006). MHB may also exert an influence over the biology of ECM fungi or of the plant through the use of small secreted proteins as the helper strain *P. fluorescens* BBc6R8 requires its type III secretion-like system (T3SS) to promote mycorrhizal root formation by *L. bicolor* S238N (Cusano et al. 2010). T3SS are molecular syringae which are used by pathogenic and symbiotic bacteria to inject effectors into the cytoplasm of eukaryotic cells and to take over the functioning of the host cell (McCann and Guttman 2008). Interestingly, Warmink and van Elsas (2008) noticed that bacteria colonizing the mycorrhizal root tips of *L. proxima* were enriched with strains harboring a T3SS-like as compared to the bulk soil. A similar pattern was described in the mycorrhizosphere of *Medicago truncatula* (Viollet et al. 2011). Altogether, this suggests a potential important role of T3SS in bacterial–fungal–plant interactions in soils. However, it remains to determine if these T3SS-like systems truly function as molecular syringe that inject effectors into host cells.

An alternative mechanism of action of the MHB which has not been considered yet by researchers is the interference of bacteria with the signaling between the roots and the ECM fungi. Indeed, some bacteria have the potential to degrade, produce or modify signaling molecules used by eukaryotic cells. For example, many bacteria colonizing the rhizosphere and the mycorrhizosphere can produce and/or degrade auxins (Frey-Klett et al. 2005; Leveau and Gerards 2008). Thus, communication during mycorhiza formation is probably much more complex than a simple dialogue between two organisms. Instead, it probably rather implies series of cross exchanges between several protagonists talking with different voices.

4.2 Do ECM Fungi Eavesdrop on Bacteria?

A less explored area in the interaction between ECM fungi and bacteria is the potential role of bacterial quorum sensing molecules as signaling agents. Many bacteria auto-regulate and coordinate their activities at the population level by a cell-density dependent mechanism called quorum sensing. To do so, bacteria use small diffusible molecules that are secreted in the environment and that are perceived by other bacteria from the same population (Waters and Bassler 2005). The mycorrhizosphere may be very "noisy" since a number of the strains of bacteria that surround ECM fungi produce quorum sensing molecules (Frey-Klett et al. unpublished). In some cases, this bacterial chit-chat can be heard or suppressed by other organisms. For example, plants such as *Arabidopsis thaliana*, barley, *Medicago truncatula* or tomato react to and interfere with bacterial quorum sensing signaling (Mathesius et al. 2003; Wheeler et al. 2006; von Rad et al. 2008; Schuhegger et al. 2006; Gotz et al. 2007; Ortiz-Castro et al. 2008). It is not known whether the ECM fungi are able to "hear" these signals and to react accordingly. But we have observed that the QS molecule 3,O-C12-HSL induces morphological and transcriptomic modifications in the ECM *L. bicolor* S238N (Frey-Klett et al. unpublished), suggesting that some of these molecules can impact the behavior of an ECM fungus. Furthermore, several ECM strains have the ability to quench bacterial quorum sensing signaling (Uroz and Heinonsalo 2008). It remains to understand if these responses have a functional role in the interactions between bacteria and fungi or if they are just an artifact, a background noise that the fungi would not pay much attention to. Should the quorum sensing quenchers produced by ECM fungi be important in a natural environment, it could be supposed that this is a strategy developed by the fungus to interfere with the growth of populations of bacteria and thus would participate to control the surrounding bacterial community.

All of these studies demonstrate that very complex and dynamic interactions occur between helper bacteria and fungi and the host plant. While MHB may not be essential for the formation of mycorrhizal root tips, their presence and the chemical dialogue between the two partners is important for increasing the fitness of the mutualistic relationship. It is very likely that the helper effect is a consequence of a combination of effects of the bacteria as suggested by the pleiotropic transcriptional response of the fungi to the presence of helper bacteria (Schrey et al. 2005; Deveau et al. 2007). However, can we consider these dynamic interactions as a form of communication, a way of exchanging information, or are they, rather, a kind of action-reaction process? To date, this question remains unsolved. Microorganisms have developed very complex mechanisms of perception of their biotic and abiotic environment, therefore, knowing the complexity of biotic interactions between microorganisms, it would not be surprising if they have also evolved some kind of communication rather than just a series of action-reaction processes. The existence of inter-kingdom perception of quorum sensing would support this hypothesis, but concrete evidences remain to be found.

5 Conclusions and Future Research

The nature of the signals released by the ectomycorrhizal symbionts, how these signals are transduced within the partners, and how these processes trigger the expression of symbiosis-regulated genes that assist in partner recognition and the formation of symbiotic tissues are only beginning to be understood. It is very interesting to note throughout the literature, however, that there is a striking similarity between the colonization methodology of pathogenic, mutualistic and 'helper' organisms. Like in the well-developed pathogen model of colonization, therefore, we can expect that new components of the transduction pathways, such as membrane and nuclear receptors, will soon be identified, facilitating an understanding of the cross-talk between organisms and between signaling networks. Dozen of symbiosis-related genes coding for regulatory transcriptional factors, hormone metabolism and secreted effector proteins have been identified in various ectomycorrhizal associations, the products of which may play a role in (1) recognition and attachment of the mycobiont onto root surfaces, (2) accommodation of the mycobiont *in planta*, (3) signaling networks, (4) organogenesis, and (5) novel symbiotic metabolism of the multitrophic ectomycorrhizal system. However, many questions concerning the differentiation of plant and fungal symbiotic structures, and MHB interactions remain unanswered. Master regulatory genes that may control morphogenesis during the symbiosis have not yet been isolated. What transcription factors regulate symbiosis regulated-gene expression, and how elicitor/signal-dependent activation of these transcription factors is achieved also remains unknown. Overall, it also remains to be seen who has the 'upper hand' in this three-way relationship if, in fact, any one organism does hold this lofty position. It would appear that ECM and AM fungi have a more developed ability to control their plant partners than the other way around while MHB have the ability to affect the biology of the fungus and possibly the plant. And yet, as we have emphasized above, this is an interaction that is mutually beneficial for all partners involved, indicating that there either must be unknown factor(s) that act to balance the relationship between all organisms involved or one or more of the partners willingly, if we may anthropomorphize, does not take advantage of the others. The answers to these questions will provide further highlights into the signaling networks and early gene regulation processes involved in ectomycorrhiza development. Eventually, such studies will lead to a better understanding of plant–microbe interactions and evolution of plant–fungus associations.

Acknowledgements We thank colleagues at INRA-Nancy (France) for their input and collaboration, specifically Annegret Kohler, Claire Veneault-Fourrey and Judith Felten whose contributions made this work possible. The authors also thank Dr Krista Plett for careful reading of the manuscript. This project was funded by the Agence Nationale de la Recherche (project FungEffector, ANR-06-BLAN-0399) and Institut National de la Recherche Agronomique. This research was also sponsored by the Genomic Science Program of the US Department of Energy, Office of Science, Biological and Environmental Research under contract DE-AC05-00OR22725 (Plant-Microbe Interface).

References

Akiyama K, Matsuzaki K, Hayashi H (2005) Plant sesquiterpenes induce hyphal branching in arbuscular mycorrhizal fungi. Nature 435:824–827

Aspray TJ, Frey-Klett P, Jones JE, Whipps JM, Garbaye J, Bending GD (2006) Mycorrhization helper bacteria: a case of specificity for altering ectomycorrhiza architecture but not ectomycorrhiza formation. Mycorrhiza 16:533–541

Besserer A, Puech-Pages V, Kiefer P, Gomez-Roldan V, Jauneau A, Roy S, Portais JC, Roux C, Becard G, Sejalon-Delmas N (2006) Strigoactones stimulate arbuscular mycorrhizal fungi by activating mitochondria. PLoS Biol 4:e226

Besserer A, Becard G, Jauneau A, Roux C, Sejalon-Delmas N (2008) GR24, a synthetic analog of strigolactones, stimulates the mitosis and growth of the arbuscular mycorrhizal fungus *Gigaspora rosea* by boosting its energy metabolism. Plant Physiol 148:402–413

Bonas U, Stall RE, Staskawicz BJ (1989) Genetic and structural characterization of the avirulence gene *avrBs3* from *Xanthomonas campestris pv. vesicatoria*. Mol Gen Genet 218:127–136

Brulé C, Frey-Klett P, Pierrat JC, Courrier S, Gérard F, Lemoin M-C, Rousselet J-L, Sommer J, Garbaye J (2001) Survival in the soil of the ectomycorrhizal fungus *Laccaria bicolor* and the effects of a mycorrhiza helper *Pseudomonas fluorescens*. Soil Biol Biochem 33:1683–1694

Charvet-Candela V, Hitchin S, Ernst D, Sandermann J-H, Marmeisse R, Gay G (2002) Characterization of an AUX/IAA cDNA upregulated in *Pinus pinaster* roots in response to colonization by the ectomycorrhizal fungus hebeloma cylindrosporum. New Phytol 154:769–777

Chen Z, Agnew JL, Cohen JD, He P, Shan L, Sheen J, Kunkel BN (2007) *Pseudomonas syringae* type III effector AvrRpt2 alters *Arabidopsis thaliana* auxin physiology. Proc Natl Acad Sci USA 104:20131–20136

Cusano AM, Burlinson P, Deveau A, Vion P, Uroz S, Preston GM, Frey-Klett P (2010) *Pseudomonas fluorescens* BBc6R8 type III secretion mutants no longer promote ectomycorrhizal symbiosis. Environ Microbiol Rep 3:203–210

Depuydt S, Hardtke C-S (2011) Hormone signalling crosstalk in plant growth regulation. Curr Biol 21:R365–R373

Deslandes L, Olivier J, Peeters N, Feng DX, Khounlotham M, Boucher C, Somssich I (2003) Physical interaction between RRS1-R, a protein conferring resistance to bacterial wilt, and PopP2, a type III effector targeted to the plant nucleus. Proc Natl Acad Sci USA 100:8024–8029

Deveau A, Palin B, Delaruelle C, Peter M, Kohler A, Pierrat JC, Sarniguet A, Garbaye J, Martin F, Frey-Klett P (2007) The mycorrhiza helper *Pseudomonas fluorescens* BBc6R8 has a specific priming effect on the growth, morphology and gene expression of the ectomycorrhizal fungus *Laccaria bicolor* S238N. New Phytol 175:743–755

Deveau A, Brulé C, Palin B, Champmartin D, Rubini P, Garbaye J, Sarniguet A, Frey-Klett P (2010) Role of fungal trehalose and bacterial thiamine in the improved survival and growth of the ectomycorrhizal fungus *Laccaria bicolor* S238N and the helper bacterium *Pseudomonas fluorescens* BBc6R8. Environ Microbiol Rep 2:560–568

Dou D, Kale SD, Wang X, Jiang RHY, Bruce NA, Arredondo PD, Zhang X, Tyler BM (2008) RXLR-mediated entry of *Phytophthora sojae* effector Avr1b into soybean cells does not require pathogen-encoded machinery. Plant Cell 20:1930–1947

Felten J, Kohler A, Morin E, Bhalerao RP, Palme K, Martin F, Ditengou FA, Legué V (2009) The ectomycorrhizal fungus *Laccaria bicolor* stimulates lateral root formation in poplar and arabidopsis through auxin transport and signaling. Plant Physiol 151:1991–2005

Felten J, Martin F, Legué V (2011) Signalling in ectomycorrhizal symbiosis. In: Baluska F, Perotto S (eds) Signalling and communication in plant symbiosis. Springer-Verlag book series. Springer, Berlin, pp 123–142

Founoune H, Duponnois R, Bâ AM, Sall S, Branget I, Lorquin J, Neyra M, Chotte J (2002) Mycorrhiza helper bacteria stimulate ectomycorrhizal symbiosis of *Accaria holoserica* with the *Pisolithus albus*. New Phytol 153:81–89

Frey P, Frey-Klett P, Garbaye J, Berge O, Heulin T (1997) Metabolic and genotypic fingerprinting of fluorescent Pseudomonads associated with the Douglas fir-*Laccaria bicolor* mycorrhizosphere. Appl Environ Microbiol 63:1852–1860

Frey-Klett P, Chavatte M, Clausse ML, Courrier S, Le Roux C, Raaijmakers J, Martinotti MG, Pierrat JC, Garbaye J (2005) Ectomycorrhizal symbiosis affects functional diversity of rhizosphere fluorescent pseudomonads. New Phytol 165:317–328

Frey-Klett P, Garbaye J, Tarkka M (2007) The mycorrhiza helper bacteria revisited. New Phytol 176:22–36

Fries N, Serck-Hanssen K, Häll DL, Theander O (1987) Abietic acid, an activator of basidiospore germination in ectomycorrhizal species of the genus Suillus (Boletaceae). Exp Mycol 11:360–363

Gogala N (1991) Regulation of mycorrhizal infection by hormonal factors produced by hosts and fungi. Experientia 47:331–340

Gomez-Roldan V, Fermas S, Brewer PB, Puech-Page V, Dun EA, Pillot JP, Letisse F, Matusova R, Danoun S, Portais JC, Bouwmeester H, Bécard G, Christine A, Beveridge CA, Rameau C, Rochanges SF (2008) Strigolactone inhibition of shoot branching. Nature 455:189–194

Gotz C, Fekete A, Gebefuegi I, Forczek ST, Fuksova K, Li X, Englmann M, Gryndler M, Hartmann A, Matucha M, Schmitt-Kopplin P, Schroder P (2007) Uptake, degradation and chiral discrimination of N-acyl-D/L-homoserine lactones by barley (*Hordeum vulgare*) and yam bean (*Pachyrhizus erosus*) plants. Anal Bioanal Chem 389:1447–1457

Hawes M-C, Gunawardena U, Miyasaka S, Zhao X (2000) The role of root border cells in plant defense. Trends Plant Sci 5:128–133

Hawes M-C, Curlango-Rivera G, Wen F, White G, VanEtten H, Xiong Z (2011) Extracellular DNA: the tip of root defenses? Int J Exp Plant Biol 180:741–745

Heupel S, Roser B, Kuhn H, Lebrun MH, Villalba F, Requena N (2010) Erl1, a novel era-like GTPase from *Magnaporthe oryzae*, is required for full root virulence and is conserved in the mutualistic symbiont *Glomus intraradices*. Mol Plant Microbe Interact 23:67–81

Hibbett DS, Matheny PB (2009) The relative ages of ectomycorrhizal mushrooms and their plant hosts estimated using Bayesian relaxed molecular clock analyses. BMC Biol 7:13

Horan DP, Chilvers GA (1990) Chemotropism: the key to ectomycorrhizal formation? New Phytol 116:297–301

Huffaker A, Pearce G, Ryan CA (2006) An endogenous peptide signal in *Arabidopsis* activates components of the innate immune response. Proc Natl Acad Sci USA 103:10098–10103

Jambois A, Dauphin A, Kawano T, Ditengou FA, Bouteau F, Legué V, Lapeyrie F (2005) Competitive antagonism between IAA and indole alkaloid hypaphorine must contribute to regulate ontogenesis. Physiol Plant 123:120–129

Johansson J, Paul LR, Finlay RD (2004) Microbial interactions in the mycorrhizosphere and their significance for sustainable agriculture. FEMS Microbiol Ecol 48:1–13

Kale SD, Gu B, Capelluto DGS, Dou D, Feldman E, Rumore A, Arredondo FD, Fudal I, Rouxel T, Lawrence CB, Shan W, Tyler BM (2010) External phosphatidylinositol-3-phosphate mediates host cell entry by eukaryotic pathogen effectors. Cell 142:284–295

Kalinova J, Radova S (2009) Effect of rutin on the growth of *Botrytis cinerea Alternaria alternata* and *Fusarium solani*. Acta Phytopathol Entomol Hung 44:39–47

Kanneganti T-D, Bai X, Tsai C-W, Win J, Meulia T, Goodin M, Kamoun S, Hogenhout SA (2007) A functional genetic assay for nuclear trafficking in plants. Plant J 50:149–158

Kapulnik Y, Delaux P-M, Resnick N, Mayzlish-Gati E, Wininger S, Bhattacharya C, Séjalon-Delmas N, Combier J-P, Bécard G, Belausov E, Beeckman T, Dor E, Hershenhorn J, Koltai H (2011) Strigolactones affect lateral root formation and root-hair elongation in Arabidopsis. Planta 233:209–216

Khan R, Straney DC (1999) Regulatory signals influencing expression of the PDA1 gene of *Nectria haematococca* MPVI in culture and during pathogenesis of pea. Mol Plant-Microbe Interact 12:733–742

Kiers ET, Duhamel M, Beesetty Y, Mensah JA, Franken O, Verbruggen E, Fellbaum CR, Kowalchuk GA, Hart MM, Bago A, Palmer TM, West SA, Vandenkoornhuyse P, Jansa J, Büking H (2011) Reciprocal rewards stabilize cooperation in the mycorrhizal symbiosis. Science 333:880–882

Kloppholz S, Kuhn H, Requena N (2011) A secreted fungal effector of *Glomus intraradices* promotes symbiotic biotrophy. Curr Biol 21:1204–1209

Kosuta S, Chabaud M, Lougnon G, Gough C, Dénarié J, Barker DG, Bécard G (2003) A diffusible factor from arbuscular mycorrhizal fungi induces symbiosis-specific MtENOD11 expression in roots of *Medicago truncatula*. Plant Physiol 131:952–962

Lagrange H, Jay-Allemand C, Lapeyrie F (2001) Rutin, the phenolglycoside from *Eucalyptus* root exudates stimulates *Pisolithus* hyphal growth at picomolor concentrations. New Phytol 150:349–355

Lehr NA, Schrey SD, Bauer R, Hampp R, Tarkka MT (2007) Suppression of plant defence response by a mycorrhiza helper bacterium. New Phytol 174:892–903

Leveau JH, Gerards S (2008) Discovery of a bacterial gene cluster for catabolism of the plant hormone indole 3-acetic acid. FEMS Microbiol Ecol 65:238–250

Ludwig-Müller J, Bendel U, Thermann P, Ruppel M, Epstein E, Hilgenberg W (1993) Concentrations of indole-3-acetic acid in plants of tolerant and susceptible varieties of Chinese cabbage infected with *Plasmodiophora brassicae* Woron. New Phytol 125:763–769

Ludwig-Müller J, Epstein E, Hilgenberg W (1996) Auxin-conjugate hydrolysis in Chinese cabbage: characterization of an amidohydrolase and its role during infection with clubroot disease. Physiol Plant 97:627–634

Maillet F, Poinsot V, André O, Puech-Pagès V, Haouy A, Gueunier M, Cromer L, Giraudet L, Formey D, Niebel A, Andres Martinez E, Driguez H, Bécard G, Denarié J (2011) Fungal lipochitooligosaccharide symbiotic signals in arbuscular mycorrhiza. Nature 469:58–63

Martin F (2007) Fair trade in the underworld: the ectomycorrhizal symbiosis. In: Howard RJ, Gow NAR (eds) Biology of the fungal cell, vol VIII, 2nd edn, The mycota. Springer, Berlin/Heidelberg, pp 291–308

Martin F, Duplessis S, Ditengou F, Lagrange H, Voiblet C, Lapeyrie F (2001) Developmental cross talking in the ectomycorrhizal symbiosis: signals and communication genes. New Phytol 151:145–154

Martin F, Aerts A, Ahrn D, Brun A, Danchin EGJ, Duchaussoy F et al (2008) The genome sequence of the basidiomycete fungus *Laccaria bicolor* provides insights into the mycorrhizal symbiosis. Nature 452:88–92

Martin F, Kohler A, Murat C, Balestrini R, Coutinho PM, Jaillon O, Montanini B et al (2010) Périgord black truffle genome uncovers evolutionary origins and mechanisms of symbiosis. Nature 464:1033–1038

Massicote H-B, Perterson R-L, Ashford A-E (1987) Ontogeny of *Eucalyptus piluliris-Pisolithus tinctorius* ectomycorrhizae. I. Light microscopy and scanning electron microscopy. Can J Bot 65:1927–1939

Mathesius U, Mulders S, Gao M, Teplitski M, Caetano-Anolles G, Rolfe BG, Bauer WD (2003) Extensive and specific responses of a eukaryote to bacterial quorum-sensing signals. Proc Natl Acad Sci USA 100:1444–1449

McCann HC, Guttman DS (2008) Evolution of the type III secretion system and its effectors in plant-microbe interactions. New Phytol 177:33–47

Niemi K, Häggman H, Sarjala T (2002) Effect of diamines on the interaction between ectomycorrhizal fungi and adventitious root formation on Scots pine in vitro. Tree Physiol 22:373–381

Ortiz-Castro R, Martinez-Trujillo M, Lopez-Bucio J (2008) N-acyl-L-homoserine lactones: a class of bacterial quorum-sensing signals alter post-embryonic root development in *Arabidopsis thaliana*. Plant Cell Environ 31:1497–1509

Paszkowski U (2006) Mutualism and parasitism: the yin and yang of plant symbioses. Curr Opin Plant Biol 9:364–370

Pearce G, Yamaguchi Y, Barona G, Ryan CA (2010) A subtilisin-like protein from soybean contains an embedded, cryptic signal that activates defense-related genes. Proc Natl Acad Sci USA 107:14921–14925

Peters NK, Frost JW, Long SR (1986) A plant flavone, luteolin, induces expression of *Rhizobium meliloti* nodulation genes. Science 233:977–980

Plett JM, Martin F (2012) Poplar root exudates contain compounds that induce the expression of MiSSP7 in *Laccaria bicolor*. Plant Signal Behav 7:12–15

Plett JM, Kemppainen M, Kale SD, Kohler A, Legué V, Brun A, Tyler BM, Pardo AG, Martin F (2011) A secreted effector protein of *Laccaria bicolor* is required for symbiosis development. Curr Biol 21:1197–1203

Reddy SM, Hitchin S, Melayah D, Pandey AK, Raffier C, Henderson J, Marmeisse R, Gay G (2006) The auxin-inducible GH3 homologue ppGH3.16 is downregulated in *Pinus pinaster* root systems on ectomycorrhizal symbiosis establishment. New Phytol 170:391–400

Redman RS, Dunigan DD, Rodriguez RJ (2001) Fungal symbiosis from mutualism to parasitism: who controls the outcome, host or invader? New Phytol 151:705–716

Requena N, Serrano E, Ocón A, Breuninger M (2007) Plant signals and fungal perception during arbuscular mycorrhiza establishment. Phytochemistry 68:33–40

Riedlinger J, Schrey S, Tarkka M, Hampp R, Kapur M, Fielder H-P (2006) Auxofuran, a novel metabolite that stimulates the growth of fly agaric, is produced by the mycorrhiza helper bacterium *Streptomyces* strain AcH 505. Appl Environ Microbiol 72:3350–3557

Schornack S, Ballvora A, Gürlebeck D, Peart J, Ganal M, Baker B, Bonas U, Lahaye T (2004) The tomato resistance protein Bs4 is a predicted non-nuclear TIR-NB-LRR protein that mediates defense responses to severely truncated derivatives of AvrBs4 and overexpressed *AvrBs3*. Plant J 37:46–60

Schrey S, Schellhammer M, Ecke M, Hampp R (2005) Mycorrhiza helper bacterium *Streptomyces* AcH 505 induces differential gene expression in the ectomycorrhizal fungus *Amanita muscaria*. New Phytol 168:205–216

Schuhegger R, Ihring A, Gantner S, Bahnweg G, Knappe C, Vogg G, Hutzler P, Schmid M, Van Breusegel F, Eberl L, Hartmann A, Langerbartels C (2006) Induction of systemic resistance in tomato by N-acyl-L-homoserine lactone-producing rhizosphere bacteria. Plant Cell Environ 29:908–918

Smith S-E, Read D-J (2008) Mycorrhizal symbiosis, 3rd edn. Academic, London

Splivallo R, Fischer U, Gobel C, Feussner I, Karlovsky P (2009) Truffles regulate plant root morphogenesis via the production of auxin and ethylene. Plant Physiol 150:2018–2029

Uroz S, Heinonsalo J (2008) Degradation of N-acyl homoserine lactone quorum sensing signal molecules by forest root-associated fungi. FEMS Microbiol Ecol 65:271–278

Uroz S, Calvaruso C, Turpault MP, Pierrat JC, Mustin C, Frey-Klett P (2007) Mycorrhizosphere effect on the genotypic and metabolic diversity of the soil bacterial communities involved in mineral weathering in a forest soil. Appl Environ Microbiol 73:3019–3027

Vanneste S, Friml J (2009) Auxin: a trigger for change in plant development. Cell 136:1005–1016

Van West P, De Bruijn I, Minor KL, Phillips AH, Robertson EJ, Wawra S, Bain J, Anderson VL, Secombes CJ (2010) The putative RxLR effector protein in SpHtp1 from the fish pathogenic oomycete *Saprolegnia parasitica* is translocated into fish cells. FEMS Microbiol Lett 310:127–137

Veneault-Fourrey C, Martin F (2011) Mutualistic interactions on a knife-edge between saprotrophy and pathogenesis. Curr Opin Plant Biol 14:444–450

Viollet A, Corberand T, Mougel C, Robin A, Lemanceau P, Mazurier S (2011) Fluorescent pseudomonads harboring type III secretion genes are enriched in the mycorrhizosphere of *Medicago truncatula*. FEMS Microbiol Ecol 75:457–467

Vivas A, Marulanda A, Ruiz-Lozano JM, Barea JM, Azcon R (2003) Influence of a Bacillus sp. on physiological activities of two arbuscular mycorrhizal fungi and on plant responses to PEG-induced drought stress. Mycorrhiza 13:249–256

von Rad U, Klein I, Dobrev P, Kottova J, Zazimalova E, Fekete A, Hartmann A, Schmitt-Kopplin P, Durner J (2008) Response of *Arabidopsis thaliana* to N-hexanoyl-DL-homoserine-lactone, a bacterial quorum sensing molecule produced in the rhizosphere. Planta 229:73–83

Wang D, Pajerowska-Mukhtar K, Culler AH, Dong X (2007) Salicylic acid inhibits pathogen growth in plants through repression of the auxin signaling pathway. Curr Biol 17:1784–1790

Warmink JA, van Elsas JD (2008) Selection of bacterial populations in the mycosphere of *Laccaria proxima*: is type III secretion involved? ISME J 2:887–900

Waters CM, Bassler BL (2005) Quorum sensing: cell-to-cell communication in bacteria. Annu Rev Cell Dev Biol 21:319–346

Wen F, Curlango-Rivera G, Hawes M-C (2007) Proteins among the polysaccharides: a new perspective on root cap slime. Plant Signal Behav 2:410–412

Wen F, White G-J, Van Etten H-D, Xiong Z, Hawes M-C (2009) Extracellular DNA is required for root tip resistance to fungal infection. Plant Physiol 151:820–829

Wheeler GL, Tait K, Taylor A, Brownlee C, Joint I (2006) Acyl-homoserine lactones modulate the settlement rate of zoospores of the marine alga *Ulva intestinalis* via a novel chemokinetic mechanism. Plant Cell Environ 29:608–618

Yamaguchi Y, Huffaker A, Bryan AC, Tax FE, Ryan CA (2010) PEPR2 is a second receptor for the Pep1 and Pep2 peptides and contributes to defense responses in *Arabidopsis*. Plant Cell 22:508–522

Yamaguchi Y, Barona G, Ryan CA, Pearce G (2011) GmPep914, an eight-amino acid peptide isolated from soybean leaves, activates defense-related genes. Plant Physiol 156:932–942

Yang B, Zhu W, Johnson LB, White FF (2000) The virulence factor AvrXa7 of *Xanthomonas oryzae pv. oryzae* is a type III secretion pathway-dependent, nuclear-localized, double-stranded DNA binding protein. Proc Natl Acad Sci USA 97:9807–9812

Zhu W, Yang B, Chittoor JM, Johnson LB, White FF (1998) *AvrXa10* contains an acidic transcriptional activation domain in the functionally conserved C terminus. Mol Plant Microbe Interact 11:824–832

Lipid-Mediated Signaling Between Fungi and Plants

Eli J. Borrego and Michael V. Kolomiets

Abstract Lipid-mediated inter-kingdom signaling in plant-fungal interactions is the exchange of molecules between plants and fungal pathogens and symbiotes. Recently these interactions were implicated in determining whether inter-organismal interactions result in parasitism, symbiosis or commencialism. Lipids constitute a very large group of structurally diverse molecules that have diverse functions in cell metabolism. One group of lipids, oxygenated lipids (oxylipins), is gaining increased interest as molecular signals that orchestrate a myriad of metabolic processes in both plants and fungi. Growing momentum implicates these metabolites as key players during the signal exchange between different interacting organisms. Recent studies have revealed oxylipins as key regulators of sporulation and secondary metabolite production while others have discovered their roles in manipulating plant metabolism and defense responses for the advantage of fungal and other pathogens. The focus of this chapter is to describe recent advances in our understanding of oxylipin-mediated signal communication between fungi and plants, highlighting pathogenic systems.

1 Introduction

Despite advancements in mechanical and chemical technology, crops continue to suffer annual losses of yield, with significant losses attributed to fungal diseases (Agrios 2005). In order to secure a growing food supply, plant–pathogen interactions must be elucidated for development of disease resistance and novel, environmentally friendly prevention strategies. Lipids have gained appreciation in the current years for their role as mediators in the interaction between plant and fungi,

E.J. Borrego • M.V. Kolomiets (✉)
Department of Plant Pathology and Microbiology, Texas A&M University, 2132 TAMU, College Station, TX 77843-2132, USA
e-mail: kolomiets@tamu.edu

G. Witzany (ed.), *Biocommunication of Fungi*, DOI 10.1007/978-94-007-4264-2_16,

contributing to both plant defense and fungal pathogenicity. A particular interest has been directed towards oxygenated fatty acids, termed oxylipins, which have potent signaling activities, endogenously and exogenously. This chapter's focus will be on the role of these molecules in the signaling exchange between plants and fungi. Lipid-mediated cross-kingdom signaling has recently been recognized in playing a central role in governing the interactions between mammals and fungi (described in Chap. 19 of this book). One of the major triggers for the biosynthesis of oxylipins in fungi, plants and mammals is the tightly-regulated alteration of redox status of the cell, intrinsically associated with the burst of reactive oxygen species (ROS) and lipid peroxidation processes (described in Chaps. 17, 18 and 19).

2 Oxylipins

Oxylipins, as a term for a group of secondary metabolites, was first introduced for the fatty acid products of enzyme driven oxygenation (Gerwick et al. 1991). Currently, this group of metabolites has been expanded to also include products of non-enzymatic lipid oxygenation (e.g., reactive oxygen species-driven reactions). Oxylipin biosynthesis begins with fatty acids liberated from diverse cellular membranes containing glycerolipids by lipases, however, evidence grows to suggests that esterified fatty acid moieties also provide substrate to oxylipin production (Andreou et al. 2009). The fatty acid tails may then be oxygenated by several mono- and di-oxygenases, which depends on species, stimulus, and subcellular co-localization of the fatty acid substrates and respective enzymes. While a complete list of plant or fungal oxylipins has yet to be generated, it is generally accepted that hundreds of distinct endogenously-produced oxylipins and oxylipin-derivatives exist. To date, despite noticeable recent advancements in characterization of selected oxylipin groups, much remains to be elucidated in this rapidly expanding field.

2.1 Phyto-oxylipins: Biosynthesis, Roles, and Perception

In plants, oxylipins are produced primarily through the so called lipoxygenase (LOX) pathway consisting of several separate pathway branches. Plant LOXs are encoded by relatively large gene families with individual enzyme isoforms often having different substrate and regio-specificities. Typical substrates of plant LOXs are polyunsaturated fatty acids linoleic (C18:2) and linolenic (C18:3) acids. Generally, plant LOXs are categorized as either a 9-LOX or 13-LOX, or a mixed regio-specificity 9/13-LOX, depending on their ability to incorporate molecular oxygen at 9- or 13- or both 9- and 13-hydrocarbon positions of the fatty acid aliphatic chain. This peroxidation either yields hydroperoxy-octadecadienoic acid (HPOD) or hydroperoxy-octadecatrienoic acid (HPOT), dependent of fatty acid substrate;

linoleic acid and linolenic acid are catalyzed into HPOD and HPOT, respectively. The 9- or 13- hydroperoxides of the LOX reaction are subsequently fluxed into at least six distinct branches of oxylipin metabolism (allene oxide synthase (AOS), divinyl ester synthase, epoxy alcohol synthase, hydroperoxide lyase (HPL), lipoxygenase (LOX), and peroxygenase) or reduced to their hydroxy derivatives (9/13- HOT/D) via yet to be characterized reductase enzyme (Feussner and Wasternack 2002). Additional contributors to oxylipin production in plants include the α-dioxygenase gene family and reactive oxygen species, which produce 2-hydroperoxides (2-HPOT/D) (Hamberg et al. 2005) and phytoprostanes (Mueller 2004), respectively.

Undeniably, the greatest knowledge of plant oxylipins come from the studies of oxylipin hormones produced by the 13-LOX reactions, especially the jasmonates (Wasternack 2007) and green leaf volatiles (Matsui 2006). Produced by the AOS sub-branch of the LOX pathway, jasmonates are a group of structurally similar metabolites that includes jasmonic acid, its precursor 12-oxo-phytodienoic acid (12-oxo-OPDA), and their derivatives produced from linolenic acid. Jasmonates are better known for their involvement in the transition from growth to defense (e.g., defense against necrotrophic fungi and insect herbivores). Besides the defensive response other physiological functions for these compounds exist, many are species-specific such as sexual organ development, male or female fertility, senescence, and many of which are based on JA proapoptotic activity (Acosta et al. 2009; Yan et al. 2012). To date, only the isoleucine conjugate of jasmonic acid (JA-Ile) has been conclusively identified to be biologically active (Fonseca et al. 2009), although other jasmonates have the ability to induce unique gene expression signature, suggesting specialization amongst diverse but related oxylipins. The other better understood group of oxylipins derived from the 13-LOX-mediated oxygenation of linolenic acid are green leaf volatiles (GLVs), produced by the HPL branch. These 6-carbon containing molecules (e.g. hexenyl acetate, leaf aldehyde, and leaf alcohol) are better known for their signaling role in plant-to-plant and plant-to-insect communication.

To date, JA-Ile is the only plant oxylipin with an identified receptor (Chico et al. 2008). Prior to induction, JA-dependent transcription factors are inhibited by JA ZIM-domain repressor proteins (JAZ). In the presence of JA-Ile, SCF^{COI1}, an E3-ubiquitin ligase, tags JAZ proteins for degradation by the 26S proteasome, releasing JA-responsive transcription factors for RNA polymerase recruitment and subsequent expression of JA-responsive genes. COI1-JAZ protein complex may represent an atypical oxylipin eukaryotic receptor, since known oxylipin receptors in animal systems are G-protein coupled receptors (GPCR) (Bos et al. 2004).

2.2 *Myco-oxylipin: Biosynthesis, Roles, and Perception*

Fungal oxylipins are produced through oxygenases similar to those found in plants and animals (Andreou et al. 2009; Brodhun and Feussner 2011). Oxygenases of particular relevance to plant-pathogen interactions include the LOX and Ppo enzyme families (Brodhun and Feussner 2011; Christensen and Kolomiets 2011). Ppo enzymes (Psi producing oxygenase) catalyze the formation of Psi (precocious sexual inducer) factors, which are oxylipin signals originally described for regulating the balance between sexual and asexual sporulation in fungi (Champe and el-Zayat 1989).

The genome of the seed-rot pathogen, *Aspergillus flavus*, houses four Ppo paralogs (Brown et al. 2009) and one LOX gene (Brown et al. 2008). Here, Ppo mutations displayed abnormal sporogenic and secondary metabolite phenotypes during in vivo assays with maize and peanut seed. Both *ΔppoA* and *ΔppoC* mutants displayed decreased conidiation and increased aflatoxin accumulation; however, *ΔppoD* behaved as wild type, suggesting a role of specific fungal psi factors during plant–pathogen interactions (Brown et al. 2009). A triple mutation strain of these enzymes was severely compromised in conidiation and produced elevated aflatoxin accumulation.

Disruption of the putative arachidonate 5-LOX from *A. flavus* affected sporulation and secondary metabolite production. The density-dependent morphological shift between sclerotia and conidiation was reduced in the knock-out strains suggesting a role for *Aflox* in fungal quorum signaling (Brown et al. 2008). However, despite no conidiation difference during infection on peanut or maize seed, the *Aflox* mutant accumulated increased levels of aflatoxin (Brown et al. 2009) (Fig. 1).

Unfortunately, to date, oxylipin receptors have not been reported in fungi. However, as in animal systems, it is likely that GPCRs are involved in oxylipin perception (Hosoi et al. 2002).

3 Plant–Fungal Interactions

Intimate associations between plants and fungi are common in nature and the interaction lies somewhere on a continuum from a positive, neutral, or negative effect to one or both partners. Among the potentially many different types of chemical exchanges existing between plant and fungi, emerging evidence strongly suggest that lipids and lipid derivatives play a key role in determining whether these interactions are parasitic, synergistic, or commensalistic. This hypothesis is based on the premise that plant- and fungal-derived oxylipins biochemically and structurally resemble one another allowing recognition by the opposing partner (Tsitsigiannis and Keller 2007; Gao and Kolomiets 2009; Brodhun and Feussner 2011; Christensen and Kolomiets 2011). With mounting evidence supporting this hypothesis, two themes have developed: (strategy 1) fungi recognize plant-derived oxylipins as growth promoting signals, and (strategy 2) fungi modulate host metabolism to promote a more hospitable environment for self-proliferation. Likely, the associating biological partners have a specific interaction, during which one or both strategies are employed, on a per case basis.

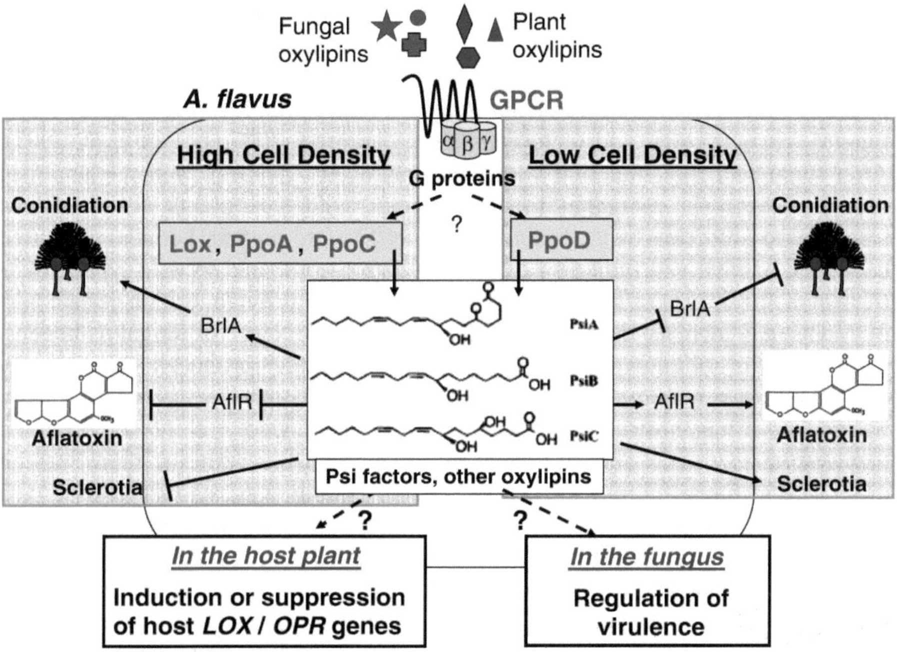

Fig. 1 A model showing the known roles of *Aspergillus flavus* Ppo and LOX genes in density dependent development and the hypothesized roles of host/fungal oxylipins in the regulation of conidiation, aflatoxin production, and sclerotia development. The model is based on the most recent findings of Brown et al. (2008, 2009). Because of the demonstrated involvement of G-proteins in the regulation of sporulation and secondary metabolite biosynthesis and the fact that some mammalian eicosanoid oxylipins are ligands for G-protein coupled receptors (*GPCRs*), it was hypothesized that fungal and plant oxylipins may serve as ligands of fungal GPCRs to initiate signaling cascades leading to mycotoxin and spore production (Brodhagen and Keller 2006). The *A. flavus* Ppo and LOX enzymes generate diverse oxylipins that function in the regulation of conidia formation (via the BrlA transcription factor), or regulation of secondary metabolism (e.g. via transcription factor AflR required for aflatoxin synthesis). Brodhagen et al. (2008) provided genetic evidence that fungal oxylipins are involved in regulation of expression of plant LOX genes, leading to possible alterations in the fungal/host interaction

3.1 Susceptible Plant–Pathogen Interaction

A patatin-like lipid acyl hydrolase (*AtPLP2)* from *Arabidopsis* is a potential producer of free PUFAs which may lead to oxylipin production. Silencing of this enzyme results in increased resistance to the fungal pathogen *Botrytis cinerea,* while overexpression increased susceptibility (La Camera et al. 2005). These results suggest a possibility that *AtPLP2* produces a lipase that liberates fatty acids from complex membranes and feeds them into the LOX or other oxylipin biosynthetic pathways to facilitate fungal pathogenesis.

The role of 9-oxylipins during disease progression in plant tissues infected by fungal pathogens was demonstrated in maize, in which a disruption of a 9-LOX

(*ZmLOX3*) increased resistance to a number of evolutionary distant fungal species including the leaf pathogens *Colletotrichum graminicola* and *Cochliobolus heterostrophus*, the root-rotting fungal pathogen *Exserohilum pedicellatum*, seed-infecting and mycotoxin-producing *Fusarium verticillioides*, and stalk-rotting fungi *F. verticillioides* and *C. graminicola* (Gao et al. 2007; Isakeit et al. 2007). After foliar and root infection with these pathogens, *ZmLox3 m*utant plants had decreased necrosis compared to near-isogenic wild-type plants. Strikingly, in response to infection with *Fusarium verticillioides*, mutant kernels supported decreased conidia and fumonisin B1 accumulation compared to wild-type kernels (Gao et al. 2007). These findings provided strong genetic evidence that certain host oxylipins may increase during infection to facilitate virulence of fungal pathogens. Currently, it is not known whether induction of ZmLOX3-produced 9-oxylipins is a result of direct manipulation of host oxylipins metabolism by the fungus, or is a by-product of misguided defense reaction of the plant. Gao et al. (2007) observed that despite reduction of conidiation and mycotoxin biosynthesis, there was no difference in fungal biomass. This unexpected finding suggests that host-derived oxylipins affect only specific developmental and secondary metabolism processes without affecting the overall ability of the fungus to colonize the seed. Although the precise mechanistic explanation of such selective effect of oxylipins on fungal pathogenicity is not available to date, the important conclusion from that study is that *ZmLOX3*-dependent product is required to stimulate secondary metabolite and asexual reproduction and not virulence per se.

13-LOX-like fungal-derived oxylipins are also implicated in promoting disease progression and pathogenicity by inducing JA-responsive genes (Thatcher et al. 2009). For example, the vascular-wilt pathogen, *Fusarium oxysporum,* requires JA perception to induce senescence for successful colonization of *Arabidopsis*. *Arabidopsis coi1* mutants were less susceptible to *F. oxysporum* infection compared to wild type or JA biosynthetic mutants, indicating that JA signaling rather than JA-production is responsible in the interaction. Strengthening this hypothesis is analytical evidence of *F. oxysporum* producing at least ten endogenous plant jasmonates (Miersch et al. 1999). Collectively, these and other studies support the notion that fungi either induce biosynthesis of host oxylipins, or produce their own oxylipins to mimic host derivatives to upregulate plant signal transduction pathways typically required for other vital physiological processes (e.g. jasmonate perception and downstream signaling is hijacked to promote senescence processes to facilitate tissue colonization as in Thatcher et al. (2009).

3.2 Resistant Plant–Pathogen Interaction

As with compatible interactions (when a plant is susceptible and a pathogen is virulent), the interacting partners produce oxylipins the cumulative action of which leads to plant host resistance. The best example of such oxylipins on the host side is JA that has long been implicated in defense against fungi with diverse life styles.

A widely known hallmark of JA signaling is increased resistance to necrotrophic pathogens (Thomma et al. 1998; Glazebrook 2005).

9-LOX mediated products also play roles in plant resistance to fungal invasion. In contrast to the enhanced resistance to *F. verticillioides*, ZmLOX3 disruption increased susceptibility to *Aspergilli spp.* When subjected to *A. flavus* and *A. nidulans,* mutant kernels had increased fungal growth, conidiation and aflatoxin accumulation compared with near-isogenic wild type (Gao et al. 2009). Similarly, a pepper 9-LOX *(CaLox1)* was shown to be involved in defense against *Colletotrichum coccodes* possibility through H_2O_2 production, lipid peroxidation, and salicylic acid accumulation (Hwang and Hwang 2010). These observations exemplify the fungal species-dependency of oxylipins in determining the outcome during plant–fungal interactions.

3.3 Phyto-oxylipins Effect on Fungi

Plant-derived oxylipins have direct effects on the reproduction and secondary metabolite production of fungi (strategy 1). Exogenous applications of plant LOX products induced dose-dependent sporogenic effects on *Aspergilli spp.*; at low concentrations the immediate LOX products, 9*S*-HPOD and 13*S*-HPOD, inhibited conidiation of *A. nidulans*, while at high concentrations conidia production was promoted (Calvo et al. 1999). The reverse was observed with regards to sexual spore production; low concentrations promoted ascospore formation, while high concentrations inhibited. Opposite effect was observed for biosynthesis of mycotoxins, low concentrations of 9*S*-HPOD induce aflatoxin production in *A. parasiticus*, whereas low concentrations of 13*S*-HPOD inhibit production (Burow et al. 1997). Expression of mycotoxin producing ketoreductases was prolonged after 9*S*-HPOD treatment, but reduced expression levels were observed after 13S-HPOD, suggesting regulation of mycotoxin synthesis by these oxylipins is at the transcription level.

In vivo studies with *A. nidulans* conidia- and mycotoxin-deficient mutant Δ*ppoAC* showed that transgenic expression of the maize 9-LOX gene, *ZmLOX3*, in this mutant reversed completely the levels of conidia and sterigmatocystin to those found in the wild-type fungus (Brodhagen et al. 2008). This study showed that plant oxylipin biosynthesis enzymes may functionally substitute for native fungal fatty acid oxygenase and their metabolites.

Plant-derived oxylipins may also serve a role in fungal survival against unfavorable environmental conditions. Sclerotia are the overwintering asexual fruiting bodies of several fungi (e.g., *A. flavus*) whose formation appears inhibited by high concentrations of 13*S*-HPOD (Calvo et al. 1999). Interestingly, sclerotia contain the sexual reproductive structure (i.e., stroma) of several *Aspergillus* teleomorphs (Horn et al. 2009a, b, 2011) suggesting a possibility for a role of 13*S*-HPOD in sexual reproduction across several species.

Volatile oxylipins play important roles in plant defense against fungal infections, however, less is known about their roles in fungal metabolism. Exogenous applications of MeJA to liquid cultures of *A. parasiticus* increased aflatoxin accumulation without affecting vegetative growth (Vergopoulou et al. 2001). In contrast, an earlier study showed that MeJA application to *A. flavus* cultures decreased aflatoxin accumulation both in media and on seed culture (Goodrich-Tanrikulu et al. 1995). After MeJA treatment, the shelf-mushroom, *Ganoderma lucidum* increased production of ganoderic acids through transcriptional regulation (Ren et al. 2010), which are compounds studied for their pharmacological benefit for human health (Shi et al. 2010). Similarly, MeJA treatment induced accumulation of the alkaloid, hypaphorine, in *Pisolithus microcarpus* (Jambois and Lapeyrie 2005). Hyphaphorine was shown to interfere with auxin activity and may be used by members of this genus to form symbiosis with plants (Reboutier et al. 2002). Furthermore, application of MeJA causes a delay in spore germination of *A. flavus* (Goodrich-Tanrikulu et al. 1995). Additionally, GLVs appear to disrupt growth and mycotoxin production of *A. flavus* (Wright et al. 2000; Boue et al. 2005; De Lucca et al. 2011). Collectively these experiments show that volatile plant-derived oxylipins are conserved signals across broad range of fungal species, however, the response to these volatiles are likely species- and volatile-specific.

3.4 Myco-oxylipins Effect on Plants

The signal exchange between plants and fungi is bidirectional; as fungal physiological and metabolic processes are regulated by phyto-oxylipins, so too do fungal oxylipin products regulate responses of their phyto-partners. This has been demonstrated best in experiments with the oxylipin-deficient *A. nidulans* Ppo mutants colonizing peanut seed. Unlike their wild-type counterpart, *ppoB*, *ppoAC*, and *ppoABC* mutants were not capable of induction of expression of two peanut 13-LOXs (PnLox2/3) in the seed upon infection (Brodhagen et al. 2008). In particular, the *ppoAC* and *ppoABC* mutant strains had the greatest reduction in their ability to induce PnLox2/3 which correlates with their decreased virulence (Tsitsigiannis and Keller 2006). This study suggests normal infection processes rely on exchange of oxylipins between parasite and host (strategy 2), or at least require intact oxylipin biosynthesis in both interacting organisms for normal pathogenicity.

An additional mechanism employed by fungi may be the direct manipulation of host lipid metabolism by the secretion of enzymes directly into plant cell. The head blight fungus, *Fusarium graminearum,* requires activity of an extracellular lipase (FGL1) for normal infection (Voigt et al. 2005). Likewise, *Magnaporthe oryzae* appears to employ its sole LOX as a biotrophy-associated secreted protein (Mosquera et al. 2009). The best explored instance of a pathogen exploiting host oxylipin-mediated signaling pathway is the case of the bacterial pathogen, *Pseudomonas syringae*, which secretes into the host coronatine, a phytotoxin functionally analogous to JA-Ile. Among the most significant implications of such oxylipin's mimicry for bacterial pathogenesis is the activation of JA synthesis and response pathways, which subsequently inhibits SA-dependent defense and results in increased susceptibility of the host to *P. syringae* (Brooks et al. 2005). As described previously, *Fusarium oxysporum* also appears to utilize host JA-signaling and subsequent gene induction for successful colonization of host (Thatcher et al. 2009).

Manipulation of host lipid metabolism is not only limited to pathogens, symbiotic fungi may also manipulate the host metabolism by secretion of signaling compounds into host cells. The biocontrol agent *Trichoderma virens*, secretes the protenaceous elicitor Sm1 to induce expression of JA and GLV biosynthetic enzymes, which in turn, is believed to trigger induced systemic resistance (Djonovic et al. 2007). Additionally, mycorrhizal fungi may also directly regulate the oxylipin biosynthetic genes such as LOX, AOS, and jasmonic acid methylesterase (Lopez-Raez et al. 2010), which may play a role in maintaining mutualistic interactions between the two organisms.

4 Perception of Partner's Oxylipins

Outside of the COI1-JAZ protein receptor complex for JA-Ile, no other oxylipin receptor has been identified in either plants or fungi, however, it is expected that for other oxylipins to have activity they too will have such targets (Christensen and Kolomiets 2011). Currently, GPCR are being implicated as oxylipin receptors in these kingdoms by recent studies. First, GPCRs are found in the genomes of both fungi and plants, and second, in the animal kingdom these act as receptors for both oxylipins (Bos et al. 2004) and fatty acids (Briscoe et al. 2003). Indeed, animal GPCRs are able to perceive oxylipins similar to those from plants (Obinata et al. 2005), adding to evidence of the oxylipin signal exchange.

5 Foresight

Various elucidations are in need for a better understanding of the oxylipin-mediated cross-kingdom signal exchange. This includes detailed characterization of oxylipin biosynthesis, evidence for oxylipins exchange/secretion, ligand-receptor pair identification, and signaling downstream targets are essential. Direct modification of interacting partner's oxylipins for signal modulation may also be a possibility during plant–fungal interactions. Modification of plant-derived lipids by insects to produce interaction signals is a well-studied phenomenon in plant–insect interactions. In this system, plant-derived fatty acids are conjugated with insect-derived amino acids within the insect herbivore and serve to the plant as a defensive signal (e.g., volicitin) (Pare et al. 1998). Plants, fungi, or both may use such mechanisms to suppress, promote, or alter the activity of oxylipin signals of their opposing partners.

6 Conclusion

Oxylipin mediated cross-kingdom signal exchange is a young but steadily growing field. Recent advancements have established that both plants and fungi are influenced by oxylipin-mediated signals produced by the opposing kingdom. Special interest has focused on the hijacking of plant oxylipin signaling for fungal proliferation. In short, fungi utilize these signals for sporulation and mycotoxin biosynthesis as well as produce their own to manipulate host metabolism. Understanding the chemical orchestra during plant–fungal interactions would allow novel developments for disease resistance and prevention.

Acknowledgements This work was supported by the NSF grant IOS-0951272 to Dr. Michael Kolomiets.

References

Acosta IF, Laparra H, Romero SP, Schmelz E, Hamberg M, Mottinger JP, Moreno MA, Dellaporta SL (2009) tasselseed1 is a lipoxygenase affecting jasmonic acid signaling in sex determination of maize. Science 323:262–265

Agrios GN (2005) Plant pathology. Elsevier Academic, Burlington

Andreou A, Brodhun F, Feussner I (2009) Biosynthesis of oxylipins in non-mammals. Prog Lipid Res 48:148–170

Bos CL, Richel DJ, Ritsema T, Peppelenbosch MP, Versteeg HH (2004) Prostanoids and prostanoid receptors in signal transduction. Int J Biochem Cell Biol 36:1187–1205

Boue SM, Shih BY, Carter-Wientjes CH, Cleveland TE (2005) Effect of soybean lipoxygenase on volatile generation and inhibition of *Aspergillus flavus* mycelial growth. J Agric Food Chem 53:4778–4783

Briscoe CP, Tadayyon M, Andrews JL, Benson WG, Chambers JK, Eilert MM, Ellis C, Elshourbagy NA, Goetz AS, Minnick DT, Murdock PR, Sauls HR Jr, Shabon U, Spinage LD, Strum JC, Szekeres PG, Tan KB, Way JM, Ignar DM, Wilson S, Muir AI (2003) The orphan G protein-coupled receptor GPR40 is activated by medium and long chain fatty acids. J Biol Chem 278:11303–11311

Brodhagen M, and Keller NP (2006) Signalling pathways connecting mycotoxin production and sporulation. Mol Plant Pathol 7:285–301

Brodhagen M, Tsitsigiannis DI, Hornung E, Goebel C, Feussner I, Keller NP (2008) Reciprocal oxylipin-mediated cross-talk in the Aspergillus-seed pathosystem. Mol Microbiol 67:378–391

Brodhun F, Feussner I (2011) Oxylipins in fungi. FEBS J 278:1047–1063

Brooks DM, Bender CL, Kunkel BN (2005) The *Pseudomonas syringae* phytotoxin coronatine promotes virulence by overcoming salicylic acid-dependent defences in *Arabidopsis thaliana*. Mol Plant Pathol 6:629–639

Brown SH, Scott JB, Bhaheetharan J, Sharpee WC, Milde L, Wilson RA, Keller NP (2009) Oxygenase coordination is required for morphological transition and the host-fungus interaction of *Aspergillus flavus*. Mol Plant Microbe Interact 22:882–894

Brown SH, Zarnowski R, Sharpee WC, Keller NP (2008) Morphological transitions governed by density dependence and lipoxygenase activity in *Aspergillus flavus*. Appl Environ Microbiol 74:5674–5685

Burow GB, Nesbitt TC, Dunlap J, Keller NP (1997) Seed lipoxygenase products modulate *Aspergillus mycotoxin* biosynthesis. Mol Plant Microbe Interact 10:380–387

Calvo AM, Hinze LL, Gardner HW, Keller NP (1999) Sporogenic effect of polyunsaturated fatty acids on development of Aspergillus spp. Appl Environ Microbiol 65:3668–3673

Champe SP, el-Zayat AA (1989) Isolation of a sexual sporulation hormone from *Aspergillus nidulans*. J Bacteriol 171:3982–3988

Chico JM, Chini A, Fonseca S, Solano R (2008) JAZ repressors set the rhythm in jasmonate signaling. Curr Opin Plant Biol 11:486–494

Christensen SA, Kolomiets MV (2011) The lipid language of plant-fungal interactions. Fungal Genet Biol 48:4–14

De Lucca AJ, Carter-Wientjes CH, Boué S, Bhatnagar D (2011) Volatile trans-2-hexenal, a soybean aldehyde, inhibits *Aspergillus flavus* growth and aflatoxin production in corn. J Food Sci 76:M381–M386

Djonovic S, Vargas WA, Kolomiets MV, Horndeski M, Wiest A, Kenerley CM (2007) A proteinaceous elicitor Sm1 from the beneficial fungus *Trichoderma virens* is required for induced systemic resistance in maize. Plant Physiol 145:875–889

Feussner I, Wasternack C (2002) The lipoxygenase pathway. Annu Rev Plant Biol 53:275–297

Fonseca S, Chini A, Hamberg M, Adie B, Porzel A, Kramell R, Miersch O, Wasternack C, Solano R (2009) (+)-7-iso-Jasmonoyl-L-isoleucine is the endogenous bioactive jasmonate. Nat Chem Biol 5:344–350

Gao XQ, Kolomiets MV (2009) Host-derived lipids and oxylipins are crucial signals in modulating mycotoxin production by fungi. Toxin Rev 28:79–88

Gao X, Shim WB, Gobel C, Kunze S, Feussner I, Meeley R, Balint-Kurti P, Kolomiets M (2007) Disruption of a maize 9-lipoxygenase results in increased resistance to fungal pathogens and reduced levels of contamination with mycotoxin fumonisin. Mol Plant Microbe Interact 20:922–933

Gao X, Brodhagen M, Isakeit T, Brown SH, Gobel C, Betran J, Feussner I, Keller NP, Kolomiets MV (2009) Inactivation of the lipoxygenase ZmLOX3 increases susceptibility of maize to Aspergillus spp. Mol Plant Microbe Interact 22:222–231

Gerwick WH, Moghaddam M, Hamberg M (1991) Oxylipin metabolism in the red alga *Gracilariopsis lemaneiformis*: mechanism of formation of vicinal dihydroxy fatty acids. Arch Biochem Biophys 290:436–444

Glazebrook J (2005) Contrasting mechanisms of defense against biotrophic and necrotrophic pathogens. Annu Rev Phytopathol 43:205–227

Goodrich-Tanrikulu M, Mahoney NE, Rodriguez SB (1995) The plant growth regulator methyl jasmonate inhibits aflatoxin production by *Aspergillus flavus*. Microbiology 141 (Pt 11):2831–2837

Hamberg M, Ponce de Leon I, Rodriguez MJ, Castresana C (2005) Alpha-dioxygenases. Biochem Biophys Res Commun 338:169–174

Horn BW, Moore GG, Carbone I (2009a) Sexual reproduction in *Aspergillus flavus*. Mycologia 101:423–429

Horn BW, Ramirez-Prado JH, Carbone I (2009b) The sexual state of *Aspergillus parasiticus*. Mycologia 101:275–280

Horn BW, Moore GG, Carbone I (2011) Sexual reproduction in aflatoxin-producing *Aspergillus nomius*. Mycologia 103:174–183

Hosoi T, Koguchi Y, Sugikawa E, Chikada A, Ogawa K, Tsuda N, Suto N, Tsunoda S, Taniguchi T, Ohnuki T (2002) Identification of a novel human eicosanoid receptor coupled to G(i/o). J Biol Chem 277:31459–31465

Hwang IS, Hwang BK (2010) The pepper 9-lipoxygenase gene CaLOX1 functions in defense and cell death responses to microbial pathogens. Plant Physiol 152:948–967

Isakeit T, Gao X, Kolomiets M (2007) Increased resistance of a maize mutant lacking the 9-lipoxygenase gene, ZmLOX3, to root rot caused by *Exserohilum pedicellatum*. J Phytopathol 155:758–760

Jambois A, Lapeyrie F (2005) Jasmonates, together with zeatin, induce hypaphorine accumulation by the ectomycorrhizal fungus *Pisolithus microcarpus*. Symbiosis 39:137–141

La Camera S, Geoffroy P, Samaha H, Ndiaye A, Rahim G, Legrand M, Heitz T (2005) A pathogen-inducible patatin-like lipid acyl hydrolase facilitates fungal and bacterial host colonization in Arabidopsis. Plant J 44:810–825

Lopez-Raez JA, Verhage A, Fernandez I, Garcia JM, Azcon-Aguilar C, Flors V, Pozo MJ (2010) Hormonal and transcriptional profiles highlight common and differential host responses to arbuscular mycorrhizal fungi and the regulation of the oxylipin pathway. J Exp Bot 61:2589–2601

Matsui K (2006) Green leaf volatiles: hydroperoxide lyase pathway of oxylipin metabolism. Curr Opin Plant Biol 9:274–280

Miersch O, Bohlmann H, Wasternack C (1999) Jasmonates and related compounds from *Fusarium oxysporum*. Phytochemistry 50:517–523

Mosquera G, Giraldo MC, Khang CH, Coughlan S, Valent B (2009) Interaction transcriptome analysis identifies magnaporthe oryzae BAS1-4 as biotrophy-associated secreted proteins in rice blast disease. Plant Cell 21:1273–1290

Mueller MJ (2004) Archetype signals in plants: the phytoprostanes. Curr Opin Plant Biol 7:441–448

Obinata H, Hattori T, Nakane S, Tatei K, Izumi T (2005) Identification of 9-hydroxyoctadecadienoic acid and other oxidized free fatty acids as ligands of the G protein-coupled receptor G2A. J Biol Chem 280:40676–40683

Pare PW, Alborn HT, Tumlinson JH (1998) Concerted biosynthesis of an insect elicitor of plant volatiles. Proc Natl Acad Sci USA 95:13971–13975

Reboutier D, Bianchi M, Brault M, Roux C, Dauphin A, Rona JP, Legue V, Lapeyrie F, Bouteau F (2002) The indolic compound hypaphorine produced by ectomycorrhizal fungus interferes with auxin action and evokes early responses in nonhost *Arabidopsis thaliana*. Mol Plant Microbe Interact 15:932–938

Ren A, Qin L, Shi L, Dong X, da Mu S, Li YX, Zhao MW (2010) Methyl jasmonate induces ganoderic acid biosynthesis in the basidiomycetous fungus *Ganoderma lucidum*. Bioresour Technol 101:6785–6790

Shi L, Ren A, Mu D, Zhao M (2010) Current progress in the study on biosynthesis and regulation of ganoderic acids. Appl Microbiol Biotechnol 88:1243–1251

Thatcher LF, Manners JM, Kazan K (2009) Fusarium oxysporum hijacks COI1-mediated jasmonate signaling to promote disease development in Arabidopsis. Plant J 58:927–939

Thomma BP, Eggermont K, Penninckx IA, Mauch-Mani B, Vogelsang R, Cammue BP, Broekaert WF (1998) Separate jasmonate-dependent and salicylate-dependent defense-response pathways in Arabidopsis are essential for resistance to distinct microbial pathogens. Proc Natl Acad Sci USA 95:15107–15111

Tsitsigiannis DI, Keller NP (2006) Oxylipins act as determinants of natural product biosynthesis and seed colonization in *Aspergillus nidulans*. Mol Microbiol 59:882–892

Tsitsigiannis DI, Keller NP (2007) Oxylipins as developmental and host-fungal communication signals. Trends Microbiol 15:109–118

Vergopoulou S, Galanopoulou D, Markaki P (2001) Methyl jasmonate stimulates aflatoxin B-1 biosynthesis by *Aspergillus parasiticus*. J Agric Food Chem 49:3494–3498

Voigt CA, Schafer W, Salomon S (2005) A secreted lipase of *Fusarium graminearum* is a virulence factor required for infection of cereals. Plant J 42:364–375

Wasternack C (2007) Jasmonates: an update on biosynthesis, signal transduction and action in plant stress response, growth and development. Ann Bot 100:681–697

Wright MS, Greene-McDowelle DM, Zeringue HJ, Bhatnagar D, Cleveland TE (2000) Effects of volatile aldehydes from Aspergillus-resistant varieties of corn on *Aspergillus parasiticus* growth and aflatoxin biosynthesis. Toxicon 38:1215–1223

Yan Y, Christensen S, Isakeit T, Engelberth J, Meeley R, Hayward A, Neil Emery RJ, Kolomiets M (2012) Disruption of OPR7 and OPR8 Reveals the Versatile Functions of JA in Maize Development and Defense. Plant Cell (in print)

Fungus Development and Reactive Oxygen: Phytopathological Aspects

Andrey A. Aver'yanov, Tatiana A. Belozerskaya, and Natalia N. Gessler

Abstract Fungi produce intracellular and extracellular reactive oxygen species (ROS) via different mechanisms. Diverse fungal metabolites (elicitors, toxins, antioxidants) modify ROS production by plants. Fungal antioxidants control signaling and destructive effects of plant and their own ROS. Disruption of pro-/antioxidant balance in host–pathogen system may disturb fungal development and hinder its virulence.

1 Introduction

Reactive oxygen species (ROS) include hydrogen peroxide (H_2O_2), superoxide anion (O_2^-) and hydroxyl ($^{\cdot}OH$) free radicals, and singlet excited oxygen ($^1O_2$), as well as their derivatives. They share some properties with nitric oxide (NO) and related compounds representing reactive nitrogen species (RNS). Compounds eliminating ROS are usually termed antioxidants.

ROS have long been treated as undesirable metabolites causing uncontrolled biodegradation. Their involvement in phagocytosis, revealed later, is controlled but again destructive. At last, ROS formed in the organism have been recognized as necessary signals. Thus, biological roles of ROS, as well as that of antioxidants are diverse.

A.A. Aver'yanov (✉)
Research Institute of Phytopathology, Russian Academy of Agricultural Sciences,
B. Vyazemy, Moscow 143050, Russia
e-mail: aaveryanov@post.ru

T.A. Belozerskaya • N.N. Gessler
A.N. Bach Institute of Biochemistry, Russian Academy of Sciences,
Leninsky pr. 33, Moscow 119071, Russia
e-mail: tabinbi@mail.ru

G. Witzany (ed.), *Biocommunication of Fungi*, DOI 10.1007/978-94-007-4264-2_17,

The concept of ROS in biology of fungi and oomycetes has passed the similar evolution and is rapidly enriching with abundant new facts. This mini-review covers mainly recent works related to ROS and antioxidants involved in interactions of phytopathogens with their hosts.

2 ROS that Impact Fungi

2.1 ROS Formation in Fungi

Reduction of dioxygen by I and III mitochondrial respiratory complexes of aerobes gives rise to superoxide radical (Aguirre et al. 2005).

One of main ROS and RNS sources in the fungal cell is the membrane-bound NADPH oxidase complex (NOX) and NO synthase (NOS). These enzymes control fungal development, differentiation and host–parasite interactions (Wang and Higgins 2005; Gessler et al. 2007; Semighini and Harris 2008; Egan et al. 2007; Li et al. 2010).

O_2^- generation by xanthine oxidase was observed in *Rhizoctonia solani, Sclerotinia sclerotiorum, Sclerotium rolfsii* and *Sclerotinia minor* (Papapostolou and Georgiou 2010).

Fungal sugar oxidases, including glucose oxidase, glyoxal oxidase and galactose oxidase function outside fungal cells to produce H_2O_2 (Sierra-Campos and Pardo 2009).

Wood-decomposing fungi produce hydroxyl radical and other ROS by enzymatic (lignin, manganese and versatile peroxidases, and laccase), and non-enzymatic (through quinone redox cycling) mechanisms. They use this very potent oxidant to metabolize cellulose and lignin (Gómez-Toribio et al. 2009). This way of breaking through host cell wall is probably not reported for phytopathogenic fungi.

Singlet oxygen along with other ROS is formed under light by photosensitizers. The best known one is cercosporin, the toxin of *Cercospora* fungi (Daub 1987).

2.2 ROS Formation in Plants Evoked by Fungi

Parasites are subjected to their own and host-originated ROS. The early oxidative burst is a part of resistance mechanism (Shetty et al. 2008). It is ascribed usually to the host. However, the contribution of the pathogen to the joint ROS pool may be significant as it was shown for rice blast caused by *Magnaporthe grisea* (Aver'yanov et al. 2007c).

In interactions with plants, fungi produce compounds inducing or suppressing host's ROS-dependent defense, which in certain situations promote disease. Microbial elicitors induce responses, including oxidative burst that often lead to disease resistance.

Microbial toxins, on the contrary, favor susceptibility but they might also elevate ROS levels in planta. Some toxins, such as cercosporin produce ROS by themselves while others stimulate the host for oxygen activation. The resulting ROS-driven plant cell death is a factor of pathogenicity of necrotrophs although not unconditionally (Daub 1987).

Toxin-stimulated oxidative burst associated with disease resistance is also known. Blast fungus toxins α-picolinic (Zhang et al. 2004a) and tenuazonic (Aver'yanov et al. 2007a) acids induce ROS production and resistance to blast in treated rice plants. Deoxynivalenol of *Fusarium* elevates H_2O_2 level in wheat leaves, followed by programmed cell death and expression of defense genes, the responses compromised by exogenous antioxidant ascorbate (Desmond et al. 2008). Presumably, the toxin-induced ROS-mediated host cell death favors necrotrophic fungal growth. But, like in the case of elicitors, toxin-induced ROS can induce antimicrobial host defenses.

Biocontrol fungi may lead to similar outcomes. The yeast antagonists, *Metschnikowia fructicola* and *Candida oleophila* generate O_2^- and induce H_2O_2 accumulation in fruit tissues (Macarisin et al. 2010).

3 ROS Involvement in Fungal Ontogenesis and Pathogenicity

3.1 Effects of Exogenous ROS on Fungal Development and Metabolism

The action of artificially added ROS helps understanding the influence of these compounds formed naturally. Indeed, ROS can kill any living cells and abort their development at any stage. Cell tolerance to oxidative stress is usually tested at millimolar or stronger H_2O_2 solutions, which is, though, beyond physiological levels. Meanwhile, much lower ROS concentrations are yet active. 10^{-2} M H_2O_2 inhibits spore germination of *Cladosporium cucumerinum*, causal agent of cucurbit scab, and *M. grisea* almost completely, and its toxicity is practically absent at about 10^{-5}–10^{-6} M (Aver'yanov et al. 2007b). However, weaker peroxide solutions (10^{-10}–10^{-12} M) inhibit again spore germination (and appressorium formation) even stronger than millimolar ones. Therefore, H_2O_2 can be bioactive at very low concentrations, presumably due to its signaling features, although the toxicity of stronger solutions is apparently imply gross damages.

It does not mean that all pathogenic fungi are uniformly vulnerable to oxidative damage. In fact, spores of the necrotroph *Botrytis cinerea* can germinate in 180 mM H_2O_2, and its mycelium tolerates even higher concentrations (Gil-ad and Mayer 1999). This may explain the ability of *B. cinerea* to colonize its hosts despite their oxidative bursts.

Oxidative stress induces multiple changes in parasite's metabolism that promote or hinder pathogenesis. In *Aspergillus flavus*, exogenous organic peroxides trigger aflatoxin synthesis (Kim et al. 2008). Although fungal elicitors induce plant oxidative burst they could, on the contrary, be secreted from *M. grisea* spores as a consequence of oxidative stress (Zakharenkova et al. 2010).

Therefore, even small amounts of ROS prove able to change significantly fungal metabolism and development.

3.2 *Regulation of Fungal Development and Pathogenicity by Endogenous ROS*

Fungi produce and often secrete ROS at various rates during different developmental stages. The necessity of these metabolites for development and pathogenesis is shown in several cases.

In suspensions of rice blast spores, extracellular O_2^- is found after their germination, and H_2O_2 is revealed after appressorium formation (Aver'yanov et al. 2007c). Cytochemical assay reveals O_2^- production in conidia cytoplasm before germination and, later, in the immature appressorium (Egan et al. 2007). Ascorbate (antioxidant) and diphenylene iodonium (inhibitor of NOX) suppress ROS formation, spore germination and appressorium formation. It also causes morphological aberrations. The activity of NOX is crucial for pathogenesis since the corresponding deletion mutants are incapable to infect rice plants (Egan et al. 2007).

In *Aspergillus parasiticus,* O_2^- formation also increases in vitro soon after conidium germination which is followed by an increase in H_2O_2 and lipoperoxide levels (Reverberi et al. 2008).

The induction of sclerotial differentiation of the phytopathogenic filamentous fungi *Rhizoctonia solani, Sclerotinia scterotiorum, S. minor,* and *Sclerotium rolfsii* involve O_2^-. It is produced more intensively in the sclerotiogenic fungi than in their non-differentiating counterpart strains (Papapostolou and Georgiou 2010).

Besides the up-regulation of fungal development by ROS, down-regulation also exists and might be responsible for quorum-effects. Spore germination of *M. grisea* (Aver'yanov and Lapikova 1990) and *C. cucumerinum* (Aver'yanov et al. 2011) is suppressed in their diluted or dense suspensions in vitro. Diffusates from both suspensions are toxic to spores present at optimal concentration suggesting that toxicants are extracellular. Their action depends on ROS because exogenous antioxidants eliminating H_2O_2, $O_2^{-\cdot}$ or $^{\cdot}OH$ restore spore germination in suspensions at extreme densities or in diffusates prepared from these suspensions. Fungitoxicity of *C. cucumerinum* diffusates correlates with their chemically assayed $O_2^{-\cdot}$ production. Thus, ROS might down-regulate fungus development in crowded or rare populations. Apparently, this action alters pathogenicity of *M. grisea* and *C. cucumerinum* because inocula at extreme concentrations produce weaker disease symptoms in their susceptible hosts than at concentrations optimal for spore germination in vitro. But the reversible inhibition would benefit the fungus by retarding its development until the population density becomes optimal (Aver'yanov et al. 2011).

Another example of ROS-controlled development is the apical dominance in fungal hyphae. The effect correlates with the localized ROS accumulation produced by NOX and activated by Rac1 in the apical region of *A. nidulans* hyphae (Semighini and Harris 2008).

Little is known about fungus-originated nitric oxide. NOS-originated NO regulates spore germination in *Colletotrichum coccodes* (Wang and Higgins 2005) and appressorium formation in *Blumeria graminis* (Prats et al. 2008).

4 Fungal Antioxidants

Antioxidants maintain the proper pro-/antioxidant balance in the organism, regulate development and protect from adverse environment. In microorganisms, antioxidant systems control stress-inducible adaptive responses (Gessler et al. 2007; Vandenbroucke et al. 2008).

In general, antioxidants, especially, those secreted by phytopathogens, help withstanding host-originated oxidative burst but roles of particular elements of antioxidant complex are not always clear.

Fungal antioxidant systems include enzymes (superoxide dismutases (SOD), catalases, glutathione- and ascorbic-dependent peroxidases and transferases) and non-enzymatic systems (melanins, carotenes, ascorbate, proline, mannitol, etc.) along with some other mechanisms (chelating ferrous ions, scavenging hydroxyl and organic free radicals, alternative pathways of mitochondrial respiratory chain) eliminating the pro-oxidant function (Gessler et al. 2007).

Some enzymatic antioxidants inactivate ROS directly but more number of auxiliary ones re-reduce substrates necessary for the inactivation. Several antioxidant enzymes are secreted (SODs, catalase, etc.). Catalases of fungi and oomycetes are located in peroxisomes (Valenciano et al. 1996), in cytosol (Schliebs et al. 2006) and in spores (Navarro and Aguirre 1998). Bifunctional catalase-peroxidases has been identified in *Neurospora crassa* (Peraza and Hansberg 2002), and *Septoria tritici* (Levy et al. 1992). Three catalase genes are characterized in *Phytophthora nicotianae* (Blackman and Hardham 2008). Secretable catalases have been identified in *M. grisea* (Skamnioti et al. 2007), *Claviceps purpurea* (Garre et al. 1998) and *B. graminis* (Zhang et al. 2004b), that have been implicated in overcoming the host defense response. *P. nicotianae* was suggested to secrete catalase-peroxidases (Blackman and Hardham 2008).

Chitosans are well known fungal elicitors but their antioxidant properties are also reported i.e. chelating ferrous ions, scavenging hydroxyl and organic free radicals (Yen et al. 2007).

5 Involvement of Antioxidants in Fungal Ontogenesis and Pathogenicity

Localization and activity of various antioxidant enzymes change in the course of development (Gessler et al. 2007).

The sclerotial differentiation in *Rhizoctonia solani*, *Sclerotinia scterotiorum*, *S. minor*, and *Sclerotium rolfsii* is associated not only with higher level of O_2^- but also lower SOD activity (Papapostolou and Georgiou 2010). This enzyme of all four species is partially secreted suggesting its protective function against host's ROS. The antioxidant down-regulation of sclerotial differentiation in *S. minor* could be realized through fungal thiol redox state. This is relevant to pathogenicity as undifferentiated hyphae are more vulnerable to degradation by soil microorganisms (Papapostolou and Georgiou 2008).

Functions of some SOD enzymes in morphogenesis are obscure. Just so the biotrophic pathogen *C. purpurea* contains copper-zinc SOD1 associated predominantly in the cell wall (Moore et al. 2002). The necessity of the enzyme for fungal development is evidenced by the slower (than in the wild type) mycelial growth of the deletion mutant in an axenic culture. However, artificial oxidative stress does not induce this gene in the wild type (WT) and inhibits the growth of the mutant not stronger than that of WT. So, the antidotal role of this SOD against host-originated O_2^- is not clear. Its involvement in pathogenicity is also uncertain. The disease index during the first days post inoculation of rye by the mutant is really lower than in case of WT but equal by the 10th day.

The secreted catalase of *M. grisea* (CATB) is significant for development since spore production of the deletion mutant is 30-fold fewer than that of WT (Skamnioti et al. 2007). Mutant conidia and appressoria are more sensitive to hyperosmotic and oxidative stresses. In the WT, the gene expression increases up to fourfold in H_2O_2 and 600-fold during barley leaf penetration. The mutant is severely less pathogenic, and its infectivity is further reduced by pre-exposure to exogenous H_2O_2. Nonetheless, there are no differences between the mutant and WT in the host's H_2O_2 production beneath appressoria and pathogen's capacity to decompose it. The authors suggest that CATB rather strengthens the fungal cell wall than detoxifies plant ROS (Skamnioti et al. 2007).

The obligate biotroph, *B. graminis* f. sp. *hordei*, elicits H_2O_2 burst in its host barley and accumulates *cat*B transcripts at sites of germ tube invasion. The H_2O_2–degrading activity is visualized at the same sites and suggests catalase contribution to pathogenicity (Zhang et al. 2004b).

In *P. nicotiana*, catalase activity and the expression of peroxisomal PnCat2 are the highest in sporulating hyphae and the lowest in germinated cysts (Blackman and Hardham 2008). The infected tobacco leaves show the increase in *PnCat2* expression. The activity of this enzyme is not found in healthy leaves or in inoculated leaves of resistant culture but rises dramatically in the inoculated susceptible one. Meantime, the expression of the host catalase gene is, in contrast, diminished upon infection, earlier in the susceptible than in the resistant cultivar. Thus, pathogen seems to up-regulate its catalase to protect itself from host-originated ROS but it down-regulates the host enzyme to promote cell death and necrotrophic colonization.

Necrotroph *B. cinerea* being extremely resistant to hydrogen peroxide degrades it rapidly (Gil-ad and Mayer 1999). This ability is adaptive because mycelia pre-exposed to H_2O_2 degrade faster the same solution added 24 h later.

The antioxidant machinery of the human pathogen *Candida albicans* is partly extracellular. A diffusate from yeast culture protects other yeast cells from agents generating either H_2O_2 or O_2^- (Westwater et al. 2005). The protection is indirect because the diffusate induces the expression of *CAT, SOD, SOD2* and *SOD4.* Farnesol is the yeast exometabolite responsible for the effect at least partially.

Other ROS-detoxifying enzymes have also been shown to be important for virulence, including the *Cryptococcus neoformans* glutathione reductase (Glr1) and the *Ustilago maydis* Yap1-induced peroxidase (Missall et al. 2006; Molina and Kahmann 2007).

Fungal antioxidant capacity depends also on siderophores. They allow iron acquisition and use, for example, for catalase biosynthesis. Besides, binding free iron prevents its involvement in $^{\cdot}OH$ formation. *M. grisea* secretes siderophores. The mutants defective in their biosynthesis have 2–3 times lower catalase activity in mycelia (Hof et al. 2009). This may explain their reduced growth rates and fewer production of conidia.

Fungal toxins may not only stimulate but also suppress oxidative burst in infected plants as it was demonstrated for oxalate secreted by *S. sclerotiorum* in interaction with tobacco or soybean (Cessna et al. 2000) and by *B. cinerea* interacting with *Arabidopsis* (L'Haridon et al. 2011).

Nitric oxide is multi-functional. In incompatible biotrophic interactions associated with hypersensitive necrosis, NO amplifies phytotoxicity of O_2^- (Deledonne et al. 1998). But little is known about NO in necrotrophic pathogenesis. In leaves of *Nicotiana benthamiana* compatibly infected with *B. cinerea*, productions in both NO and O_2^- are induced but with different consequences. The inhibitor of NOX slightly reduces disease symptoms while the inhibitor of NOS increases them. So, NO participates in defense against this necrotroph whereas ROS favor infection. In general, hypersensitive cell death is important for *B. cinerea* virulence but may also be responsible for resistance, depending on particular conditions (Asai and Yoshioka 2009).

Since NO may be fungitoxic, this suggests fungal anti-RNS activity. Human fungal pathogen *Candida albicans* faces both ROS and NO generated by macrophages and commensal bacteria. Exposing *C. albicans* to NO upregulates the content of flavohemoglobin Yhblp, which converts NO to harmless nitrate. To date, mechanisms of fungal NO sensing and signal transduction are uncertain. In *C. albicans,* the protein Cta4p appears to be involved in initiating NO response. Deletion of this gene causes a small but significant decrease in virulence (Chiranand et al. 2008).

6 Signaling by ROS and Antioxidants in Fungi

The delicate balance between ROS generation and elimination is maintained by many complex mechanisms. Unfortunately, data on redox signaling through Yap1 and its homologs in pathogens are controversial (Molina and Kahmann 2007; Yang et al. 2009; Temme and Tudzynski 2009). Only certain proteins of redox regulated pathways are unraveled up to date (Gessler et al. 2007; Jain et al. 2009; Forman et al. 2010; Roetzer et al. 2011).

Along with redox signaling, histidine kinase phosphorelay pathway mediates pathogen's adaptations to environmental signals (Nathues et al. 2007; Fassler and West 2011). ROS signal transduction in fungi might follow some other pathways common to living systems: the cyclic AMP (cAMP) signaling pathway, Ca^{2-}/calcineurin signaling pathway, protein kinase C (PKC)/Mpk1 (also known as Slt2) mitogen-activated protein kinase (MAPK) pathway, and stress-activated Hog1 MAPK pathway (Aguirre et al. 2006; Segmüller et al. 2007; Bahn 2008; Schumacher et al. 2008; Brown et al. 2009).

Thus, many ROS-transduction mechanisms might regulate a set of genes responsible for oxidative stress response. Future investigations will reveal specialized or synergistic regulatory interactions between these signaling pathways for diverse physiological functions of fungi.

7 Conclusions

Mountain of data witnesses that phytopathogenic fungi and oomycetes naturally produce and scavenge ROS as well as their host-plants do. In general, both opposing activities are indispensable for microbial development and tolerance to harsh environment including plant defense responses. However, particular host–pathogen interactions not always display simple patterns, especially as to specific genes and their products responsible for certain effects.

References

Aguirre J, Rios-Momberg M, Hewitt D, Hansberg W (2005) Reactive oxygen species and development in microbial eukaryotes. Trends Microbiol 13:111–118

Aguirre J, Hansberg W, Navarro R (2006) Fungal responses to reactive oxygen species. Med Mycol 44:S101–S109

Asai S, Yoshioka H (2009) Nitric oxide is a partner of reactive oxygen species participates in disease resistance to necrotrophic pathogen *Botrytis cinerea* in *Nicotiana benthaniana*. Mol Plant Microbe Interact 22:619–629

Aver'yanov AA, Lapikova VP (1990) Activated oxygen as a possible factor in the autoinhibition of spore germination of the fungus *Pyricularia oryzae*. Biokhimia (Moscow) 55:1397–1402

Aver'yanov AA, Lapikova VP, Lebrun M-H (2007a) Tenuazonic acid, toxin of rice blast fungus, induces disease resistance and reactive oxygen production in plants. Russ J Plant Physiol 54:749–754

Aver'yanov AA, Lapikova VP, Pasechnik TD, VlV K, Baker CJ (2007b) Suppression of early stages of fungus development by hydrogen peroxide at low concentrations. Plant Pathol J 6:242–247

Aver'yanov AA, Pasechnik TD, Lapikova VP, Gaivoronskaya LM, VlV K, Baker CJ (2007c) Possible contribution of blast spores to the oxidative burst in the infection droplet on rice leaf. Acta Phytopathol Entomol Hung 42:305–319

Aver'yanov AA, Lapikova VP, Pasechnik TD, Zakharenkova TS, Baker CJ (2011) Self-inhibition of spore germination via reactive oxygen in the fungus *Cladosporium cucumerinum*, causal agent of cucurbit scab. Eur J Plant Pathol 130:541–555

Bahn YS (2008) Master and commander in fungal pathogens: the two-component system and the HOG signaling pathway. Eukaryot Cell 7:2017–2036

Blackman LM, Hardham AR (2008) Regulation of catalase activity and gene expression during *Phytophthora nicotianae* development and infection of tobacco. Mol Plant Pathol 9:495–510

Brown A, Haynes K, Quinn J (2009) Nitrosative and oxidative stress responses in fungal pathogenicity. Curr Opin Microbiol 12:384–391

Cessna SG, Sears VE, Dickman MB, Low PS (2000) Oxalic acid, a pathogenicity factor for *Sclerotinia sclerotiorum*, suppresses the oxidative burst of the host plant. Plant Cell 12:2191–2199

Chiranand W, McLeod I, Zhou H, Lynn JJ, Vega LA, Myers H, Yates JR, Lorenz MC, Gustin MC (2008) *CTA4* transcription factor mediates induction of nitrosative stress response in *Candida albicans*. Eukaryot Cell 7:268–278

Daub ME (1987) The fungal photosensitizer cercosporin and its role in plant disease. In: Heitz JR, Downum KR (eds) Light activated pesticides. American Chemical Society, Washington, DC, pp 271–280

Deledonne M, Xia Y, Dixon R, Lamb C (1998) Nitric oxide functions as a signal in plant disease resistance. Nature 394:685–688

Desmond OJ, Manners JM, Stephens AE, Maclean DJ, Schenk PM, Gardiner DM, Munn AL, Kazan K (2008) The *Fusarium* mycotoxin deoxynivalenol elicits hydrogen peroxide production, programmed cell death and defence responses in wheat. Mol Plant Pathol 9:435–445

Egan MJ, Zheng-Yi W, Jones MA, Smirnoff N, Talbot NJ (2007) Generation of reactive oxygen species by fungal NADPH oxidases is required for rice blast disease. Proc Natl Acad Sci USA 104:11772–11777

Fassler JS, West AH (2011) Fungal Skn7 stress responses and their relationship to virulence. Eukaryot Cell 10:156–167

Forman HJ, Maiorino M, Ursini F (2010) Signalling functions of reactive oxygen species. Biochemistry 49:835–842

Garre V, Tenberge KB, Eising R (1998) Secretion of a fungal extracellular catalase by *Claviceps purpurea* during infection of rye: putative role in pathogenicity and suppression of host defense. Phytopathology 88:744–753

Gessler NN, Aver'yanov AA, Belozerskaya TA (2007) Reactive oxygen species in regulation of fungal development. Biochemistry (Moscow) 72:1091–1109

Gil-ad NL, Mayer AM (1999) Evidence for rapid breakdown of hydrogen peroxide by *Botrytis cinerea*. FEMS Microbiol Lett 176:455–461

Gómez-Toribio V, García-Martín AB, Martínez MJ, Martínez AT, Guillén F (2009) Induction of extracellular hydroxyl radical production by white-rot fungi through quinone redox cycling. Appl Environ Microbiol 75:3944–3953

Hof C, Eisfeld K, Antelo L, Foster AJ, Anke H (2009) Siderophore synthesis in *Magnaporthe grisea* is essential for vegetative growth, conidiation and resistance to oxidative stress. Fungal Genet Biol 46:321–332

Jain C, Yun M, Politz SM, Rao RP (2009) A pathogenesis assay using *Saccharomyces cerevisiae* and *Caenorhabditis elegans* reveals novel roles for yeast AP-1, Yap1, and host dual oxidase BLI-3 in fungal pathogenesis. Eukaryot Cell 8:1218–1227

Kim JH, Yu J, Mahoney N, Chan KL, Molyneux RJ, Varga J, Bhatnagar D, Cleveland TE, Nierman WC, Campbell BC (2008) Elucidation of the functional genomics of antioxidant-based inhibition of aflatoxin biosynthesis. Int J Food Microbiol 122:49–60

L'Haridon F, Besson-Bard A, Binda M, Serrano M, Abou-Mansour E, Balet F, Schoonbeek H-J, Hess S, Ricardo M, Léon J, Lamotte O, Métraux J-P (2011) A permeable cuticle is associated with the release of reactive oxygen species and induction of innate immunity. PLoS Pathog 7: e1002148

Levy E, Eyal Z, Hochman A (1992) Purification and characterization of catalase-peroxidase from the fungus *Septoria tritici*. Arch Biochem Biophys 296:321–327

Li B, Fu Y, Jiang D, Xie J, Cheng J, Li G, Hamid MI, Yi X (2010) Cyclic GMP as a second messenger in the nitric oxide-mediated conidiation of the mycoparasite *Coniothyrium minitans*. Appl Environ Microbiol 76:2830–2836

Macarisin D, Droby S, Bauchanc G, Wisniewski M (2010) Superoxide anion and hydrogen peroxide in the yeast antagonist–fruit interaction: a new role for reactive oxygen species in postharvest biocontrol? Postharvest Biol Technol 58:184–202

Missall TA, Pusateri ME, Donlin MJ, Chambers KT, Corbett JA, Lodge JK (2006) Posttranslational, translational, and transcriptional responses to nitric oxide stress in *Cryptococcus neoformans*: implications for virulence. Eukaryot Cell 5:518–529

Molina L, Kahmann R (2007) An *Ustilago maydis* gene involved in H_2O_2 detoxification is required for virulence. Plant Cell 19:2293–2309

Moore S, De Vries OMH, Tudzynski P (2002) The major Cu, Zn SOD of the pathogen *Claviceps purpurea* is not essential for pathogenicity. Mol Plant Pathol 3:9–22

Nathues E, Jörgens C, Lorentz N, Tudzynski P (2007) The histidine kinase CpHK2 has impact on spore germination, oxidative stress and fungicide resistance, and virulence of the ergot fungus *Claviceps purpurea*. Mol Plant Pathol 8:653–665

Navarro R, Aguirre J (1998) Posttranscriptional control mediates cell type-specific localization of catalase during *Aspergillus nidulans* development. J Bacteriol 180:5733–5738

Papapostolou I, Georgiou CD (2008) Differentiation of *Sclerotinia minor* depends on thiol redox state and oxidative stress. Can J Microbiol 54:28–36

Papapostolou I, Georgiou CD (2010) Superoxide radical is involved in the sclerotial differentiation of filamentous phytopathogenic fungi: identification of a fungal xanthine oxidase. Fungal Biol 114:387–395

Peraza L, Hansberg W (2002) Neurospora crassa catalases, singlet oxygen and cell differentiation. Biol Chem 383:569–575

Prats E, Carver TLW, Mur LAJ (2008) Pathogen-derived nitric oxide influences formation of the appressorium infection structure in the phytopathogenic fungus *Blumeria graminis*. Res Microbiol 159:476–480

Reverberi M, Zjalic S, Ricelli A, Punelli F, Camera E, Fabbri C, Picardo M, Fanelli C, Fabbri AA (2008) Modulation of antioxidant defense in *Aspergillus parasiticus* is involved in aflatoxin biosynthesis: a role for the Ap*yapA* gene. Eukaryot Cell 7:988–1000

Roetzer A, Klopf E, Gratz N, Marcet-Houben M, Hiller E, Rupp S, Toni G, Kovarik P, Schüller C (2011) Regulation of *Candida glabrata* oxidative stress resistance is adapted to host environment. FEBS Lett 585:319–327

Schliebs W, Würtz CH, Kunau W-H, Veenhuis M, Rottensteiner H (2006) A eukaryote without catalase-containing microbodies: *Neurospora crassa* exhibits a unique cellular distribution of its four catalases. Eukaryot Cell 5:1490–1502

Schumacher J, de Larrinoa IF, Tudzynski B (2008) Calcineurin-responsive zinc finger transcription factor CRZ1 of *Botrytis cinerea* is required for growth, development, and full virulence on bean plants. Eukaryot Cell 7:584–601

Segmüller N, Ellendorf U, Tudzynski B, Paul T (2007) BcSAK1, a stress-activated mitogen-activated protein kinase, is involved in vegetative differentiation and pathogenicity in *Botrytis cinerea*. Eukaryot Cell 6:211–221

Semighini CP, Harris SD (2008) Regulation of apical dominance in *Aspergillus nidulans* hyphae by reactive oxygen species. Genetics 179:1919–1932

Shetty NP, Jorgensen HJL, Jensen JD, Collinge DB, Shetty HS (2008) Roles of reactive oxygen species in interactions between plants and pathogens. Eur J Plant Pathol 121:267–280

Sierra-Campos E, Pardo JP (2009) The relationship between the antioxidant system and the virulence in *Ustilago maydis*, a fungal pathogen. Rev Latinoam Microbiol 51:7–17

Skamnioti P, Henderson C, Zhang Z, Robinson Z, Gurr SJ (2007) A novel role for catalase B in the maintenance of fungal cell-wall integrity during host invasion in the rice blast fungus *Magnaporthe grisea*. Mol Plant Microbe Interact 20:568–580

Temme N, Tudzynski P (2009) Does *Botrytis cinerea* ignore H_2O_2-induced oxidative stress during interaction? Characterization of *Botrytis* activator protein 1. Mol Plant Microbe Interact 22:987–998

Valenciano S, Lucas JR, Pedregosa A, Monistrol IF, Laborda F (1996) Induction of β-oxidation enzymes and microbody proliferation in *Aspergillus nidulans*. Arch Microbiol 166:336–341

Vandenbroucke K, Robbens S, Vandepoele K, Inzé D, Van de Peer Y, Van Breusegem F (2008) Hydrogen peroxide-induced gene expression across kingdoms: a comparative analysis. Mol Biol Evol 25:507–516

Wang J, Higgins VJ (2005) Nitric oxide has a regulatory effect in the germination of conidia of *Colletotrichum coccodes*. Fungal Genet Biol 42:284–292

Westwater C, Balish E, Schofield DA (2005) *Candida albicans*-conditioned medium protects yeast cells from oxidative stress: a possible link between quorum sensing and oxidative stress resistance. Eukaryot Cell 4:1654–1661

Yang SL, Lin C-H, Chung K-R (2009) Coordinate control of oxidative stress tolerance, vegetative growth, and fungal pathogenicity via the AP1 pathway in the rough lemon pathotype of *Alternaria alternata*. Physiol Mol Plant Pathol 74:100–110

Yen M-T, Tseng Y-H, Li R-C, Mau J-L (2007) Antioxidant properties of fungal chitosan from shiitake stipes. LWT 40:255–261

Zakharenkova TS, Aver'yanov AA, Pasechnik TD, Lapikova VP, Baker CJ (2010) Release of elicitors from rice blast spores under the action of reactive oxygen species. Russ J Plant Physiol 57:615–619

Zhang H, Zhang X, Mao B, Li Q, He Z (2004a) Alpha-picolinic acid, a fungal toxin and mammal apoptosis-inducing agent, elicits hypersensitive-like response and enhances disease resistance in rice. Cell Res 14:27–33

Zhang Z, Henderson C, Gurr SJ (2004b) *Blumeria graminis* secretes an extracellular catalase during infection of barley: potential role in suppression of host defence. Mol Plant Pathol 5:537–547

Oxidative Stress and Oxylipins in Plant-Fungus Interaction

Massimo Reverberi, Anna A. Fabbri, and Corrado Fanelli

Abstract Considering the huge quantity of studies related to the implication of oxidative stress in the biological processes, we could say that reactive oxygen species (ROS) are actually "everywhere" around us and play different important roles in the life of all organisms. Reactive species can affect cell physiology both in a quantitative and in a qualitative manner, that is, not only the amount of ROS accumulated in the cell, but also the kind of ROS produced affect cell biology. Indeed, ROS are formed at the plant-fungus interface during their interaction and they influence both the plant and the pathogen by altering the respective metabolism. Here we hypothesise that peroxisomes play an important role in both challenging organisms in modulating the ROS signal and transform it in oxylipins "words", e.g. jasmonates for plants and hormone-like substances (i.e. psi factors) for fungi. Plant secretes oxylipins to alter fungal metabolism and differentiation, and the fungus use these lipid signals for switching secondary metabolism on. On the other hand, fungi are able to secrete Mn-lipoxygenase and Jasmonates into the plant cell for driving the host metabolism toward their own parasitic needs.

1 Oxidative Stress in Plant and Fungi: A *"Primum Movens"* in Their Interaction?

1.1 ROS Generation in Plants and Fungi: When and Where

ROS are produced by all living organisms, either constitutively as by-products of several metabolic processes and at the early stages of plant-microorganism

M. Reverberi (✉) • A.A. Fabbri • C. Fanelli
Department of Environmental Biology, University "Sapienza",
L.go Cristina di Svezia 24, 00165 Rome, Italy
e-mail: massimo.reverberi@uniroma1.it

G. Witzany (ed.), *Biocommunication of Fungi*, DOI 10.1007/978-94-007-4264-2_18,

interactions or, in a more controlled manner, by a complex interplay between oxidant and antioxidant activities. ROS are known to play a dual role depending on their accumulation levels. On one hand, high intracellular concentration of ROS can cause extensive cell injury or death; therefore, the levels of ROS need to be tightly regulated to avoid cell damage (Neill et al. 2002; Mittler et al. 2004; Kotchoni and Gachomo 2006). On the other hand, when modulated, ROS play a pivotal role in signal transduction network of stress-inducible genes (Bartels 2001; Mittler et al. 2004; Davletova et al. 2005). It is emerging that a balanced amount of ROS is crucial for many different metabolic processes (Bartels 2001; Kotchoni and Gachomo 2006; Davletova et al. 2005). ROS are physiologically produced during aerobic phase of photosynthesis and photorespiration (Asada and Takahashi 1987; Kotchoni and Gachomo 2006). During cellular metabolism, oxygen molecules are often converted into several intermediates such as anion superoxide O_2^-, hydrogen peroxide H_2O_2, hydroxyl radical $HO^\cdot$, which often leak out from electron transport chain (Banerjee et al. 2003). Different enzymes have been implicated in the generation of ROS and among these, NADPH oxidases (Nox) play an important role in the production of superoxide radicals during the oxidative burst triggered to defend cells from pathogen invasion. These enzymes share common structural features and are evolutionarily of ancient origin and thus ubiquitous in multicellular eukaryotes (Bedard and Krause 2007; Bedard et al. 2007). Nox are integral membrane proteins and oxidize NADPH in the cell as well as reducing molecular oxygen into superoxide radicals into the apoplast (Sumimoto 2008), which is quickly dismutated into H_2O_2 either spontaneously or by superoxide dismutase (SOD) enzymes. In plants, Nox form a small multigenic family which is involved in diverse events including innate immunity development. Other sources of ROS include cell-wall-bound peroxidase, chloroplasts, and mitochondria (Davletova et al. 2005). The chloroplast is considered to be a focal point of plant ROS metabolism in fact, it is a major producer of O_2^- and H_2O_2 and contains also a large array of ROS-scavenging mechanisms (Davletova et al. 2005). In particular, light-driven production of ROS inside the plant occurs by uncoupling or by inhibiting the photosynthesis and the photorespiration associated with chloroplast and peroxisomes (Karpinski et al. 2003). The mitochondria are another intracellular source of ROS (Tiwari et al. 2002). The mitochondrial electron transport chain can produce significant quantities of ROS, primarily because of the presence of the ubisemiquinone radical, which can transfer a single electron to oxygen and produce O_2^- (Overmyer et al. 2003). Owing to the fact that ROS are toxic and in many cases short-lived, the activity of these oxidases is tightly regulated both temporally and spatially. The battery of ROS scavenging systems present in the cells, including ascorbate peroxidases, glutathione cycle, superoxide dismutases and catalases, ensures rapid turnover of the ROS to maintain ROS homeostasis (Torres and Dangl 2005). To prevent oxidative damage due to excessive ROS accumulation, especially in chloroplasts, these organelles contain multiple ROS scavenging systems. At least four different isozymes of SOD (one Cu, Zn-SOD, three Fe-SODs) and enzymes of the ascorbate-glutathione cycle capable of reducing oxidized ascorbic acid and glutathione are present in plastids (Asada and Takahashi 1987). However, Davletova et al. (2005) demonstrated that in the absence of the cytosolic H_2O_2-scavenging enzyme ascorbate peroxidase APX1, the entire chloroplastic H_2O_2-scavenging system of *Arabidopsis thaliana* collapses suggesting that the accumulation of ROS in other cellular compartments must be tightly regulated to avoid oxidation of macromolecules. Nevertheless, high levels of ROS from metabolic origins and/or from down-regulation of ROS-scavenging systems can also accumulate in different compartments of the plant cell. This compartmentalization could contribute to the specific functions attributed to ROS (Torres 2010).

Even in fungi, ROS can be generated by Nox and can play an important role in different aspects of fungal development, such as growth and differentiation (Lambeth 2004; Lara-Ortiz et al. 2003). In particular, in *Neurospora crassa*, the onset of the transition from conidia to germination is affected by the cell redox imbalance generated by the formation of singlet oxygen (Lledias et al. 1999). Three different subfamilies of Nox are found in the fungal kingdom (Aguirre et al. 2005). The ROS production catalysed by Nox has also been involved in a defence role against other fungi and during the interaction with the plant (Haedens et al. 2005; Silar 2005). Thus, ROS production by Nox represents an universal signaling system among multi-cellular organisms. In fungi, apart from the respiratory chain, O_2^- appears in reactions involving xanthine oxidase, microsomal monooxygenases, lipoxygenase, cyclooxygenase, and as a result of auto-oxidation of thiols, flavins, quinones, catecholamines, and reduction of the xenobiotic cycle (Georgiou et al. 2006). Differently from plants, ROS are also originated by cytosolic peroxidases, by glucose oxidase, which is a flavin enzyme not typical of plants but usually present in fungi, and by glyoxal oxidases which are proteins associated with plasma membrane. Glucose oxidase, which can be excreted to the extracellular medium, oxidizes glucose to gluconic acid leading to H_2O_2 formation whereas glyoxal oxidase oxidizes small aldehydes and generates H_2O_2 (Gessler et al. 2007).

The efficiency of the cell in maintaining ROS at safe levels mainly depends on the effectiveness of its antioxidant system (Jayashree and Subramanyam 2000; Passi et al. 2005). In fungi, as well as in animal cells, some transcription factors are able to act as sensors of oxidants in the cell (Nguyen et al. 2003; Pinkus et al. 1996). In yeast and filamentous fungi, it has been shown that oxidative stress-related transcription factors such as Yap1, Skn7, Hsf1-2, and Msn2-4 are differentially activated by oxidative stimuli provided by peroxides, diamide, and free radical generators (Moye-Rowley 2003; Reverberi et al. 2008), as well as by antioxidant treatment (Kim et al. 2005). In particular, Yap1 is a nuclear factor localized in the cytoplasm (where it interacts with the export receptor Crm1) which, under oxidative conditions, migrates into the nucleus, where it binds with responsive elements (YRE–TGACTC) of many antioxidant-related genes (*gst*, *sod1*, *sod2*, *cta1*, *ctt*, *trr*, and *txl*). Yap1 does not directly sense ROS but indirectly, i.e. through the activity of Gpx3, which acts as hydroperoxidase and peroxiredoxin (Apel and Hirt 2004).

1.2 The Role of the Peroxisome in ROS Regulation

Several studies indicate that the peroxisomes are crucial knots in the metabolism of ROS, RNS and in β-oxidation with concomitant production of intra- and inter-cellular signaling molecules. Since these molecules are produced during normal cellular metabolism, their role in signaling largely depends on the balance between synthesis, utilisation and degradation (Corpas et al. 2001; Nyathi and Baker 2006), playing a key role in redox-regulated responses to pathogen attack and abiotic stresses. Peroxisomes are ubiquitous organelles with an oxidative-type metabolism which fulfils a range of metabolic functions (Grant and Loake 2000; Neill et al. 2002).

In plants, in addition to the role of leaf peroxisomes in the photorespiratory C2 cycle, specialised plant peroxisomes are also involved in β-oxidation of fatty acids and the glyoxylate cycle in oilseeds as well as ureide metabolism in root nodules of leguminosae (del Rio et al. 2003). In fungi, peroxisomes are involved in the β-oxidation (together with mitochondria) of long-chain fatty acids (Maggio-Hall and Keller 2004), oxalate synthesis and the metabolism of methanol, amines and alkanes, in antibiotics synthesis (Kiel et al. 2009) and probably in the first steps of mycotoxin synthesis (Maggio-Hall et al. 2005). A peculiarity of peroxisomes is their metabolic plasticity, since the enzymatic content of peroxisomes can vary depending on the organism, cell/ tissue-type, and environmental conditions (Bosch et al. 1992; Hayashi et al. 2000). Since peroxisomes contain a set of antioxidant enzymes and molecules, they also contribute to the reduction of oxidative stress (Willekens et al. 1995; del Rio et al. 2002). Nevertheless, as an inevitable consequence of their oxidative-type metabolism, peroxisomes are simultaneously a significant source of ROS. Superoxide anion is generated by membrane proteins via NAD(P)H oxidation and by xanthine oxidase in the peroxisomal matrix (del Rio et al. 2002). The production of H_2O_2 takes place via flavin-containing oxidases and O_2^- dismutation both in matrix and in peroxisomal membrane (Corpas et al. 2001; del Rio et al. 2002; Igamberdiev and Lea 2002). In leaf peroxisomes, O_2^- dismutation is catalysed by Mn-SOD present in the organelle matrix (Palma et al. 2009) and in peroxisomes of oilseed plants it is catalysed by a Cu, Zn-SOD in the matrix and by a membrane-bound Mn-SOD on the cytosolic side of the peroxisomal membrane (Sandalio et al. 1997; Corpas et al. 1998). Hydrogen peroxide produced in peroxisomes is mostly decomposed by CAT isoenzymes localised in the matrix, usually in the form of organised structures called the crystalline core (Willekens et al. 1995; Igamberdiev and Lea 2002). Similarly, in filamentous fungi, as *Penicillium chrysogenum,* peroxisomes have both ROS-detoxifying and ROS-producing enzymes such as catalase/peroxidase KatG, CatC, peroxiredoxin and hydrogen-peroxide-producing oxidases (amine-, formate-oxidase) (Kiel et al. 2009). Further, the Pmp20 family members are peroxisomal antioxidant proteins which have a more important role in decomposing lipid hydroperoxides (Horiguchi et al. 2001). Thus, peroxisomes are cellular organelles where ROS are both generated and detoxified and they appear to be instrumental in redox-homeostasis-mediated defence against abiotic and biotic stresses. Several stresses that generate H_2O_2 as a signaling molecule result in peroxisome proliferation *via* the up-regulation of the peroxins *pex* genes required for the biogenesis of these organelles and import of proteins. Wounding and pathogen attack, which also generate H_2O_2 as a signaling molecule, were demonstrated to induce *pex1* expression in plants (López-Huertas et al. 2000). Further, the transcript level of *ospex5l* in rice seeds was markedly increased by diverse stress signaling molecules including rice blast fungus, fungal elicitor, H_2O_2, ABA, JA, and SA, suggesting that the defence system in plants may be regulated by a complex network of defence signaling pathways including those initiated by plant hormones (Lee et al. 2007). Peroxisomes are prompted to proliferate in *A. nidulans* when the fungus is grown onto viable maize seeds. Organelles proliferation is related to sexual reproduction and to toxin formation (Maggio-Hall et al. 2005).

1.3 ROS in Plant-Fungus Interaction

Considering the important role played by ROS in the plant and fungal metabolism as signaling their involvement in plant-fungus interaction is not so surprising.

The diffusible nature of O_2^- and H_2O_2 make them ideal second messengers for signaling within the cell, and in the case of H_2O_2, which can pass through the cell membrane, inter-cellular signaling. In plant-microbe interactions, ROS participate in plant defence system in a number of ways including acting as signaling agents (Lamb and Dixon 1997; Kotchoni and Gachomo 2006) or by causing reinforcement of cell wall through oxidative cross-linking (Mellersh et al. 2002). H_2O_2 most likely acts within a pathway involving transcription/translation and the expression of wall-associated responses such as the accumulation of fungal inhibiting compounds (Aist and Brushnell 1991). Under pathogen attack, ROS metabolism is regulated by a network which involves at least 152 genes in *A. thaliana* (Mittler et al. 2004). Interestingly, pathogens have developed ways to alter ROS accumulation or signaling to modify plant defences. In both symbiotic and pathogenic relations, the transient production of ROS is detected in the early events of plant-microorganism interactions and ROS production appears as the only common feature of the plant responses. ROS production is typically apoplastic (Levine et al. 1994) and biphasic, with a first unspecific, transitory phase that usually takes place within minutes of the interaction with the pathogen, and a second sustained phase that occurs hours after pathogen attack and that is usually associated with the establishment of the defences and the hypersensitive response (HR; Grant and Loake 2000). However, three phases of ROS accumulation have been observed in some cases, e.g., for *Blumeria graminis* f. sp. *Hordei* infecting barley (Hückelhoven and Kogel 2003) and *Septoria tritici* infecting wheat (Shetty et al. 2003). During the first minutes of interaction between plants and microorganisms, a molecular dialogue involving several signal molecules, takes place at the interface leading to physical interaction. As previously stated, ROS play a signaling function by mediating defence gene activation, by establishing additional defences, by redox control of transcription factors and by interacting with other signaling components as phosphorylation cascades (Kovtun et al. 2000; Mou et al. 2003; Kanda et al. 2010).

1.3.1 ROS in Plant-Fungus Beneficial Interactions

The oxidative burst seems to differ in intensity and length between plant-pathogen and plant-symbiote interactions. This difference could act as a specific signal predefining the host response to the microbe. During mycorrhizal fungus and plant interaction, a dialogue similar to that occurring in plant-Rhizobia interaction (Oldroyd and Downie 2008) is observed. In plant-bacteria interaction, the nodulation factor (NF) production is responsible for ROS production in the infection threads because ROS formation did not occur when *Medicago truncatula* plants were inoculated with rhizobia unable to produce NFs (Ramu et al. 2002). Similarly, ROS play an important role in mycorrhizal symbiosis too: in the *M. truncatula-Glomus intraradices* interaction, an H_2O_2 accumulation is observed in plant cells and this accumulation is hypothesized to be a consequence of activation of a plant Nox in response to the fungus (Salzer et al. 1999). This event is analogous to what occurs during the defence-related HR. In this symbiotic interaction, H_2O_2 accumulation is mostly observed surrounding the arbuscular structures of the fungus. This suggests that ROS play a role in the control of fungal proliferation within the plant (Fester and Hause 2005). Evidence for the participation of ROS and antioxidant systems in the ectomycorrhizal symbiosis has also been found between the fungus *Pisolithus tinctorius* and the plant *Castanea sativa* where the combined action of the antioxidant enzymes SOD and CAT play an important role during the interaction (Baptista et al. 2007). In a genetic screen to identify fungal symbiotic genes that control the mutualistic symbiotic interaction between *Epichloe festucae* and perennial ryegrass, Tanaka et al. (2006) found that an *E. festucae noxA* mutant strain changed the interaction from mutualistic to antagonistic. In wild-type associations *E. festucae* grows systemically in the intercellular spaces as infrequently branched hyphae parallel to the axis of the leaf (Christensen et al. 2002; Tan et al. 2001). Growth of the hyphae is synchronized with that of the host throughout the life cycle of the grass. Inactivation of *noxA* resulted in unregulated hyphal growth in the meristems and in mature tissue of leaves. Plants infected with the *E. festucae noxA* mutant lose apical dominance, become severely stunted and undergo precocious senescence (Tanaka et al. 2006). These studies demonstrate that not only plant-related but also fungal-produced ROS is an important signaling mechanism to maintain the interaction at mutualistic level.

1.3.2 ROS in Pathogenic Interactions

The microbe must produce ROS scavenging enzymes in order to successfully infect the plant or down-regulate the plant ROS producing systems. Fungal pathogens have also developed ways to sense and modify ROS accumulation in host plants

(Torres et al. 2010). In *Fusarium pseudograminearum*, the production of deoxynivalenol (DON) triggers the formation of H_2O_2 in plants (Desmond et al. 2008), but it has been established that H_2O_2 present in the interaction *milieu* is, in turn, a potent inducer of DON biosynthesis (Ponts et al. 2009). DON can inhibit the expression of some proteins involved in plant defence mechanisms and it is also phytotoxic, causing chlorosis, necrosis and apoptosis *in planta* (Lemmens et al. 2005). *Magnaporthe grisea*, a fungus which is the causal agent of rice blast disease, has to overcome the plants innate immunity in order to infect it. The massive production of ROS during the early stages of interaction is part of the plants innate immunity response. To overcome this line of defence, *M. grisea* must be able to counter the oxidative burst by producing ROS scavenging enzymes. Fungal Nox-catalysed production of ROS is also an important signaling mechanism for inducing disease in plant. Both *nox1* and *nox2* mutants of *M. grisea* are non-pathogenic due to a defect in appressorium formation. Δ*nox1* and Δ*nox2* mutants are incapable of causing plant disease because of an inability to bring about appressorium-mediated cuticle penetration. The initiation of rice blast disease, therefore, requires production of O_2^- by the invading pathogen or its presence in the leaf surface (Egan et al. 2007). In *Magnaporthe oryzae* the defence suppressor 1 (DES1) is a novel pathogenicity gene which regulates counter-defences against plant basal defence. DES1 controls the ferrous ions availability thus altering peroxidase activity necessary for scavenging extracellular ROS within host cells (Chi et al. 2009). ROS production also accumulates within the penetration peg that is formed by *Alternaria alternata* during host invasion (Shinogi et al. 2003). In *Claviceps purpurea*, deletion of *nox1* affected spore germination, resistance to oxidative stress and virulence, in fact the *nox1*/*nox2* double mutant shows a loss-of-pathogenicity phenotype. In this pathogen, the deletion of CPTF1, a creB-like transcription factor controlling the expression of CAT, results in a mutant that induces a plant oxidative burst in rye (Nathues et al. 2004). Another example is shown by the biotrophic fungus *Ustilago maydis*. In this pathogen, Yap1, is required for virulence and responsible for preventing the accumulation of H_2O_2 produced by the plant Nox near the hyphae during early stages of biotrophic growth, allowing the fungus to cope with early plant defences (Molina and Kahmann 2007). The modulation of antioxidant responses by the use of transcription factors to modify the host oxidative burst could represent a general strategy for the pathogen to cope with the early plant defences. Thus, successful pathogens have to overcome or suppress the host defence mechanisms, for example, by the secretion of SOD and CAT which convert the ROS into less reactive molecules. In some pathogenic fungi, MnSOD was reported to be involved in fungal virulence since the ability of defending against ROS is a determinant for pathogenicity (Fréalle et al. 2005). Anyway, ROS produced by the plant can in turn deeply affect fungal metabolism. Resistant plants can influence *M. grisea* as they respond to the infection by overproducing and secreting ROS and RNS, both capable of inhibiting cytochrome oxidase in this pathogenic fungus (Gessler et al. 2007).

Thus, ROS signal can affect plant and pathogen metabolism both by acting directly on primary metabolism and also by affecting protein function reacting with sulphur-containing groups, such as the invariant cysteine residues found in protein tyrosine phosphatases and cysteine-rich regions of transcription factors (Finkel 2003; Liu et al. 2005). MAP kinase pathways are also regulated by ROS through the signaling intermediates, thioredoxin and glutaredoxin, which serve as both sensors and transducers of oxidative stress (Finkel 2000; Fujino et al. 2006). The functional MLO protein [Mutation-induced recessive alleles *(mlo)*] of the barley *mlo* locus has been proposed to be a central negative regulator of defence mechanisms and cell death in plants. Loss of MLO function leads to unspecific *Blumeria graminis* f.sp. *Hordei* resistance in barley where induces spontaneous cell death and chlorophyll degradation (Piffanelli et al. 2002). Powdery mildew-resistant *mlo* mutants (*mlo5*) accumulate H_2O_2 at sites of pathogen attack more frequently, earlier and apparently to higher concentrations (Huckelhoven et al. 1999; Piffanelli et al. 2002). Additionally, *mlo* expression is triggered by pathogen attack and by oxidative stress, suggesting that MLO is both a putative sensor and an effector of the cellular redox status (Piffanelli et al. 2002; Kim et al. 2002). MLO, by monitoring Ca^{2+} and ROS activities, antagonizes non-specific defence against *B. graminis* (Hückelhoven and Kogel 2003).

Furthermore, the oxidative stressors present at the plant-fungus interface can trigger the activation of plant and fungal LOX and, in turn, can lead to the synthesis and release of oxylipins (Reverberi et al. 2007). On the other hand, ROS can generate lipid derivatives by non-enzymatic oxygenation that can produce membrane damage or can function as signaling molecules. The free fatty acid hydroperoxides which are produced enzymatically can also be substrates of Fenton-like reactions, leading, similarly, to the production of alkoxy radicals (Halliwell and Gutteridge 2007) and thus enhancing the free radical mediated lipid peroxidation.

2 Oxylipin Biosynthesis in Plant and Fungi

2.1 Are Peroxisomes a Site for Oxylipin Synthesis?

Some authors (Nyathi and Baker 2006) report that the functions of peroxisomes in plants include not only regulation of ROS and catabolism of fatty acids, but also action on a range of other substrates that give rise to signaling molecules, including indole acetic acid (IAA), salicylic acid and the oxylipin-related jasmonates. In the canonical pathway, 12-oxo-phytodienoic acid (OPDA) is synthesized in the plastid and transferred into peroxisome where it undergoes reduction to 3-oxo-2-(20-pentenyl)-cyclopentane-1-octanoic acid and acyl chain shortening by three rounds of β-oxidation to yield jasmonic acid. The oxylipin profiling of the defective ABC transporter *cts* mutant showed a dramatic reduction in both basal and wound-induced jasmonic acid in leaves. This suggests that CTS is involved in importing OPDA (or its CoA ester) into peroxisomes in parallel with a second, probably passive, transport pathway (Nyathi and Baker 2006). In yeast and fungi oxylipins can be produced into peroxisome in different manners. The non-enzymatic generation of oxylipins is indicated by the presence into peroxisome of the alcohol oxidase which contains FAD (or flavin semiquinone) and peroxisomal catalase which contains heme as a cofactor. In stress conditions, these cofactors or iron molecules can be released from the protein and can catalyze the formation of hydroxyl radicals from H_2O_2, which then attack peroxisomal membrane, resulting in the generation of lipid hydroperoxides (Horiguchi et al. 2001). In the interaction between maize seeds and *Aspergillus flavus*, peroxisome hyper-proliferation modify the ROS and oxylipins formation at the interface. This modification alters plant defences and fungal metabolism (Reverberi M., 2012, personal communication). More recently, many antioxidant and ROS-generating enzymes, such as a putative dioxygenase, a formate oxidase (Fod1) and a putative flavin-containing monooxygenase have been found also into fungal peroxisomes (*P. chrysogenum;* Kiel et al. 2009).

This study can provide the basis for a more direct approach for uncovering the real contribution of peroxisome to the creation of an oxylipin signature.

2.2 Oxylipins Biosynthesis-Related Pathways

Oxylipins represent a class of compounds of several hundred of distinct oxidized lipophilic molecules that are derived from lipid oxidation (Andreou and Feussner 2009). The initial oxidation of unsaturated fatty acids may either occur by chemical oxidation or by enzymatic peroxidation by dioxygenases (DOX), lipoxygenases (LOX) or CYP74 enzymes (Schneider et al. 2007; Stumpe and Feussner 2006). A large variety of oxylipin classes are generated by an array of alternative reactions further converting hydroperoxy fatty acids. The structural diversity of oxylipins is further increased by their occurrence either as free fatty acid derivatives or as esters in complex lipids. Oxylipins exhibit crucial biological activities as signals of intra- and inter-cellular communication in plants, vertebrates, invertebrates and fungi. Furthermore, oxylipins may perform various biological roles ranging from antimicrobials to hormone-like functions (Göbel and Feussner 2009). Plant oxylipins are derived from the oxidation of the most abundant polyunsaturated fatty acids (PUFAs) in plants: linoleic acid [LA, 18:2(n − 6)], α-linolenic acid [ALA, 18:3(n − 3)] and roughanic acid [16:3(n − 3)] in the plastids (Mosblech et al. 2009). Fungal oxylipins are primarily derived from oleic (C18:1), linoleic and linolenic acids (Tsitsigiannis and Keller 2007). Oxylipins formed in plants include hydroperoxy-, hydroxy-, oxo- and epoxy-fatty acids, divinyl ethers, volatile aldehydes and the plant hormones, jasmonates (Grechkin 1998). The first enzymes, linoleate diol synthases, involved in oxylipin biosynthesis in fungi are homologous to the mammalian prostaglandin H synthases or cyclooxygenases (COX) (Hornsten et al. 1999). In fungi exist other potential routes of oxylipin biosynthesis, indicated by the presence of lipoxygenases and glutathione transferases (Cristea et al. 2003; Oakley 2005). Additionally, through a non-enzymatic process that is initiated by oxidative stress and free radical-catalyzed mechanisms, linolenic acid can be converted to an array of prostaglandin-like compounds called phytoprostanes in all aerobic PUFA-containing organisms. In plants, oxylipins are involved in different events: they stimulate signals implicated in the onset of plant defences against pathogens and pests, have antimicrobial effects, provide building units of physical barriers against pathogen invasion, regulate plant cell death and have a major role in the formation of phytohormones and in senescence (La Camera et al. 2004). Also in fungi, by-products of LOX and DOX enzymatic activities play a relevant role in different events, that is by modulating morphogenesis and secondary metabolism (e.g., mycotoxins biosynthesis) (Oliw 2002; Tsitsigiannis and Keller 2007).

2.3 Oxylipins in Plant-Fungi Interactions

The oxyilipins and the PUFA produced by plants can control, by a cross-kingdom cross-talk, the synthesis of toxins by pathogenic fungi and, *vice versa,* fungi may alter plant-defence responses by releasing plant-like oxylipins (e.g. jasmonic acid) into the host cell (Christensen and Kolomiets 2010). One of the first example of an oxylipin-based cross-talk between the plant and fungal-like organisms is provided by the function of arachidonic acid in the interaction between *Phythophthora infestans* and potato tuber. In fact, arachidonic acid is released by the oomycete plasma membrane probably through the action of secreted plant lipases. This early PAMP is then translocated into plant cell and therein transformed by a potato typical 5-arachidonate-LOX into 19-S-hydroperoxyeicosatetraenoic acid (19-S-HPETE); this, *ante litteram*, oxylipin, acting as an elicitor, is able to trigger HR in resistant potato (Castoria et al. 1992). In turn, plant volatile organic compounds, such as methyl salicylate and oxylipins (e.g. green leaf volatiles), trigger sporulation (Hountondji et al. 2006) and modulate toxin biosynthesis in pathogenic fungi (Gao and Kolomiets 2009), apparently by replacing fungal endogenous oxylipins (Mita et al. 2007; Tsitsigiannis and Keller 2007; Reverberi et al. 2010a). But how endogenous oxylipins can affect fungal development and secondary metabolism? Probably, as suggested by Tsitsigiannis and Keller (2007) oxylipins act in an autochrine/parachrine fashion for modulating, through a GPCR/FadA/PkaA pathway of signaling, these events. Changes in the lifestyle of fungi are often due to the "perception" of signals coming from the host, which trigger morphological transitions (Noverr and Huffnagle 2004) or the production of virulence factors (Desmond et al. 2008) and toxic compounds (Lillehoj 1991). Oxylipin perception in plants, apart considering the well described Ja-Ile/Jaz1/COI1 system (Gfeller et al. 2010), is rather elusive. Even in plants GPCR could have a role in oxylipin perception even if a unique receptor class has not yet described. In humans, the peroxisome proliferators activated receptors – PPAR – are activated following PUFA or oxylipin binding to their ligand binding domain (LBD); then PPAR migrate into the nucleus, where, by recognising PPRE (PPAR responsive elements), control the transcription of many genes involved into lipid metabolism and inflammatory responses (Dinarello 2010). Recently, the *Xenopus laevis* PPARα (xPPARα) nuclear receptor has been inserted in *Nicotiana tabacum*. It has been demonstrated that xPPARα is functional in plants and its expression in tobacco leads to changes in general lipid metabolism and peroxisomal proliferation as reported in animal cells. This may indicate that there is an endogenous ligand in tobacco cells able to activate xPPARα and, in turn, an endogenous FA-related receptor (Nila et al. 2006).

Intriguingly, since the oxylipins are formed following ROS formation-related to biotic and abiotic stresses, they could represent a common language between organisms devoid of an oral language. For instance, plants do not sense insects only by the physical injury caused by herbivory. Plant-insect interactions are mediated by complex chemical signaling involving volicitin *inter alia*. Volicitin, N-17(S)-hydroxylinolenoyl-L glutamine, was isolated from oral secretions of beet army worm and is composed of 17(S)-hydroxy linolenic acid and L-glutamine. Volicitin triggers the release of a specific blend of volatiles by the plant and this attracts parasitic insects to the herbivore (Weber 2002). An oxylipin-mediated cross-talk was demonstrated to occur in the interaction between *A. nidulans* and maize. In this scenario (i.e. the relationship between fungus and seed), the mycotoxigenic fungus secretes oxylipins, which, acting as PAMP, elicit a defence response by the seed; this response is also based on the activation of the plant LOX pathway and the production of lipoperoxides (Fabbri et al. 1983; Gao et al. 2009). In turn, the lipid hydroperoxides released by the seed, probably mimicking fungal oxylipins, induce several responses in the pathogenic fungus, including the activation of secondary metabolism and changes in morphogenesis (Brodhagen et al. 2008). In fact, hydroperoxides formed from linoleic and linolenic acid (HPOD/TE) by the action of maize LOX, differentially influence toxin biosynthesis in *A. flavus* and *A. nidulans*, i.e. 9S-HPODE stimulates, whereas 13S-HPODE inhibits, their biosynthesis. Actually, the plant-produced oxylipins which show a regulatory effect on fungal metabolism and differentiation are diverse (Gao and Kolomiets 2009 and references therein). In fact, one of the extracellular signals known to regulate both asexual and sexual spore development in fungi is a mixture of oxylipins collectively called psi – precocious sexual inducer – factors that are proposed as hormone precursors which repress conidiation and promote precocious sexual development (Champe and el-Zayat 1989). *A. nidulans* Δ*ppoA*Δ*ppoC* and Δ*ppoA*Δ*ppoB*Δ*ppoC* (psi producing oxidases – ppo) mutants are unable to produce sterigmatocystin *in vitro* or *in planta*. The psi factors chemically resembles plant 8-oxylipins phytoprostane B1, 13S- and 9S-HPOD/TE. Thus plant oxylipins can affect fungal development on the basis of a metabolic "fake". The effect of the disruption of an arachidonate 15-*lox* (Ao*loxA*) in *A. ochraceus* shows a pleiotropic effect: the oxylipin-defective mutant displays altered development, does not trigger 9-oxylipins in contaminated wheat seeds and consequently the seeds do not support OTA biosynthesis (Reverberi et al. 2010). During the storage of wheat seeds, *A. ochraceus* initially colonises the germ and then differentiates conidia. The ΔAoloxA strain presents delayed conidia formation, copious sclerotia production, and hyphae distribution patterns that involve the whole seed's surface, i.e. the hyphal growth it is not limited to the germ as in the WT strain. *In vitro*, the ΔAoloxA strain shows that oxylipin biosynthetic pathways switch from 13-HPODE to a prevalent formation of 7,8- and 8,13-DiHODE. It has been shown in *M. grisea* and other Aspergilli (i.e., *A. fumigatus*) that 7,8-DiHODE and 5,8-DiHODE, which are derived from the fatty acid dioxygenase enzyme PpoA (Cristea et al. 2003; Garscha et al. 2007), regulated the ratio between asexual to sexual spores or sclerotia formation (Tsitsigiannis and Keller 2007; Horowitz Brown et al. 2008). *In vitro*, a large number of sclerotia, which are considered to be vestigial sexual cleistothecia (Yager 1992; Geiser et al. 1996), are formed by the ΔAoloxA mutant, possibly due to an increase of such diol formation. *In vivo*, LOX activity of wheat seeds is stimulated after contact with *A. ochraceus* WT; during this period a steady-state increase of 9-/13-HODE ratio was detected. This reaction occurs in response to *A. ochraceus* contamination and it is similar to that observed in the *A. nidulans*–maize interaction, which is likely to be mediated by fungal oxylipins. In fact, in both uninfected or ΔAoloxA strain-inoculated seeds, a lower LOX activity is detected and the ratio between 9- and 13-HODE is maintained constantly around the unitary value. In wheat seeds, the expression of PR1 mRNA, which is an index of the onset of plant defence responses (Dixon et al. 1994), is enhanced after contamination with the WT and not with the oxylipin-defective mutant. In fact, 9-HPODE and PR1 can exert an antimicrobial effect against *A. ochraceus* as suggested by the results obtained in other OTA-producing fungi such as *Aspergillus carbonarius* (Mita et al. 2007). It can be inferred that the oxylipins formed by the fungus elicit a plant defence response through the formation of plant oxylipins. Furthermore, as already observed in *A. flavus* (Burow et al. 1997), the increase of 9- HPODE seems to positively correlate with OTA biosynthesis. In contrast, levels of 13-HPODE comparable or higher than 9-HPODE abrogated the effects on the OTA formation. How 9-HPODE promotes OTA biosynthetic genes in this fungus has not yet elucidated even if receptors expressed on the fungal membrane can respond differently to 9- and 13-HPODE produced by plants (Tsitsigiannis and Keller 2007). Intriguingly, PPAR, as previously described, are differentially activated by 9 and 13-HODE (Zuo et al. 2006). In filamentous fungi and in plants, where no obvious PPAR-like factors are reported, the peroxin Pex11 shows extensive amino acid sequence similarity to the ligand-binding domain (LBD) of PPAR (Barnett et al. 2000). By analogy, Pex11 might contain a binding site for fatty acids (van Roermund et al. 2000) which enables it to “perceive” oxylipins and, in turn, to activate peroxisome proliferation and subsequently its extremely various metabolism.

3 Concluding Remarks

Plant and fungi communicate at various levels and “words”. Firstly, following a chemical oxylipin-based contact, plant responds to the fungus by producing ROS, also at peroxisomal level. Also fungi produce ROS subsequently or concomitantly to the host perception. The ROS produced at the interface act as signals able to trigger attack and counter-attack responses by both actors. The chronic ROS synthesis, typical in biotrophic steps of infection, can prompt the non-enzymatic lipid peroxidation, which, in turn, can trigger the modulated enzymatic ones. Thus a new set of oxylipins is produced and released in both direction.

In this scenario the "generic" ROS words can be converted in "specific" oxylipin language to modulate symbiotic and pathogenic interactions between plants and fungi.

References

Aguirre J, Rios-Momberg M, Hewitt D, Hansberg W (2005) Reactive oxygen species and development in microbial eukaryotes. Trends Microbiol 25:111–118

Aist JR, Brushnell WR (1991) Invasion of plants by powdery mildew fungi, and cellular mechanisms of resistance. In: Cole GT, Hoch HC (eds) The fungal spore and disease interaction in plants and animals. Plenum Press, New York, pp 321–345

Andreou AZ, Feussner I (2009) Lipoxygenases – structure and reaction mechanism. Phytochemistry 70:1485–1503

Apel K, Hirt H (2004) Reactive oxygen species: metabolism, oxidative stress, and signal transduction. Annu Rev Plant Biol 55:373–399

Asada K, Takahashi M (1987) Production and scavenging of active oxygen in photosynthesis. In: Kyle DJ, Osmond CB, Arntzen CJ (eds) Photoinhibition. Elsevier, Amsterdam, pp 227–287

Banerjee AK, Mandal A, Chanda D, Chakraborti S (2003) Oxidant, antioxidant and physical exercise. Mol Cell Biochem 253:307–312

Baptista P, Martins A, Pais MS, Tavares RM, Lino-Neto T (2007) Involvement of reactive oxygen species during early stages of ectomycorrhiza establishment between *Castanea sativa* and *Pisolithus tinctorius*. Mycorrhiza 17:185–193

Barnett P, Tabak HF, Hettema EH (2000) Nuclear hormone receptors arose from pre-existing protein modules during evolution. Trends Biochem Sci 25:227–228

Bartels D (2001) Targeting detoxification pathways: an efficient approach to obtain plants with multiple stress tolerance? Trends Plant Sci 6:284–286

Bedard K, Krause KH (2007) The NOX family of ROS-generating NADPH oxidases: physiology and pathophysiology. Physiol Rev 87:245–313

Bedard K, Lardy B, Krause KH (2007) NOX family NADPH oxidases: not just in mammals. Biochimie 89:1107–1112

Bosch VDH, Schutgens RBH, Wanders RJA, Tager JM (1992) Biochemistry of peroxisomes. Annu Rev Biochem 61:157–197

Brodhagen M, Tsitsigiannis DI, Hornung E, Goebel C, Feussner I, Keller NP (2008) Reciprocal oxylipin-mediated cross-talk in the Aspergillus-seed pathosystem. Mol Microbiol 67 (2):378–391

Burow GB, Nesbitt TC, Dunlap J, Keller NP (1997) Seed lipoxygenase products modulate Aspergillus mycotoxin biosynthesis. Mol Plant Microbe Interact 10:380–387

Castoria R, Fanelli C, Fabbri AA, Passi S (1992) Metabolism of arachidonic acid involved in its eliciting activity in potato tuber. Physiol Mol Plant Pathol 41:127–137

Champe SP, el-Zayat AA (1989) Isolation of a sexual sporulation hormone from *Aspergillus nidulans*. J Bacteriol 171:3982–3988

Chi M-H, Park S-Y, Kim S, Lee Y-H (2009) A novel pathogenicity gene is required in the Rice Blast Fungus to suppress the basal defences of the host. PLoS Pathog 5:e1000401

Christensen SA, Kolomiets MV (2010) The lipid language of plant-fungal interactions. Fungal Genet Biol 48(1):4–14

Christensen MJ, Bennett RJ, Schmid J (2002) Growth of Epichloe/Neotyphodium and p-endophytes in leaves of Lolium and Festuca grasses. Mycol Res 106:93–106

Corpas FJ, Barroso JB, Sandalio LM, DiStefano S, Palma JM, Lupianez JA, del Rio LA (1998) A dehydrogenase-mediated recycling system of NADPH in plant peroxisomes. Biochem J 330:777–784

Corpas FJ, Barroso JB, del Rìo LA (2001) Peroxisomes as a source of reactive oxygen species and nitric oxide signal molecules in plant cells. Trends Plant Sci 6:145–150

Cristea M, Osbourn AE, Oliw EH (2003) Linoleate diol synthase of the rice blast fungus *Magnaporthe grisea*. Lipids 38:1275–1280

Davletova S, Rizhsky L, Liang H, Shenggiang Z, Oliver DJ, Coutu J, Shulaev V, Schlauch K, Mittler R (2005) Cytosolic ascorbate peroxidase 1 is a central component of the reactive oxygen gene network of Arabidopsis. Plant Cell 17:268–281

del Rio LA, Corpas FJ, Sandalio LM, Palma JM, Gomez M, Barroso JB (2002) Reactive oxygen species, antioxidant systems and nitric oxide in peroxisomes. J Exp Bot 53:1255–1272

del Rio LA, Corpas FJ, Sandalio LM, Palma JM, Barroso JB (2003) Plant peroxisomes, reactive oxygen metabolism and nitric oxide. IUBMB Life 55:71–81

Desmond OJ, Manners JM, Stephens AE, Maclean DJ, Schenk PM, Gardiner DM, Munn AL, Kazan K (2008) The *Fusarium* mycotoxin deoxynivalenol elicits hydrogen peroxide production, programmed cell death and defence responses in wheat. Mol Plant Pathol 9(4):435–445

Dinarello CA (2010) Anti-inflammatory agents: present and future. Cell 140(6):935–950

Dixon RA, Harrison MJ, Lamb CJ (1994) Early events in the activation of plant defence responses. Annu Rev Phytopathol 32:479–501

Egan MJ, Wang Z-Y, Jones MA, Smirnoff N, Talbot NJ (2007) Generation of reactive oxygen species by fungal NADPH oxidases is required for rice blast disease. Proc Natl Acad Sci USA 104(28):11772–11777

Fabbri AA, Fanelli C, Panfili G, Passi S, Fasella P (1983) Lipoperoxidation and aflatoxin biosynthesis by *Aspergillus parasiticus* and *A. flavus*. J Gen Microbiol 129:3447–3452

Fester T, Hause G (2005) Accumulation of reactive oxygen species in arbuscular mycorrhizal roots. Mycorrhiza 15:373–379

Finkel T (2000) Redox-dependent signal transduction. FEBS Lett 476:52–54

Finkel T (2003) Oxidant signals and oxidative stress. Curr Opin Cell Biol 15:247–254

Fréalle E, Noel C, Viscogliosi E, Camus D, Dei-Cas E, Delhaes L (2005) Manganese superoxide dismutase in pathogenic fungi: an issue with pathophysiological and phylogenetic involvements. FEMS Immunol Med Microbiol 45:411–422

Fujino G, Noguchi T, Takeda K, Ichijo H (2006) Thioredoxin and protein kinases in redox signaling. Semin Cancer Biol 16:427–435

Gao XQ, Kolomiets MV (2009) Host-derived lipids and oxylipins are crucial signals in modulating mycotoxin production by fungi. Toxin Rev 28:79–88

Gao XQ, Brodhagen M, Isakeit T, Brown SH, Göbel C, Betran J, Feussner I, Keller NP, Kolomiets MV (2009) Inactivation of the lipoxygenase ZmLOX3 increases susceptibility of maize to Aspergillus spp. Mol Plant Microbe Interact 22(2):222–231

Garscha U, Jerneren F, Chung DW, Keller NP, Hamberg M, Oliw EH (2007) Identification of dioxygenases required for *Aspergillus* development. J Biol Chem 282:34707–34718

Geiser DM, Timberlake WE, Arnold ML (1996) Loss of meiosis in *Aspergillus*. Mol Biol Evol 13:809–817

Georgiou CD, Patsoukis N, Papapostolou I, Zervoudakis G (2006) Sclerotial metamorphosis in filamentous fungi is induced by oxidative stress. Integr Comp Biol 46:1–22

Gessler NN, Averyanov AA, Belozerskaya TA (2007) Reactive oxygen species in regulation of fungal development. Biochemistry (Moscow) 72(10):1091–1109

Gfeller A, Liechti R, Farmer EE (2010) Arabidopsis jasmonate signaling pathway. Sci Signal 3 (109):cm4

Göbel C, Feussner I (2009) Methods for the analysis of oxylipins in plants. Phytochemistry 70:1485–1503

Grant JJ, Loake GJ (2000) Role of reactive oxygen intermediates and cognate redox signaling in disease. Plant Physiol 124:21–29

Grechkin A (1998) Recent developments in biochemistry of the plant lipoxygenase pathway. Prog Lipid Res 37:317–352

Haedens V, Malagnac F, Silar P (2005) Genetic control of an epigenetic cell degeneration syndrome in *Podospora anserina*. Fungal Genet Biol 42:564–577

Halliwell B, Gutteridge JMC (2007) In: Halliwell B, Gutteridge JMC (eds) Free radicals in biology and medicine. Oxford University Press, Oxford, p 851

Hayashi M, Toriyama K, Kondo M, Kato A, Mano S, De Bellis L, Hayashi-Ishimaru Y, Yamaguchi K, Hayashi H, Nishimura M (2000) Functional transformation of plant peroxisomes. Cell Biochem Biophys 32:295–304

Horiguchi H, Yurimoto H, Kato N, Sakai Y (2001) Antioxidant system within yeast peroxisome: biochemical and physiological characterization of cbpmp20 in the methylotrophic yeast *Candida boidinii*. J Biol Chem 276(17):14279–14288

Hornsten L, Su C, Osbourn AE, Garosi P, Hellman U, Wernstedt C, Oliw EH (1999) Cloning of linoleate diol synthase reveals homology with prostaglandin H synthases. J Biol Chem 274:28219–28224

Horowitz Brown S, Zarnowski R, Sharpee WC, Keller NP (2008) Morphological transitions governed by density dependence and lipoxygenase activity in *Aspergillus flavus*. Appl Environ Microbiol 74(18):5674–5685

Hountondji FCC, Hanna R, Sabelis MW (2006) Does methyl salicylate, a component of herbivore-induced plant odour, promote sporulation of the mite-pathogenic fungus *Neozygites tanajoae*? Exp Appl Acarol 39:63–74

Hückelhoven R, Kogel K-H (2003) Reactive oxygen intermediates in plant microbe interactions: who is who in powdery mildew resistance? Planta 216:891–902

Huckelhoven R, Fodor J, Preis C, Kogel K-H (1999) Hypersensitive cell death and papilla formation in barley attacked by the powdery mildew fungus are associated with H_2O_2 but not with salicylic acid accumulation. Plant Physiol 119:1251–1260

Igamberdiev AU, Lea PJ (2002) The role of peroxisomes in the integration of metabolism and evolutionary diversity of photosynthetic organisms. Phytochemistry 60:651

Jayashree T, Subramanyam C (2000) Oxidative stress as a prerequisite for aflatoxin production by *Aspergillus parasiticus*. Free Radic Biol Med 29:981–985

Kanda AK, Andrio E, Marino D, Pauly N, Dunand C (2010) Reactive oxygen species during plant-microorganism early interactions. J Integr Plant Biol 52(2):195–204

Karpinski S, Gabrys H, Mateo A, Karpinska B, Mullineaux PM (2003) Light perception in plant disease defence signaling. Curr Opin Plant Biol 6:390–396

Kiel JAKW, van den Berg MA, Fusetti F, Poolman B, Bovenberg RAL, Veenhuis M, van der Klei IJ (2009) Matching the proteome to the genome: the microbody of penicillin-producing *Penicillium chrysogenum* cells. Funct Integr Genomics 9:167–184

Kim MC, Panstruga R, Elliott C, Muller J, Devoto A, Yoon HW, Park HC, Cho MJ, Schulze-Lefert P (2002) Calmodulin interacts with MLO protein to regulate defence against mildew in barley. Nature 416:447–451

Kim JH, Campbell BC, Yu J, Mahoney N, Chan KL, Molyneux RJ, Bhatnagar D, Cleveland TE (2005) Examination of fungal stress response genes using *Saccharomyces cerevisiae* as a model system: targeting genes affecting aflatoxin biosynthesis by *Aspergillus flavus* link. Appl Microbiol Biotechnol 67:807–815

Kotchoni SW, Gachomo EW (2006) The reactive oxygen species network pathways: an essential prerequisite for perception of pathogen attack and the acquired disease resistance in plants. J Biosci 31(3):389–404

Kovtun Y, Chiu W-L, Tena G, Sheen J (2000) Functional analysis of oxidative stress-activated mitogen-activated protein kinase cascade in plants. Proc Natl Acad Sci USA 97:2940–2945

La Camera S, Gouzerh G, Dhondt S, Hoffmann L, Fritig B, Legrand M, Heitz T (2004) Metabolic reprogramming in plant innate immunity: the contributions of phenylpropanoid and oxylipin pathways. Immunol Rev 198:267–284

Lamb C, Dixon RA (1997) The oxidative burst in plant disease resistance. Annu Rev Plant Physiol Plant Mol Biol 48:251–275

Lambeth JD (2004) Nox enzymes and the biology of reactive oxygen. Nat Rev Immunol 4:181–189

Lara-Ortiz T, Riveros-Rosas H, Aguirre J (2003) Reactive oxygen species generated by microbial NADPH oxidase NOXA regulate sexual development in *Aspergillus nidulans*. Mol Microbiol 50:1241–1255

Lee JR, Park S-C, Kim M-H, Jung JH, Shin MR, Lee DH, Cheon MG, Park Y, Hahm KS, Lee SY (2007) Antifungal activity of rice Pex5p, a receptor for peroxisomal matrix proteins. Biochem Biophys Res Commun 359:941–946

Lemmens M, Sholz U, Berthiller F, Dall'Asta C, Koutnik A, Schuhmacher R, Adam G, Buerstmayr H, Mesterhazy A, Krska R, Ruckenbauer P (2005) The ability to detoxify the mycotoxin deoxynivalenol colocalizes with a major quantitative trait locus for *Fusarium* head resistance in wheat. Mol Plant Microbe Interact 18(12):1318–1324

Levine A, Tenhaken R, Dixon RA, Lamb CJ (1994) H_2O_2 from the oxidative burst orchestrates the plant hypersensitive disease resistance response. Cell 79:583–593

Lillehoj EB (1991) Aflatoxin: an ecologically elicited genetic activation signal. In: Smith JE, Henderson RS (eds) Mycotoxins and animal foods. CRC Press, Boca Raton, pp 2–30

Liu H, Colavitti R, Rovira II, Finkel T (2005) Redox-dependent transcriptional regulation. Circ Res 97:967–974

Lledias F, Rangel P, Hansberg W (1999) Singlet oxygen is part of a hyperoxidant state generated during spore germination. Free Radic Biol Med 26:1396–1404

López-Huertas E, Charlton WL, Johnson B, Graham IA, Baker A (2000) Stress induces peroxisome biogenesis genes. EMBO J 19:6770–6777

Maggio-Hall LA, Keller NP (2004) Mitochondrial β-oxidation in *Aspergillus nidulans*. Mol Microbiol 54(5):1173–1185

Maggio-Hall LA, Wilson RA, Keller NP (2005) Fundamental contribution of beta-oxidation to polyketide mycotoxin production in planta. Mol Plant Microbe Interact 18(8):783–793

Mellersh DG, Foulds IV, Higgins VJ, Heath CM (2002) H_2O_2 plays different roles in determining penetration failure in three diverse plant-fungal interactions. Plant J 29:257–268

Mita G, Fasano P, De Domenico S, Perrone G, Epifani F, Iannacone R, Casey R, Santino A (2007) 9-lipoxygenase metabolism is involved in the almond/*Aspergillus carbonarius* interaction. J Exp Bot 58(7):1803–1811

Mittler R, Vanderauwera S, Gollery M, Van Breusegem F (2004) The reactive oxygen gene network of plants. Trends Plant Sci 9:490–498

Molina L, Kahmann R (2007) An *Ustilago maydis* gene involved in H_2O_2 detoxification is required for virulence. Plant Cell 19:2293–2309

Mosblech A, Feussner I, Heilmann I (2009) Oxylipins: structurally diverse metabolites from fatty acid oxidation. Plant Physiol Biochem 47:511–517

Mou Z, Fan W, Dong X (2003) Inducers of plant systemic acquired resistance regulate NPR1 function through redox changes. Cell 113:935–944

Moye-Rowley WS (2003) Regulation of transcriptional response to oxidative stress in fungi: similarities and differences. Eukaryot Cell 2:381–389

Nathues E, Joshi S, Tenberge KB, von den Driesch M, Oeser B, Baumer N, Mihlan M, Tudzynski P (2004) CPTF1, a CREB-like transcription factor, is involved in the oxidative stress response in the phytopathogen Claviceps purpurea and modulates ROS level in its host Secale cereale. Mol Plant Microbe Interact 17:383–393

Neill S, Desikan R, Hancock J (2002) Hydrogen peroxide signaling. Curr Opin Plant Biol 5:388–395

Nguyen T, Sherratt PJ, Pickett CB (2003) Regulatory mechanisms controlling gene expression mediated by the antioxidant response element. Annu Rev Pharmacol Toxicol 43:233–260

Nila AG, Sandalio LM, Lòpez MG, Gòmez M, del Rìo L, Gòmez-Lim MA (2006) Expression of a peroxisome proliferator-activated receptor gene (xPPARα) from *Xenopus laevis* in tobacco (*Nicotiana tabacum*) plants. Planta 224:569–581

Noverr MC, Huffnagle GB (2004) Regulation of *Candida albicans* morphogenesis by fatty acid metabolites. Infect Immun 72:6206–6210

Nyathi Y, Baker A (2006) Plant peroxisomes as a source of signaling molecules. Biochim Biophys Acta 1763:1478–1495

Oakley AJ (2005) Glutathione transferases: new functions. Curr Opin Struct Biol 15:716–723

Oldroyd GE, Downie JA (2008) Coordinating nodule morphogenesis with rhizobial infection in legumes. Annu Rev Plant Biol 59:519–546

Oliw EH (2002) Plant and fungal lipoxygenases. Prostaglandins Other Lipid Mediat 68–69:313–323

Overmyer K, Brosché M, Kangasjarvi J (2003) Reactive oxygen species and hormonal control of cell death. Trends Plant Sci 8(7):335–342

Palma JM, Corpas FJ, del Río LA (2009) Proteome of plant peroxisomes: new perspectives on the role of these organelles in cell biology. Proteomics 9:2301–2312

Passi S, Ricci R, Aleo E, Cocchi M (2005) Oxidative stress, aging and aging-related diseases. Progr Nutr 7:3–22

Piffanelli P, Zhou F, Casais C, Orme J, Jarosch B, Schaffrath U, Collins NC, Panstruga R, Schulze-Lefert P (2002) The barley MLO modulator of defence and cell death is responsive to biotic and abiotic stress stimuli. Plant Physiol 129:1076–1085

Pinkus R, Weiner LM, Daniel V (1996) Role of oxidants and antioxidants in the induction of AP-1, NF-kB, and glutathione S-transferase gene expression. J Biol Chem 271:13422–13429

Ponts N, Couedelo L, Pinson-Gadais L, Verdal-Bonnin M-N, Barreau C, Richard-Forget F (2009) Fusarium response to oxidative stress by H_2O_2 is trichothecene chemotype-dependent. FEMS Microbiol Lett 293(2):255–262

Ramu SK, Peng H-M, Cook DR (2002) Nod factor induction of reactive oxygen species production is correlated with expression of the early nodulin gene rip1 in *Medicago truncatula*. Mol Plant Microbe Interact 15:522–528

Reverberi M, Zjalic S, Punelli F, Ricelli A, Fabbri AA, Fanelli C (2007) Apyap1 affects aflatoxin biosynthesis during *Aspergillus parasiticus* growth in maize seeds. Food Addit Contam 24 (10):1070–1075

Reverberi M, Zjalic S, Ricelli A, Punelli F, Camera E, Fabbri C, Picardo M, Fanelli C, Fabbri AA (2008) Modulation of antioxidant defence in *Aspergillus parasiticus* is involved in aflatoxin biosynthesis: a role for the ApyapA gene. Eukaryot Cell 7(6):988–1000

Reverberi M, Punelli F, Scarpari M, Camera E, Zjalic S, Ricelli A, Fanelli C, Fabbri AA (2010) Lipoperoxidation affects ochratoxin A biosynthesis in *Aspergillus ochraceus* and its interaction with wheat seeds. Appl Microbiol Biotechnol 85:1935–1946

Salzer P, Corbiere H, Boller T (1999) Hydrogen peroxide accumulation in *Medicago truncatula* roots colonized by the arbuscular mycorrhiza-forming fungus *Glomus intraradices*. Planta 208:319–325

Sandalio LM, López-Huertas E, Bueno P, del Río LA (1997) Immunocytochemical localization of copper, zinc superoxide dismutase in peroxisomes from watermelon (*Citrullus vulgaris* Schrad.) cotyledons. Free Radic Res 26:187–194

Schneider C, Pratt DA, Porter NA, Brash AR (2007) Control of oxygenation in lipoxygenase and cyclooxygenase catalysis. Chem Biol 14:473–488

Shetty NP, Kristensen BK, Newman M-A, Møller K, Gregersen PL, Jørgensen HJL (2003) Association of hydrogen peroxide with restriction of *Septoria tritici* in resistant wheat. Physiol Mol Plant Pathol 62:333–346

Shinogi T, Suzuki T, Kurihara T, Narusaka Y, Park P (2003) Microscopic detection of reactive oxygen species generation in the compatible and incompatible interactions of *Alternaria alternata* Japanese pear pathotype and host plants. J Gen Plant Pathol 69:7–16

Silar P (2005) Peroxide accumulation and cell death in filamentous fungi induced by contact with a contestant. Mycol Res 109:137–149
Stumpe M, Feussner I (2006) Formation of oxylipins by CYP74 enzymes. Phytochem Rev 5:347–357
Sumimoto H (2008) Structure, regulation and evolution of NOX-family NADPH oxidases that produce reactive oxygen species. FEBS J 275:3249–3277
Tan YY, Spiering MJ, Scott V, Lane GA, Christensen MJ, Schmid J (2001) In planta regulation of extension of an endophytic fungus and maintenance of high metabolic rates in its mycelium in the absence of apical extension. Appl Environ Microbiol 67:5377–5383
Tanaka A, Christensen MJ, Takemoto D, Park P, Scott B (2006) Reactive oxygen species play a role in regulating a fungusperennial ryegrass mutualistic interaction. Plant Cell 18:1052–1066
Tiwari BS, Belenghi B, Levine A (2002) Oxidative stress increased respiration and generation of reactive oxygen species, resulting in ATP depletion, opening of mitochondrial permeability transition, and programmed cell death. Plant Physiol 128:1271–1281
Torres MA (2010) ROS in biotic interactions. Physiol Plantarum 128:414–429
Torres MA, Dangl JL (2005) Functions of the respiratory burst oxidase in biotic interactions, abiotic stress and development. Curr Opin Plant Biol 8:397–403
Tsitsigiannis DI, Keller NP (2007) Oxylipins as developmental and host–fungal communication signals. Trends Microbiol 15:109–118
van Roermund CWT, Tabak HF, van den Berg M, Wanders RJA, Hettema EH (2000) Pex11p plays a primary role in medium-chain fatty acid oxidation, a process that affects peroxisome number and size in *Saccharomyces cerevisiae*. J Cell Biol 150(3):489–498
Weber H (2002) Fatty acid-derived signals in plants. Trends Plant Sci 7(5):217–224
Willekens H, Inzé D, Montagu M, Camp W (1995) Catalases in plants. Mol Breed 1:207–228
Yager LN (1992) Early developmental events during asexual and sexual sporulation in *Aspergillus nidulans*. In: Bennett JW, Klich MA (eds) Aspergillus – biology and industrial applications. Butterworth-Heinemann, Boston, pp 19–41
Zuo X, Wu Y, Morris JS, Stimmel JB, Leesnitzer LM, Fischer SM, Lippman SM, Shureiqi I (2006) Oxidative metabolism of linoleic acid modulates PPAR-beta/delta suppression of PPAR-gamma activity. Oncogene 23–25(8):1225–1241

Oxylipins in Fungal-Mammalian Interactions

Katharyn J. Affeldt and Nancy P. Keller

Abstract Oxylipins, or oxygenated fatty acids, are potent signaling molecules that assist in orchestrating fungal-mammalian interactions. These molecules are generated by several oxygenases, chiefly lipoxygenases, cyclooxygenases and dioxygenases. The fungal and mammalian oxygenases share many conserved domains resulting in the production of similar and even identical oxylipins. Mammalian oxylipins (more commonly called eicosinoids) are part of the immune response whereas fungal oxylipins direct growth and development of the producing organism. Importantly, oxylipins from both Kingdoms can be perceived and recognized by each organism, impacting fungal-mammalian interactions.

1 Introduction

With the advancement of medicine, many people are able to survive cancer, replace a non-functional organ, and live with HIV for many years. Unfortunately, these life-saving treatments require suppression of the immune system, and fungal pathogens that a healthy host can evade become the death knell for immunocompromised patients. These pathogens include *Candida albicans*, *Cryptococcus neoformans*, and *Aspergillus fumigatus* (see Shapiro et al. 2011 for review). Some fungal pathogens, such as *C. albicans*, *Blastomyces dermatitidis*, and *Histoplasma capsulatum* can also infect immunocompetent individuals. A greater understanding of fungal pathogens is imperative as there are currently not many effective antifungals, and most are damaging to the host as well.

K.J. Affeldt • N.P. Keller (✉)
Department of Medical Microbiology and Immunology,
University of Wisconsin, Madison, WI, USA
e-mail: npkeller@wisc.edu

G. Witzany (ed.), *Biocommunication of Fungi*, DOI 10.1007/978-94-007-4264-2_19,

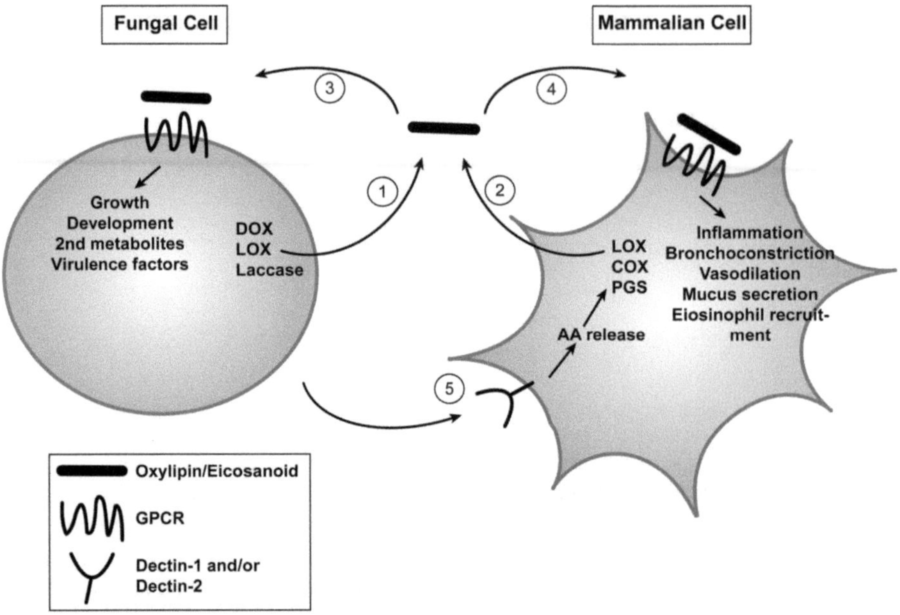

Fig. 1 A hypothetical model for oxylipin-mediated signaling between fungi and mammals. Fungi and mammals engage in oxylipin signaling to direct the outcomes of their interaction. Oxylipins produced by either organism can directly affect either organism, or the presence of one organism can induce oxylipin production in the other. In this model, both fungal (*1*) and mammalian (*2*) cells produce oxylipins, which include eicosanoids such as prostaglandins and leukotrienes, via a variety of oxygenases. These include fungal dioxygenases (*DOX*), lipoxygenases (*LOX*), and laccase, and mammalian LOX, cyclooxygenases (*COX*) and prostaglandin synthases (*PGS*). Oxylipin signals are then perceived by fungal G-protein coupled receptors (*GPCRs*) to impact growth, development, secondary metabolism, and production of various virulence factors (*3*). Oxylipins are also recognized by mammalian GPCRs as part of the immune response, which can include inflammation, bronchoconstriction, vasodilation, mucus production, and eosinophil recruitment (*4*). Because of structural similarities between fungal and mammalian oxylipins, these molecules can be produced and sensed by both organisms, allowing for crosstalk to occur and, thus, impacting the fate of the interaction. Fungi also influence oxylipin production in mammalian cells via the mammalian receptor Dectin-1 and/or Dectin-2, which recognize β-glucans and mannans (fungal cell wall components) respectively (*5*). This stimulates arachidonic acid (AA) release and mammalian oxylipin production (*2*)

Oxylipins, or oxygenated fatty acids, are potent signaling molecules that assist in orchestrating fungal-mammalian interactions. On the fungal side, oxylipins direct growth and development; on the mammalian side, oxylipins are part of the immune response. By transmitting and receiving oxylipin signals, the pathogen and the host engage in a conversation that impacts responses of both organisms to each other. This review will discuss the roles of fungal oxylipins, the effects of fungi on mammalian oxylipins, and how oxylipin signals are perceived in the context of fungal-mammalian interactions. Figure 1 presents a schematic of a hypothetical model involving fungal mammalian signaling events.

2 Oxylipins and Enzymes

Mammals and fungi both produce endogenous oxylipins chiefly through the expression of various oxygenases (Table 1). Here our focus will be on shared oxylipins produced or recognized by fungi and mammals alike, while chapters 16, 17 and 18 will discuss more thoroughly plant/fungal oxylipin-driven and lipid-driven communication, respectively. A more thorough understanding of oxylipins as communication molecules will be achieved through consideration of all three chapters.

The most common mammalian oxylipins are derived from arachidonic acid (20:4) and are generically termed eicosanoids. These include prostaglandins, prostacyclins, thromboxanes and leukotrienes (Table 1). The first step in prostaglandin synthesis requires a cyclooxygenase (COX) whose product prostaglandin H_2 (PGH_2) is converted to the various prostaglandins by different prostaglandin synthases (see Ricciotti and Fitzgerald 2011 for review). Prostaglandins contain a characteristic ring within the fatty acid hydrocarbon chain. Prostaglandin production occurs in almost every human cell type, and they are important for maintaining homeostasis and causing inflammation. Prostacyclins and thromboxanes are also derived from PGH_2 and are formed by various isomerases. Prostacyclins cause vasodilation and prevent blood clot formation. Thromboxanes have the opposite functions, causing vasoconstriction and platelet aggregation. Leukotrienes are formed via a lipoxygenase pathway (see Singh et al. 2010 for review). Leukotrienes are structurally distinct from prostaglandins as they do not have a ring structure but instead contain four double bonds, three of which are conjugated. The leukotrienes include LTB_4, which recruits neutrophils, as well as the cysteine leukotrienes (CysLT) that play an important role in respiratory diseases such as asthma and allergic rhinitis. Leukotrienes cause increased mucus secretion, bronchoconstriction, increased vascular permeability, and eosinophil recruitment, among other responses. Mammals also synthesize several oxylipins from 18-carbon polyunsaturated acids, including some that overlap with plant metabolites (Table 1).

Fungal oxylipins are derived from dioxygenases, lipoxygenases and laccases (see Brodhun and Feussner 2011 for review). *A. fumigatus* encodes three dioxygenases and two lipoxygenases. The dioxygenases were identified based on their homology to *Aspergillus nidulans* dioxygenases PpoA, PpoB, and PpoC. The Ppo enzymes contain two heme-binding domains, the first being required for oxygenation, and the second conferring isomerase activity, resulting in hydroperoxy and dihydroxy fatty acid products. Both domains are functional in *A. fumigatus* PpoA as this enzyme makes both types of products (Garscha et al. 2007; Hoffmann et al. 2011). PpoC, however, lacks a critical heme-coordinating residue in its isomerase domain, rendering it nonfunctional. As such, it only produces hydroperoxy fatty acids (Garscha et al. 2007; Garscha and Oliw 2009) (see Table 1). The functions of PpoB and the two lipoxygenases are unknown, as a *ΔppoB* mutant does not exhibit an altered oxylipin profile (Garscha et al. 2007), and the lipoxygenases have not been characterized. Studies using ELISA detection methodology suggest that at least PpoC is also able to produce a variety of eicosanoids (Tsitsigiannis et al. 2005b).

Table 1 Mammals and fungi produce oxilipins

Oxylipin	Organism	Enzyme(s)	Reference(s)
PGH_2 (prostanoid precursor)	*H. sapiens*	COX-1, COX-2	Ricciotti and Fitzgerald (2011)
Prostacyclins (PGI_2)	*H. sapiens*	Prostacyclin synthase	Ricciotti and Fitzgerald (2011)
Thromboxanes (TXA_2)	*H. sapiens*	Thromboxane synthase	Ricciotti and Fitzgerald (2011)
Prostaglandins (PGE_2, PGD_2, $PGF_{2\alpha}$)	*H. sapiens*	Prostaglandin synthases	Ricciotti and Fitzgerald (2011)
	A. fumigatus	PpoA, PpoB, PpoC	Noverr et al. (2002), Tsitsigiannis et al. (2005a, b)
	B. dermatitidis	Unknown	Noverr et al. (2002)
	C. albicans	Unknown	Noverr et al. (2002)
	C. neoformans	Laccase (Lac1, Lac2)	Noverr et al. (2002), Erb-Downward et al. (2008)
	H. capsulatum	Unknown	Noverr et al. (2002)
Prostaglandins (PGE_X)	*P. brasiliensis*	Unknown	Biondo et al. (2010)
Leukotrienes (CysLT, LTB_4)	*H. sapiens*	Lipoxygenase	Singh et al. (2010)
	A. fumigatus	Unknown	Noverr et al. (2002)
	B. dermatitidis	Unknown	Noverr et al. (2002)
	C. albicans	Unknown	Noverr et al. (2002)
	C. neoformans	Unknown	Noverr et al. (2002)
	H. capsulatum	Unknown	Noverr et al. (2002)
Farnesol	*C. albicans*	Unknown	Singh and Del Poeta (2011)
Tyrosol	*C. albicans*	Unknown	Singh and Del Poeta (2011)
(8*R*,11*S*)-DiHODE	*A. fumigatus*	Unknown	Garscha et al. (2007)
(5*S*,8*R*)-DiHODE	*A. fumigatus*	PpoA	Garscha et al. (2007), Hoffmann et al. (2011)
(8*R*)-H(p)ODE	*A. fumigatus*	PpoA, PpoC	Garscha et al. (2007), Hoffmann et al. (2011)
(10*R*)-H(p)ODE	*A. fumigatus*	PpoC	Garscha et al. (2007)
(5/8/9/11/12/15 *R*/*S*)-HETE	*H. sapiens*	5-, 12-, and 15-LOX[a]	Obinata and Izumi (2009)
(9/13 *R*/*S*)-H(p)ODE	*H. sapiens*	12/15-LOX[a]	Obinata and Izumi (2009)

[a]No human enzyme identified for 11-HETE or 9-H(p)ODE

As shown in Table 1, *C. neoformans* produces prostaglandins, and this production is blocked by cyclooxygenase inhibitors despite the lack of cyclooxygenase homologs in the *C. neoformans* genome. Lipoxygenase inhibitors also prevent prostaglandin production, though there are no lipoxygenase homologs in the genome either (Erb-Downward et al. 2008). Interestingly, several lines of evidence point to a laccase as having a role in prostaglandin production (Erb-Downward et al. 2008). Cryptococcal lysates immunodepleted of laccase are almost completely deficient in prostaglandin production, as is a *LAC1* deletion mutant. Deletion of a second laccase, *LAC2*, has no effect on prostaglandin levels. Cyclooxygenases convert arachidonic acid to PGH_2, with PGG_2 being the first stable intermediate of this reaction. Prostaglandin synthases then convert PGH_2 to the various prostaglandins. Recombinant Lac1 is unable to synthesize prostaglandins from exogenous arachidonic acid or PGH_2, indicating that it does not act as a cyclooxygenase or a prostaglandin synthase, but it is able to convert exogenous PGG_2 to the prostaglandin PGE_2. This reaction also generates other related compounds, including 15-keto-PGE_2, revealing laccase to be part of a unique chemical synthesis pathway.

3 Fungal Development

3.1 Morphogenesis and Secondary Metabolism

Fungal oxylipins, including the prostaglandins, are very important for fungal development and morphogenesis. Affinity-purified PGE_X from both *C. albicans* and *C. neoformans*, as well as commercial PGE_2, enhance germ tube formation in *C. albicans*, which is the first step in the yeast-to-hyphae transition (Noverr et al. 2001). The yeast-to-hyphae transition is important as strains unable to undergo this transition are reduced in virulence (see Shapiro et al. 2011 for review). An array of commercial fatty acids and eicosanoids, including leukotrienes, thromboxanes, and additional prostaglandins were tested for their effects on germ tube formation, and only PGE_2 and the thromboxane TxB_2 accelerated germ tube formation. Interestingly, the short-chain fatty acid butyric acid had an inhibitory effect on germination (Noverr and Huffnagle 2004).

Aspergillus oxylipins have been studied extensively as influencers of development, mediating the balance of asexual to sexual spores, biosynthesis of secondary metabolites and germination (see Tsitsigiannis and Keller 2007 for review). This was first noted by studies where addition of plant and later mammalian oxylipins to fungal cultures resulted in abberant spore development and secondary metabolism (Burow et al. 1997; Calvo et al. 1999; Tsitsigiannis et al. 2005b). *Aspergillus nidulans* oxylipins were first studied for their role in sexual development and were collectively called psi factor (precocious sexual inducer) (Champe et al. 1987). The Ppo enzymes (for psi-producing oxygenase) were linked to psi factor formation, and in *A. nidulans*, Δ*ppoA* and Δ*ppoB* result in a shift toward asexual development, while Δ*ppoC* shifts toward sexual development (Tsitsigiannis et al. 2004a, b, 2005a). Disrupting the *ppo* genes also impacts secondary metabolism. A Δ*ppoABC* triple mutant produces many metabolites not seen in the wild type, though sterigmatocystin, which is produced by the wild type, is undetectable in the mutant (Tsitsigiannis and Keller 2006). In *A. flavus*, a strain depleted for all five of its oxygenases shows increased production of aflatoxin (Georgianna et al. 2010). An *A. fumigatus* Δ*ppoC* mutant displays abnormal spore morphology (larger and oval-shaped), decreased sporulation, and altered germination compared to wild type (Dagenais et al. 2008).

3.2 *Quorum Sensing*

Quorum sensing is a density-dependent phenomenon that leads to a coordinated response from the population, such as biofilm formation by *Psuedomonas aeruginosa* and bioluminescence by *Vibrio fischeri*. As the population grows, cells secrete an inducer molecule until it surpasses a certain threshold, activates its receptor, and initiates gene transcription. *C. albicans* was the first eukaryote for which a quorum sensing phenomenon was discovered, and its main quorum-sensing molecule is the oxylipin farnesol (see Nickerson et al. 2006; Singh and Del Poeta 2011 for reviews). Farnesol inhibits germ tube formation, causing low-density populations to grow as hyphae, while high-density populations grow as yeast. However, farnesol does not inhibit hyphae elongation from cells that are already in the hyphal form. Tyrosol also acts as a quorum sensing molecule, though it has the opposite effect of farnesol: it promotes the yeast-to-hyphae transition (Chen et al. 2004). Interestingly, farnesol is always dominant to tyrosol when in direct competition, even if tyrosol is in excess (Nickerson et al. 2006). Farnesol and tyrosol's mechanisms of action are not understood, though farnesol does activate expression of many genes. *C. albicans*' quorum sensing phenomenon is important for regulating biofilm formation since biofilms are composed of both yeast for attachment and hyphae for expansion. *C. albicans* biofilms also secrete prostaglandins, and they do this at a much higher level than planktonic cells, though their role in this context is not understood (Alem and Douglas 2005).

In *Aspergillus flavus,* there is a density-dependent shift in which, at low population densities, production of conidia (asexual spores) is low while production of sclerotia (overwintering structures) is high. The opposite is seen at high densities. However, individual oxygenase mutants and, particularly the mutant with all five oxygenases depleted by RNAi mentioned earlier, disrupt this conidia-to-sclerotia balance (Horowitz Brown et al. 2008). This implies that oxylipins may act as quorum-sensing molecules to regulate density-dependent development. Volatile eight-carbon oxylipins have also been implicated in preventing germination at high population densities and stimulating production of asexual spores in *A. nidulans* (Herrero-Garcia et al. 2011), suggesting another link between oxylipins and quorum sensing.

4 Mammalian/Fungal Interactions

4.1 Fungal Virulence

Because of the importance of fungal oxylipins in *C. albicans* dimorphism and biofilm formation, processes that are key to colonization of host tissue and internal medical devices, these metabolites are thought to play a role in virulence. Supporting this hypothesis is evidence that *C. albicans* and *C. neoformans* prostaglandins act similarly to mammalian prostaglandins (Noverr et al. 2001). Affinity-purified PGE_X from both fungal species and commercially available PGE_2 decrease proliferation, decrease TNF-α production, and increase IL-10 production when applied to mitogen-stimulated murine splenocytes. *C. albicans* PGE_X also decreases IL-8 production in lung epithelial cells, like PGE_2. This suggests that fungal oxylipins are able to mimic mammalian eicosanoids, which would allow for crosstalk between fungal and mammalian cells.

A. fumigatus prostaglandins may also be involved in virulence. Mutants with RNAi-depleted ppoA, B and C genes (e.g. cyclooxygenase like dioxygenases, Table 1) produce fewer prostaglandins *in vitro* but are hypervirulent compared to the wild type in a murine model (Tsitsigiannis et al. 2005b). It is hypothesized that fungal prostaglandins may be important for stimulating the host immune system, and without this signal, the immune response is diminished or delayed. However, a Δ*ppoC* mutant was more readily phagocytosed and killed by murine alveolar macrophages than the wild type (Dagenais et al. 2008), indicating that oxylipin signaling is complex.

Laccase, the activity found necessary for prostaglandin production in *C. neoformans*, has been independently studied for its roles in melanin production and virulence in this fungus. Deletion of *LAC1* results in loss of melanin production and attenuated virulence in a murine model (Salas et al. 1996). Here the decreased virulence was attributed to melanin loss, however, at the time, the role of *LAC1* in prostaglandin production was not known, so it remains unclear whether putative prostaglandin loss in the mutant contributes to the decreased virulence. Other strains defective in melanin production have also been found to be less virulent (Kwon-Chung et al. 1982; Kwon-Chung and Rhodes 1986), hence the focus on melanin in the above study.

An additional *C. neoformans* virulence factor is the secreted phospholipase *PLB1* (Noverr et al. 2003). Phospholipases cleave fatty acids from phospholipids, and these fatty acids may include the prostaglandin precursor arachidonic acid. A *plb1* mutant grown in a nutrient-rich medium without added arachidonic acid is decreased in prostaglandins compared to the wild type. Furthermore, a *plb1* mutant is only able to synthesize prostaglandins if free arachidonic acid, and not acrachidonic acid conjugated to phosphatidylcholine, is present. The wild type and a *plb1* complement can use either substrate to generate oxylipins. Interestingly, the *plb1* mutant is avirulent, and complementation with *PLB1* restores virulence in a mouse model. Researchers found that infections of mice with the *plb1* mutant results in a decrease in prostaglandin and leukotriene production in mouse lungs post-infection, though it was not determined whether the oxylipins produced during infection with the wild type or *plb1* complement are of host or pathogen origin. However, it does suggest another layer of complexity to the fungal-host relationship in which fungal oxylipins not only mimic those of mammalian cells, but the pathogen may even be scavenging mammalian cells for oxylipin substrates.

4.2 Host Reactions

Recognition of a pathogen by a mammalian host triggers host prostaglandin and leukotriene synthesis, which leads to cytokine production and immune cell recruitment. However, just as it is possible for fungal oxylipins to mimic host signals, the reverse situation could also occur. Based on the effects of various oxylipins on fungal development and virulence discussed above, production of host oxylipins could directly impact the fungal pathogen in addition to having an indirect effect via the immune response.

Upon exposure of rabbit alveolar macrophages to *C. albicans*, arachidonic acid is released, and 55% of this is converted into prostaglandins and leukotrienes (Castro et al. 1994). *C. albicans* induces expression of mammalian cyclooxygenase COX-2 mRNA and protein, as well as prostaglandins in a variety of mammalian cell types (Filler et al. 1996; Deva et al. 2003; Suram et al. 2006; Lee et al. 2009; Smeekens et al. 2010). A pulse-chase experiment further demonstrated that *C. albicans* hyphae in particular induce host prostaglandin production. Mouse bone marrow dendritic cells were incubated with *C. albicans* yeast or hyphae for 2 h, when the fungicide amphotericin B was added. Eighteen hours later, prostaglandin levels were high in cells that had been exposed to hyphae, but not yeast (Kundu and Noverr 2011). This is interesting in light of *C. albicans* dimorphism playing a key role in infection.

Several studies have indicated that fungal cell wall components are what trigger arachidonic acid release and subsequent oxylipin production in mammalian cells. Both soluble α-mannan and β-glucan block arachidonic acid release in rabbit alveolar macrophages and human peripheral blood monocytes when coincubated with *C. albicans* (Castro et al. 1994, 1996). However, Suram et al. (2006) reported that only β-glucan and not α-mannan blocks arachidonic acid release from *C. albicans*-stimulated mouse peritoneal macrophages. Additionally, murine macrophages expressing the β-glucan receptor Dectin-1 show an increase in arachidonic acid release, COX2 protein expression, and prostaglandin PGD_2 over cells containing the vector control or the mannose receptor SIGNR1 (Suram et al. 2006). However, Suram et al. (2010) later reported that D-mannose and *C. albicans* mannans did block *C. albicans*-stimulated arachidonic acid release in those cells, as did antibodies to the mannan receptor Dectin-2. Furthermore, cells expressing Dectin-2 release arachidonic acid when stimulated with *C. albicans*, though not to the degree of cells expressing Dectin-1 (Suram et al. 2010). Human peripheral blood mononuclear cells

(PBMC) produce PGE_2 in response to stimulation with *C. albicans*, and the only fungal component that was shown to stimulate PGE_2 production in these cells is mannan, not β-glucan or chitin (Schmeekens et al. 2010). Together, these studies demonstrate that different cell types may be recognizing different components of the fungal cell wall, depending on whether they express the appropriate receptors, including Dectin-1 and Dectin-2. That recognition ultimately leads to prostaglandin production. Blocking PGE_2 production in PBMCs with a chemical inhibitor altered the cytokine profile, indicating that fungus-induced prostaglandins are important in directing downstream immune responses (Schmeekens et al. 2010).

Like *C. albicans*, *Aspergillus fumigatus* also induces expression of prostaglandin-associated genes in mammalian cells. Immature human dendritic cells (iDCs) incubated with *A. fumigatus* are upregulated in several genes involved in prostaglandin synthesis and recognition compared to unstimulated cells. *A. fumigatus* also stimulates cytokine production in iDCs, and this is reduced when Dectin-1 is silenced or blocked by anti-Dectin-1 antibodies in these cells (Mezger et al. 2008). To directly link *A. fumigatus*-stimulated oxylipin production with Dectin-1-mediated recognition will require more extensive study, though it is an intriguing possibility. It should be noted that prostaglandins are not the only oxylipins affected by fungal infection. *A. fumigatus* infection of mouse bone marrow dendritic cells causes a burst of CysLTs to be released by the cells, though no further research into the source of these CysLTs was reported (Barrett et al. 2009).

Further study is also needed to understand the impacts of fungal induction of mammalian oxylipin production. Though mammalian oxylipins are part of the immune response, this does not necessarily mean that increasing them will benefit the host. As mentioned, mouse bone marrow dendritic cells pulsed with *C. albicans* hyphae have high levels of prostaglandins while yeast-pulsed cells do not. Immunizing mice with yeast-pulsed cells causes a reduction in fungal burden during subsequent *C. albicans* infection. This is a result of the pulsed cells inducing T helper cell (Th1)-mediated protection and clearance. However, when fungal or commercial prostaglandin PGE_2 is added to the pulsed dendritic cells, the protective effect of these cells is lost, and hyphae-pulsed cells have no immunizing effect on mice either (Kundu and Noverr 2011).

5 Mechanism of Interaction: GPCRs

In order for oxylipin-mediated cross-talk to occur between fungi and mammals, there needs to be appropriate receptors in each organism to recognize and transmit these signals. G protein-coupled receptors (GPCR) have been well characterized as mammalian eicosanoid receptors. GPCRs are transmembrane proteins that pass the membrane seven times with the N-terminus in the cytoplasm and the C-terminus inside the cell. They are called GPCRs because their activation causes nucleotide exchange of GTP for GDP on the Gα subunit of a heterotrimeric G protein, resulting in dissociation of the complex and activation of downstream pathways.

CysLTs bind to two classes of GPCR: $CysLT_1$ and $CysLT_2$ receptors (see Singh et al. 2010 for review). Though the various CysLTs can bind to both classes of receptors, they have differing affinities for the different receptors. It has been suggested that additional CysLT receptors exist for two reasons. First, a dual antagonist that blocks binding to both $CysLT_1$ and $CysLT_2$ does not prevent the cellular response to CysLTs in some cases. Second, the CysLT LTE_4 can initiate signaling in cells lacking both classes of receptors. Only certain cell types express the CysLT receptors, and the expression of these proteins is regulated by cytokines.

Like the leukotrienes, prostaglandins are also perceived by GPCRs, though the prostaglandin receptors are comprised of eight different groups (see Ricciotti and Fitzgerald 2011 for review). This variety in receptor type is expanded further by alternative splicing, causing one receptor type in particular to have eight different variants that differ in their C-terminal tails. The various receptors activate different Gα subunits, and some can even activate multiple Gα subunits. This leads to signaling through a range of different pathways. Prostaglandin receptor types are tissue specific, and several prostaglandins are recognized by more than one receptor type.

Additional oxylipins beyond just the eicosanoids are recognized by mammalian GPCRs. Specifically, the human GPCR G2A responds to a range of oxylipins with 9-hydroxyoctadecadienoic acid (9-HODE, a linoleic acid derivative) eliciting the strongest response (Obinata et al. 2005). Cells were transfected with G2A, exposed to a panel of oxylipins, and monitored for production of the second messenger calcium. The results indicate that G2A is not highly specific for a single oxylipin but instead can be activated to varying degrees by a plethora of related oxylipins. Because 9-HODE showed the strongest response, it was used in a subsequent experiment to further demonstrate G2A activation. Using a radiolabeled non-hydrolyzable form of GTP, researchers showed that increasing amounts of 9-HODE causes an increase in GTP bound to the membrane fraction of cells containing G2A.

Like mammals, fungi sense extracellular signals via GPCRs (see Xue et al. 2008 for review). However, until recently, no fungal GPCR had been implicated in oxylipin recognition. We have found that *A. nidulans* responds to purified oxylipins with a dose-dependent cAMP burst and this response is absent in the GPCR GprD. Like calcium, cAMP is a second messenger that acts downstream of many GPCRs. Interestingly, the putative *A. fumigatus* homolog of GprD (63% identity), also called GprD, is required for virulence in this species (Gehrke et al. 2010). Though the ligand for *A. fumigatus* GprD is not known, its homology with *A. nidulans* GprD supports the hypothesis that it also perceives oxylipins. This is intriguing as it would suggest that *A. fumigatus* recognizes exogenous, and potentially host-derived, oxylipins that are critical for its ability to persist in the host via a GPCR, and that this receptor could be a target of future drug development. Future studies in this area will be critical for gaining a more complete understanding of how hosts and pathogens receive each other's signals and what the outcome is of those responses.

6 Conclusion

Fungal pathogens are able to infect hosts due to a variety of virulence factors, but the study of oxylipins is intriguing because they act as signaling molecules that may control these virulence factors, the host response, or both. Furthermore, oxylipins are produced and perceived by both the pathogen and the host, allowing both organisms to communicate using a common language (Fig. 1). Whether this ultimately benefits the fungus or its mammalian host depends on the individual interaction. However, a greater understanding of how oxylipin-mediated signaling proceeds, the receptors that are involved, and the downstream signaling that is initiated as a result will be important for identifying novel ways of combating these pathogens.

References

Alem MAS, Douglas LJ (2005) Prostaglandin production during growth of *Candida albicans* biofilms. J Med Microbiol 54:1001–1005. doi:10.1099/jmm.0.46172-0

Barrett NA, Maekawa A, Rahman OM et al (2009) Dectin-2 recognition of house dust mite triggers cysteinyl leukotriene generation by dendritic cells. J Immunol 182:1119–1128

Biondo GA, Dias-Melicio LA, Bordon-Graciani AP et al (2010) *Paracoccidioides brasiliensis* uses endogenous and exogenous arachidonic acid for PGE x production. Mycopathologia 170:123–130. doi:10.1007/s11046-010-9301-x

Brodhun F, Feussner I (2011) Oxylipins in fungi. FEBS J 278:1047–1063. doi:10.1111/j.1742-4658.2011.08027.x

Brown SH, Scott JB, Bhaheetharan J et al (2009) Oxygenase coordination is required for morphological transition and the host-fungus interaction of *Aspergillus flavus*. Mol Plant Microbe Interact 22:882–894

Burow GB, Nesbitt TC, Dunlap J, Keller NP (1997) Seed lipoxygenase products modulate *Aspergillus* mycotoxin biosynthesis. Mol Plant Microbe Interact 10:380–387. doi:10.1094/MPMI.1997.10.3.380

Calvo AM, Hinze LL, Gardner HW, Keller NP (1999) Sporogenic effect of polyunsaturated fatty acids on development of *Aspergillus* spp. Appl Environ Microbiol 65:3668–3673

Castro M, Ralston NV, Morgenthaler TI et al (1994) *Candida albicans* stimulates arachidonic acid liberation from alveolar macrophages through alpha-mannan and beta-glucan cell wall components. Infect Immun 62:3138–3145

Castro M, Bjoraker JA, Rohrbach MS, Limper AH (1996) *Candida albicans* induces the release of inflammatory mediators from human peripheral blood monocytes. Inflammation 20:107–122

Champe SP, Rao P, Chang A (1987) An endogenous inducer of sexual development in *Aspergillus nidulans*. J Gen Microbiol 133:1383–1387. doi:10.1099/00221287-133-5-1383

Chen H, Fujita M, Feng Q et al (2004) Tyrosol is a quorum-sensing molecule in *Candida albicans*. Proc Natl Acad Sci USA 101:5048–5052. doi:10.1073/pnas.0401416101

Dagenais TRT, Chung D, Giles SS et al (2008) Defects in conidiophore development and conidium-macrophage interactions in a dioxygenase mutant of *Aspergillus fumigatus*. Infect Immun 76:3214–3220. doi:10.1128/IAI.00009-08

Deva R, Shankaranarayanan P, Ciccoli R, Nigam S (2003) *Candida albicans* induces selectively transcriptional activation of cyclooxygenase-2 in HeLa cells: pivotal roles of Toll-like receptors, p38 mitogen-activated protein kinase, and NF-kappa B. J Immunol 171:3047–3055

Erb-Downward JR, Noggle RM, Williamson PR, Huffnagle GB (2008) The role of laccase in prostaglandin production by *Cryptococcus neoformans*. Mol Microbiol 68:1428–1437. doi:10.1111/j.1365-2958.2008.06245.x

Filler SG, Pfunder AS, Spellberg BJ et al (1996) *Candida albicans* stimulates cytokine production and leukocyte adhesion molecule expression by endothelial cells. Infect Immun 64:2609–2617

Garscha U, Oliw EH (2009) Leucine/valine residues direct oxygenation of linoleic acid by (10R)- and (8R)-dioxygenases: expression and site-directed mutagenesis oF (10R)-dioxygenase with epoxyalcohol synthase activity. J Biol Chem 284:13755–13765. doi:10.1074/jbc.M808665200

Garscha U, Jernerén F, Chung D et al (2007) Identification of dioxygenases required for *Aspergillus* development. J Biol Chem 282:34707–34718. doi:10.1074/jbc.M705366200

Gehrke A, Heinekamp T, Jacobsen ID, Brakhage AA (2010) Heptahelical receptors GprC and GprD of *Aspergillus fumigatus* are essential regulators of colony growth, hyphal morphogenesis, and virulence. Appl Environ Microbiol 76:3989–3998. doi:10.1128/AEM.00052-10

Georgianna DR, Fedorova ND, Burroughs JL, Dolezal AL, Bok JW, Horowitz-Brown S, Woloshuk CP, Yu J, Keller NP, Payne GA (2010) Beyond aflatoxin: four distinct expression patterns and functional roles associated with Aspergillus flavus secondary metabolism gene clusters. Mol Plant Pathol 11:213–226

Herrero-Garcia E, Garzia A, Cordobés S et al (2011) 8-Carbon oxylipins inhibit germination and growth, and stimulate aerial conidiation in *Aspergillus nidulans*. Fungal Biol 115:393–400. doi:10.1016/j.funbio.2011.02.005

Hoffmann I, Jernerén F, Garscha U, Oliw EH (2011) Expression of 5,8-LDS of *Aspergillus fumigatus* and its dioxygenase domain. A comparison with 7,8-LDS, 10-dioxygenase, and cyclooxygenase. Arch Biochem Biophys 506:216–222. doi:10.1016/j.abb.2010.11.022

Horowitz Brown S, Zarnowski R, Sharpee WC, Keller NP (2008) Morphological transitions governed by density dependence and lipoxygenase activity in *Aspergillus flavus*. Appl Environ Microbiol 74:5674–5685. doi:10.1128/AEM.00565-08

Kundu G, Noverr MC (2011) Exposure to host or fungal PGE_2 abrogates protection following immunization with Candida-pulsed dendritic cells. Med Mycol 49:380–394. doi:10.3109/13693786.2010.532514

Kwon-Chung KJ, Rhodes JC (1986) Encapsulation and melanin formation as indicators of virulence in *Cryptococcus neoformans*. Infect Immun 51:218–223

Kwon-Chung KJ, Polacheck I, Popkin TJ (1982) Melanin-lacking mutants of *Cryptococcus neoformans* and their virulence for mice. J Bacteriol 150:1414–1421

Lee H-S, Lee C-S, Yang C-J et al (2009) *Candida albicans* induces cyclo-oxygenase 2 expression and prostaglandin E2 production in synovial fibroblasts through an extracellular-regulated kinase 1/2 dependent pathway. Arthritis Res Ther 11:R48. doi:10.1186/ar2661

Mezger M, Kneitz S, Wozniok I et al (2008) Proinflammatory response of immature human dendritic cells is mediated by dectin-1 after exposure to *Aspergillus fumigatus* germ tubes. J Infect Dis 197:924–931. doi:10.1086/528694

Nickerson KW, Atkin AL, Hornby JM (2006) Quorum sensing in dimorphic fungi: farnesol and beyond. Appl Environ Microbiol 72:3805–3813. doi:10.1128/AEM.02765-05

Noverr MC, Huffnagle GB (2004) Regulation of *Candida albicans* morphogenesis by fatty acid metabolites. Infect Immun 72:6206–6210. doi:10.1128/IAI.72.11.6206-6210.2004

Noverr MC, Phare SM, Toews GB et al (2001) Pathogenic yeasts *Cryptococcus neoformans* and *Candida albicans* produce immunomodulatory prostaglandins. Infect Immun 69:2957–2963. doi:10.1128/IAI.69.5.2957-2963.2001

Noverr MC, Toews GB, Huffnagle GB (2002) Production of prostaglandins and leukotrienes by pathogenic fungi. Infect Immun 70:400–402

Noverr MC, Cox GM, Perfect JR, Huffnagle GB (2003) Role of PLB1 in pulmonary inflammation and cryptococcal eicosanoid production. Infect Immun 71:1538–1547

Obinata H, Izumi T (2009) G2A as a receptor for oxidized free fatty acids. Prostaglandins Other Lipid Mediat 89:66–72. doi:16/j.prostaglandins.2008.11.002

Obinata H, Hattori T, Nakane S et al (2005) Identification of 9-hydroxyoctadecadienoic acid and other oxidized free fatty acids as ligands of the G protein-coupled receptor G2A. J Biol Chem 280:40676–40683. doi:10.1074/jbc.M507787200

Ricciotti E, FitzGerald GA (2011) Prostaglandins and inflammation. Arterioscler Thromb Vasc Biol 31:986–1000. doi:10.1161/ATVBAHA.110.207449

Salas SD, Bennett JE, Kwon-Chung KJ et al (1996) Effect of the laccase gene CNLAC1, on virulence of *Cryptococcus neoformans*. J Exp Med 184:377–386

Shapiro RS, Robbins N, Cowen LE (2011) Regulatory circuitry governing fungal development, drug resistance, and disease. Microbiol Mol Biol Rev 75:213–267. doi:10.1128/MMBR.00045-10

Singh A, Del Poeta M (2011) Lipid signalling in pathogenic fungi. Cell Microbiol 13:177–185. doi:10.1111/j.1462-5822.2010.01550.x

Singh RK, Gupta S, Dastidar S, Ray A (2010) Cysteinyl leukotrienes and their receptors: molecular and functional characteristics. Pharmacology 85:336–349. doi:10.1159/000312669

Schmeekens SP, van de Veerdonk FL, van der Meer JWM et al (2010) The Candida Th17 response is dependent on mannan- and beta-glucan-induced prostaglandin E2. Int Immunol 22:889–895. doi:10.1093/intimm/dxq442

Suram S, Brown GD, Ghosh M et al (2006) Regulation of cytosolic phospholipase A2 activation and cyclooxygenase 2 expression in macrophages by the beta-glucan receptor. J Biol Chem 281:5506–5514. doi:10.1074/jbc.M509824200

Suram S, Gangelhoff TA, Taylor PR et al (2010) Pathways regulating cytosolic phospholipase A2 activation and eicosanoid production in macrophages by *Candida albicans*. J Biol Chem 285:30676–30685. doi:10.1074/jbc.M110.143800

Tsitsigiannis DI, Keller NP (2006) Oxylipins act as determinants of natural product biosynthesis and seed colonization in *Aspergillus nidulans*. Mol Microbiol 59:882–892

Tsitsigiannis DI, Keller NP (2007) Oxylipins as developmental and host-fungal communication signals. Trends Microbiol 15:109–118. doi:10.1016/j.tim.2007.01.005

Tsitsigiannis DI, Kowieski TM, Zarnowski R, Keller NP (2004a) Endogenous lipogenic regulators of spore balance in *Aspergillus nidulans*. Eukaryot Cell 3:1398

Tsitsigiannis DI, Zarnowski R, Keller NP (2004b) The lipid body protein, PpoA, coordinates sexual and asexual sporulation in *Aspergillus nidulans*. J Biol Chem 279:11344

Tsitsigiannis DI, Kowieski TM, Zarnowski R, Keller NP (2005a) Three putative oxylipin biosynthetic genes integrate sexual and asexual development in *Aspergillus nidulans*. Microbiology 151:1809

Tsitsigiannis DI, Bok J-W, Andes D et al (2005b) Aspergillus cyclooxygenase-like enzymes are associated with prostaglandin production and virulence. Infect Immun 73:4548–4559. doi:10.1128/IAI.73.8.4548-4559.2005

Xue C, Hsueh Y, Heitman J (2008) Magnificent seven: roles of G protein-coupled receptors in extracellular sensing in fungi. FEMS Microbiol Rev 32:1010–1032. doi:10.1111/j.1574-6976.2008.00131.x

Chemical Signals That Mediate Insect-Fungal Interactions

Drion G. Boucias, Verena-Ulrike Lietze, and Peter Teal

Abstract This chapter reviews the diverse types of chemical communication that mediate interactions between fungi and insects. Both life forms utilize complex chemosensory systems to process environmental cues. As they coexist in terrestrial and aquatic habitats, the consequences of their interactions range from beneficial effects for both partners to adverse ones or even death for one partner. Plants, as an additional life form, also can be involved in these multifaceted relationships. Depending on the volatility of the produced compound(s), chemical cues generally are perceived either by contact or from a distance. Responses of insects to such cues may operate on various levels, including metabolism, immune defense, and behavior. Furthermore, insects may perceive compounds as attractive, repellent, deterrent, or neutral. Fungi, on the other hand, show various metabolic responses to insect-borne chemicals that come into contact with their cell wall. While numerous published reviews and textbooks cover insect-fungal symbiosis, we here present examples of communication systems in which one or more components regulating the signaling events have been identified. These examples include fungal pathogens of insects and plants, mycophagous and fungivorous insects, mutualistic symbioses between fungi and insects, fungal kairomones utilized by insects, and floral and other scent mimicry exploited by fungi.

D.G. Boucias (✉) • V.-U. Lietze
Entomology and Nematology Department, University of Florida,
970 Natural Area Drive, Gainesville, FL 32611-0620, USA
e-mail: pathos@ufl.edu

P. Teal
Center for Medical, Agricultural and Veterinary Entomology, USDA-ARS,
1700 SW 23 Drive, Gainesville, FL 32604, USA

G. Witzany (ed.), *Biocommunication of Fungi*, DOI 10.1007/978-94-007-4264-2_20,

1 Introduction

Insects and fungi represent two of the most diverse and numerous groups of eukaryotes that co-exist in various terrestrial and aquatic habitats. In natural systems, hundreds of thousands of potential interactions exist between insect and fungal species. Members within the Kingdom Fungi represent relatively simple sessile eukaryotes with diverse metabolic pathways. Insects, more advanced, mobile eukaryotes, are renowned for their chemosensory systems capable of detecting and distinguishing between minute amounts of chemical signatures emanating from environmental sources. In terms of communication, the insect possesses an array of chemical receptors linked to a central nervous system programmed to detect both short- and long-distance chemical cues.

The ability to perceive, discriminate, and react to a multitude of chemical cues by chemoreception greatly influences the fitness and survival of insects. Insects rely on this process to identify nutritional food sources (mycophagy), to avoid intoxication (mycotoxins), and to communicate and interact on various levels with other organisms (fungi) in their environment (Silbering and Benton 2010). Two categories of chemical cues appertain: odor is produced by compounds that are volatile at ambient temperatures and is perceived at relatively low concentrations by olfactory or odorant receptors (ORs), whereas taste is produced by non-volatile compounds and is perceived at comparatively higher concentrations by gustatory receptors (GRs). In insects, ORs are expressed in the two olfactory organs, the antenna and maxillary palp; GRs are expressed in contact chemosensors on the proboscis, legs, wings, and genitalia. The fungi, although lacking formal chemoreceptors, can detect and respond readily to selective chemical inputs using various cell signaling systems.

The interactions between these groups vary; at the extremes, there are groups of fungi (mycopathogens) and insects (mycophagous feeders) that utilize each other as their primary source of nutrition. Within these groups there are specialists and generalists that may attack a select group or a broad range of host species, respectively. Between these extremes exist many non-lethal associations in which the two groups co-exist in more balanced ways. For example, many insects harbor fungal mutualists that provide nutrition or defense to their respective hosts. In general, these fungi provide a nutritional supplement to the insect or assist the insect in the assimilation of plant biomass (Douglas 2009), and, in return, the insect provides a suitable habitat and nutrient pools. Over the past three decades, numerous reviews and textbooks have been published on insect-fungal symbiosis (Barbosa et al. 1991; Gibson and Hunter 2010; Mueller et al. 2001; Schwemmler and Gassner 1989; Vega and Blackwell 2005; Wheeler and Blackwell 1984). It is beyond the scope of this chapter to review in detail the potential bidirectional signaling events that regulate these relationships.

In this review, we cover various aspects of the communication that occurs between fungi and insects and present examples of systems in which one or more components underlying the signaling events have been delineated.

2 Insect-Mycopathogen Interactions

Insect mycopathogens, represented throughout the Kingdom Mycota, produce infectious propagules that can breach the cuticle barrier, develop within the insect, and cause a lethal mycosis (Boucias and Pendland 1998). These fungi, like many animal mycopathogens, are dimorphic, producing thin-walled, budding hyphal bodies in the insect hemocoel that synchronously give rise to a tissue-invasive mycelial phenotype. Many insect mycopathogens exhibit a hemibiotrophic lifestyle; biotrophic growth in the living host leads to the host's death, followed by necrotrophic development of the pathogen, leading to the production of infectious and/or resilient propagules of the fungus. The interaction between the pathogenic fungi and insects involves a developmental program with four major phases: contact of the infectious propagule to the host cuticle; development of the penetrant germ tube; vegetative development in the nutrient-rich hemocoel; and differentiation of vegetative cells into reproductive structures. Successful pathogenesis derives from the ability of the fungus to receive host cues and respond by producing appropriate signals that overcome host barriers and defense mechanisms. A unique feature of such fungi is their ability to breach the cuticular barrier to gain access to the hemocoel. A properly orchestrated developmental program terminates in the conversion of insect tissue into fungal biomass, giving rise to numerous infectious propagules. If signaling is inappropriate, as occurs with contact onto non-hosts, the invasive program is aborted, and few, if any, infectious spores are produced by the challenged insect host.

2.1 *Conidiospore/Cuticle Interactions*

The initial event, the attachment of the propagule to the insect cuticle, involves both contact and consolidation phases (Fargues 1984). In general, infectious propagules produced by entomopathogenic fungi can be classified as hydrophobic or hydrophilic conidiospores. Hydrophobic conidiospores are characterized by an outer coat comprised of hydrophobins that form a well-defined rodlet layer (Boucias et al. 1988; Charnley 2003). Members of many insect mycopathogens, including species of *Beauveria, Metarhizium, Nomuraea*, and *Paecilomyces,* operate as r-strategists, synchronously producing numerous hydrophobic conidiospores on the surface of mycosed insects. These propagules, passively dispersed by wind and water, coat both plant and soil surfaces. Insects moving over these substrates randomly contact conidia that bind to the epicuticle surface through nonspecific electrostatic and hydrophobic interactions (Boucias et al. 1988; Holder and Keyhani 2005). In most cases, propagules adhere to the surface of both host and non-host cuticle surfaces. Modification in the hydrophobicity of the insect cuticle can lead to decreased levels of conidial adhesion. For example, the booklouse, *Liposcelis bostrychophila,* an insect resistant to many mycopathogens, has an abnormally high level of amphiphilic, fatty-acid amides on the cuticle that reduce conidial adhesion (Howard and Lord 2003).

Unlike fungi that produce a plethora of aerially dispersed hydrophobic conidiospores, the second group of fungi produces relatively few hydrophilic conidia designed to adhere to and to infect healthy insects that contact the conidiospores bound to phialides. Insect mycopathogens from the genus *Hirsutella*, operating as k-strategists, exemplify the properties of this group. Furthermore, the production of conidiospores by the entomopathogenic *Hirsutella* on cadavers occurs in pulses over relatively long periods of time. Despite producing a limited number of stationary infectious propagules, *Hirsutella* species cause annual epizootics in their host populations. Random contact between mycosed cadavers and healthy conspecifics may occur, but it is an unlikely explanation of the infection rates observed in natural populations. More likely, infected insects produce visual or chemical cues that attract healthy conspecifics.

2.2 Germination and Penetration of the Cuticle Barrier

The consolidation stage involves the hydrated conidiospore receiving a proper suite of cuticle cues. Germination of many mycopathogens requires nutrient inputs that are derived from the hydrolysis of cuticular components. Following attachment, enzymes released from the conidiospore wall act on the substrate, leading to the release of signals that trigger germination. The conidiospore possesses a broad range of surface-bound enzymes that allow it to "taste" and/or "smell" the compounds released from the substrate. For example, enzymes of the peroxisomal β-oxidation system, associated with the conidial cell wall of *Beauveria bassiana*, metabolize the epicuticle hydrocarbons (Crespo et al. 2000; Lecuona et al. 1991; Pedrini et al. 2007) and provide signals to proceed with penetrant germ tube formation. These alkane-degrading enzymes play key roles in gaining ingress through the epicuticle barrier and cause the emission of volatile organic compounds (VOCs) (Crespo et al. 2008). Detailed studies by St. Leger et al. (1998, 1999) demonstrated that the mycopathogen *Metarhizium anisopliae* responds to low levels of amino acids by producing ammonia, resulting in the alkalization of the microenvironment. The increase in pH is essential for the cuticle-degrading subtilisins that act on the proteins which form the scaffold of the laminate chitin layers underlying the epicuticle. Regulation of pH in the microhabitat of the conidiospore/germ tube also is influenced by the expression of carboxylate transporter genes that sequester organic acids, alkalizing the penetration arena while provisioning the cell with a carbon source (Jin et al. 2010).

Insect mycopathogens, like other eukaryotes, possess signal transduction pathways (mitogen-activated protein kinase [MAPK], cAMP-dependent protein kinase A) to receive environmental signals (*e.g.*, cuticular signals) and to translate them into differential gene expression (Gao et al. 2011). Functional pathways are critical for infection; Zhang et al. (2010) reported that *B. bassiana* MAPK mutants had reduced levels of adhesion to the host cuticle and were avirulent when topically applied to host insects. In addition to the above-mentioned pathways, several

Table 1 Insect-derived chemistries that impact fungal development

Insect chemistry	Source	Insect group	Activity
p-hydroxycinnamaldehyde	Whole body extract	Sawfly *Acantholyda parki*	Inhibits *Candida albicans*
β-alanyl-tyrosine	Whole body methanol extract	Grey flesh fly *Neobellieria bullata*	Inhibits yeast and fungal growth
Insect defensins (heliomycin, drosomycin, termicin, Alo3)	Synthesized by fat body, released in hemolymph	Various insects (dipterans, termites, moths)	Inhibit germination, perforate cell wall
Proline-rich peptides, metchnikowin	Fat body	*Drosophila*	Fungitoxic
Phenoloxidase cascade and associated quinones	Hemocytes	Most insects	Fungistatic to fungitoxic
PAMPs (βGRPs, gram-negative binding protein, lectins)	Hemocytes, fat body	Many insects have homologues	Recognition of fungi as non-self
Venom alkaloids	Poison gland	Fire ant *Solenopsis invicta*	Fungistatic
Amphiphilic, fatty-acid amides	Secreted over the cuticle	Booklouse, *Liposcelis bostrychophila*	Reduce conidial adhesion
Aldehydes, (E)-2-decenal	Metathoracic gland	Stink bugs, *Nezara viridula*	Inhibit germination, serve as general defensive secretion
Metapleural gland secretions	Metapleural gland	Leaf cutter ant	Fungitoxic

mycopathogens contain the seven-transmembrane protein 1 (STM1) gene, which is reported to link low nitrogen levels to cell differentiation (Gao et al. 2011).

It is important to realize that the insect cuticle is a dynamic structure; its composition and structure are altered continuously during development. Certain insect species or particular life stages of a host may contain appropriate cues that trigger germ tube formation and allow for successful ingress, whereas other insects or life stages of the same insect either lack the proper cues or produce inhibitors that block consolidation (Bogus et al. 2010; Golebiowski et al. 2008) (Table 1). Early work by Kerwin (1984) demonstrated that the adults of the lesser house fly, *Fannia canicularis*, contain sufficient oleic acid to induce penetrant germ tube formation, whereas the free fatty acid on the pupal stage stimulates the production of secondary discharge conidia rather than penetrant germ tubes of *Erynia variabilis*. Many insects secrete mycostatic lipids onto the epicuticle; their presence and their levels vary according to the physiological state of the host (Gross et al. 1998; Howard and Lord 2003; Kerwin 1982). In most cases, the cuticular compounds possessing mycostatic activity play multifunctional roles. The resistance of the stinkbug *Nezara viridula* to infection by either *M. anisopliae* or *B. bassiana* operates at the cuticle level; conidiospores applied to this insect do not germinate on its cuticle (Sosa-Gomez et al. 1997). Subsequent extraction of the cuticle produced a nonpolar fraction containing the "alarm secretion" n-dodecanol that is highly inhibitory to the germination of certain mycopathogens. A second example is the venom alkaloids emitted by the fire ant *Solenopsis invicta*; these compounds, deposited throughout the mound, actively inhibit conidial germination (Storey et al. 1991). Additionally, the metapleural gland secretions of the leaf-cutter ant *Acromyrmex octospinosus* contain dozens of compounds (Ortius-Lechner et al. 2000) that have potent antibacterial and antifungal properties (Poulsen et al. 2002; Veal et al. 1992). These secretions work along with bacterial associates in maintaining the fungal gardens. However, bioassays have shown that the fungal mutualist is sensitive to the secretions (Bot et al. 2002), suggesting that their release is spatially and temporally regulated. In general, the physiological levels of such antifungal agents vary with the life stage and therefore may dictate stage-specific susceptibility of insects to mycopathogens.

In addition to the cuticle signals that influence fungal development, signals emitted by conidiospores modulate host responses. In certain cases, insects detect and avoid habitats contaminated with entomopathogenic fungi. For example, Ormond et al. (2011) demonstrated that the ladybird beetle, *Coccinella septempuctata,* a generalist predator, avoided leaf and soil substrates and mycosed insects contaminated with *B. bassiana*. A similar avoidance behavior was exhibited by the bug *Anthocoris nemorum* (Meyling and Pell 2006). The signals mediating repellency to *Beauveria* are unknown but likely involve the emitted fungal VOC(s). Ormond et al. (2011) proposed that resources committed to detection and avoidance of entomopathogens are more efficacious for and less costly to host fitness than are those required for an effective immune response to an active infection.

An alternative to avoidance behavior is found in social insects that exhibit extensive grooming behaviors. In their colonies, healthy conspecifics continuously clean the cuticle surface of nestmates. Removal of surface-associated conidia can be extremely efficient, resulting in their nearly complete removal within hours after surface treatment (Boucias et al. 1996; Oi and Pereira 1993). In many ant species, groomed conidiospores accumulate in the infrabuccal cavity, where they are bathed in gland secretions containing mycostatic agents (Febvay et al. 1984). With termites, groomed conidiospores are ingested and, when deposited, are coated in antagonistic bacteria that inhibit germination (Boucias et al. 1996). Disruption of grooming either by pre-treatment with sublethal dosages of insecticides or by solitary confinement after treatment decreases conidiospore removal and increases susceptibility to mycosis by more than 10,000-fold (Boucias et al. 1996). Yanagawa et al. (2009) demonstrated that the termites *Coptotermes* use their chemoreceptors to detect condiospores attached to the cuticle. Surgical removal of antennae reduces the removal of conidiospores and significantly increases mortality rates. Electroantennogram (EAG) responses and feeding studies suggest that these termites can detect and discriminate between mycopathogens by using a combination of contact and odor chemoreceptors. These findings support earlier reports that termites exhibit an immediate defensive response when exposed to *M. anisopliae* spores (Rosengaus et al. 1999; Staples and Milner 2000) or to spore-bearing nestmates (Myles 2002).

In addition to volatile cues, insects use other systems to detect mycopathogens in the environment. In the higher termite *Nasutitermes corniger*, salivary secretions and cuticular washings have been reported to contain gram-negative bacteria-binding proteins (GNBPs) that exhibit β-1,3-glucanase activity (Bulmer et al. 2009). These molecules serve as remote, nest-embedded sensors which detect fungal propagules and act in concert with fungitoxic peptides such as termiticin (Lamberty et al. 2001) to inactivate fungal elements within the nest.

Whether or not the dissolved organic compounds released by the attachment and consolidation of conidiospores elicit localized defense reactions is unclear. The production of penetrant germ tubes on non-hosts can elicit the cuticle prophenoloxidase cascade, resulting in a localized melanization reaction similar to that observed in cuticle wounding. The toxic quinones produced by this reaction may be lethal to the invading fungus; if induced at high levels, this reaction can cause premature molting or, in extreme cases, result in host mortality. In part, the activation of the prophenoloxidase system may be regulated by host protease inhibitors presumed to be synthesized in the hemolymph and localized in the cuticle (Bogus et al. 2007; Boucias and Pendland 1987; Yoshida et al. 1990). These inhibitors also have been shown to inhibit fungal proteases and block germination.

2.3 Ingress into the Hemocoel: Transit from a Nutrient-Poor into a Nutrient-Rich Environment

The insect cuticle is the major barrier to fungal infection. Typically, to initiate an infection, numerous conidiospores must be applied to insects. Alternatively, much lower inoculum levels typically cause a synchronous, 100% mycosis when injected into the hemocoel of host insects, demonstrating that only a select few germ tubes successfully breach the cuticle barrier to enter the hemocoel. The vast majority of conidia making contact with the cuticle cannot penetrate the primary defense barrier. Many insect mycopathogens are dimorphic and can multiply by hyphal-elongation or by budding growth programs. Penetration of the cuticle requires the hyphal-elongation program in which cuticle-degrading enzymes are selectively released at the apical tip. Replication in the hemolymph typically involves a yeast-like budding program, in which thin-walled hyphal bodies are formed that absorb available amino acids and sugars, continuously drawing upon host nutrient reserves.

In the hemocoel, the major obstacles to fungal development are innate host defenses designed to recognize and respond to non-self (Gillespie et al. 2000). The injection of *in vitro* fungi cells into the hemocoel elicits an immediate up-regulation of the Toll signaling pathway, resulting in the activation of the hemocyte-mediated phenoloxidase cascade (Roh et al. 2009). The major fungal elicitors are the β-1,3-glucans produced by many fungi, including insect mycopathogens (Latgé et al. 1988). Immunocytochemical studies have localized β-1,3-glucans as surface cell wall components (Pendland and Boucias 1992). These glucose polymers serve as pathogen-associated molecular pattern (PAMP) molecules and are recognized by the β-1,3-glucan recognition receptors (*e.g.*, lepidopteran βGRPs, gram-negative binding protein 3) that activate the serine protease cascade that in turn activates the Toll pathway (Gottar et al. 2006; Ochiai and Ashida 2000; Roh et al. 2009). Whether insect fungi produce effector proteins, such as the LysM domain-containing effector protein Ecp6, to suppress PAMP mediated responses is unknown. In addition to fungal wall components, virulence components such as the cuticle-degrading proteases (*e.g.*, PR1A) expressed as transgenes in *Drosophila* can activate the Persephone protease that results in activation of the Toll pathway (Gottar et al. 2006). Although the cuticle contains prophenoloxidase that is likely activated by the fungal proteases, it is unknown if the cuticle-degrading enzymes under *in vivo* conditions can up-regulate the internal innate defense response prior to fungal ingress into the hemocoel.

Mycopathogens possess multiple mechanisms designed to evade, block, or overcome innate defense systems. Early studies revealed that the cell wall surface components present in *in vitro* cells were distinct from *in vivo* fungal cell phenotypes. These observations led to the discovery that certain mycopathogens utilize antigenic mimicry and cell wall masking to evade host cellular defenses (Pendland and Boucias 1993, 2000). For example, monoclonal antibodies (MAbs) raised against the epitopes on the cell walls of the fungal entomopathogen *Nomuraea rileyi* cross-react to molecules in hemolymph, hemocytes, and fat body basement membranes (ECM) from host lepidopteran larvae (Pendland and Boucias 1998). Polyclonal antibodies raised against insect hemolymph and cell lysates and against *in vitro*-produced cells all cross-react with both insect and insect mycopathogen antigens but do not bind to non-insect-pathogenic fungi (Pendland and Boucias 2000). The presence of shared epitopes between surface components on the *N. rileyi* cells and insect hemocytes allows the fungal cells to evade host immune response via molecular mimicry, *i.e.*, they are recognized as "self" rather than "non-self" by the hemocytes. Similarly, in *M. anisopliae*, the hyphal bodies, upon gaining ingress into the hemocoel, up-regulate the *Metarhizium* collagen-like (*MCl1*) gene that encodes for a collagen-type protein that coats the hyphal body surface, hiding or masking the cell wall β-1,3-glucan elicitor complex (Wang and St. Leger 2006; Wang et al. 2005). Mutant disruption of the highly expressed *MCl1* gene results in the activation of the innate defense response and rapid phagocytosis of the fungal elements by circulating hemocytes.

An alternative to masking the cell wall elicitors is selective down-regulation of their synthesis. In several cases, *in vivo* cells lack chemical cues responsible for non-self recognition. The best examples are species of *Entomophaga* (Zygomycetes). These fungi produce wall-less protoplasts in the insect hemocoel that escape immune recognition and replicate in the hemolymph (Beauvais et al. 1989). The production of unsaturated fatty acids by these protoplasts inhibits β-1,3-glucan synthase activity (Mackichan et al. 1995), resulting in the production of motile protoplasts that move by cytoplasmic streaming using pseudopodial extensions, a unique motility mechanism. A second example is *B. bassiana*, a mycopathogen that produces *in vitro* cell phenotypes that possess a well-developed cell wall and are readily phagocytosed.

Free-floating *in vivo* hyphal body cells extracted from infected hemolymph possess thin cell walls that lack cell wall surface-associated galactose residues (Pendland and Boucias 1993). The host insect *Spodoptera exigua* produces hemolymph-borne galactose-binding opsonins that mediate the phagocytosis of non-self material containing exposed galactose (Boucias and Pendland 1993). The lack of galactose residues on the *in vivo* cells provides a means to evade opsonin-mediated phagocytosis. Follow-up studies by Tartar et al. (2005) demonstrated that, in addition to lacking galactose residues, growth *in vivo* caused a three-fold down-regulation in transcription of chitin synthase and glucan synthase genes. These *in vivo* cells produced in three different insects all possess thin walls with chitin and β-1,3-glucan being localized in the interior of the cell walls. These thin-walled cells, although lacking exposed galactose residues, are coated by compounds analogous to the collagen coating detected on the *in vivo M. anisopliae* cells.

The "masked" hyphal bodies/protoplasts bathed in the insect hemolymph absorb the available nutrients and multiply exponentially. In many cases, the hemolymph becomes turbid with fungal cells that far outnumber host hemocytes. Vegetative growth of certain fungi suppresses the cellular defense system (Hung and Boucias 1992; Vilcinskas et al. 1997). In the case of *B. bassiana,* the vegetative cells produce high-molecular-weight toxins that act as potent immunosuppressive compounds (Mazet et al. 1994). In addition, a host of secondary metabolites, including the cyclic peptides cyclosporin and destruxin, play a key role in suppression of host systems during the late stage of vegetative development (Huxham et al. 1989; Sloman and Reynolds 1993; Vey 1985; Vilcinskas et al. 1997). At a particular threshold, a synchronous switch in the developmental program occurs. With *N. rileyi*, at 4 days post-vegetative growth, several key events occur. Within 1–2 h, all hyphal bodies convert from budding to the hyphal growth program. The cell wall surface chemistry is modified such that the apical growing tips all contain basement-membrane-binding domains, providing a signal to attach to and invade the somatic tissues (Pendland and Boucias 1993). The signal underlying this switch has not been identified, but cell-free hemolymph collected at this interval induces the switch at low concentrations under nutrient-rich *in vitro* conditions. Likely, a quorum-sensing chemistry such as that of the oxylipins (Tsitsigiannis and Keller 2007) is involved in this synchronous switch in cell phenotype. The resulting tissue-invasive phenotype releases enzymes and secondary metabolites that kill the host and digest and assimilate somatic tissues. The result is the production of a mycelial biomass that, provided proper moisture and temperature, differentiates into conidiospore-bearing structures. If the dimorphic program is not coordinated, an unregulated phenoloxidase cascade may be activated that results in host death and termination of the fungal developmental program.

Remarkably, the exponential vegetative growth of mycopathogens in the hemocoel does not dramatically impair host function. Insects with a major portion of their hemolymph composed of fungal cells continue to feed, move, and develop until the end of the vegetative stage. However, the presence of fungal cells may elicit signals perceived by other insects. Swansson et al. (2009) demonstrated that vegetative development of *Ascosphaera apis* induced honey bee larvae to emit a unique blend of volatiles, of which phenethyl acetate induced hygienic behavior in adult bees. In some cases, insect parasitoids have been shown to detect and reject fungal-infected hosts after probing with the ovipositor (Fransen and Vanlenteren 1993). It should be noted that in other cases, beneficial parasitoids and predators do not differentiate healthy from infected hosts and thus suffer subsequent infection (Lord 2001) or fitness costs (Roy et al. 2008).

Fungal development, using unknown signals, can modulate the feeding behavior, thermotropic, phototropic and geotropic preferences, and social behaviors of insects (Roy et al. 2006). In certain cases, fungal infection induces insects to thermoregulate, to seek habitats with temperatures unfavorable for fungal development. This behaviorally induced fever does not cure the insect, but it does slow fungal development. At the late stages of disease, several species exhibit the "summit disease" response, similar to the "Wipfelsucht" originally described for baculovirus-infected insects. In these cases, the insect climbs to an elevated position, attaches to foliage, and dies; depending on the situation, this response may aid in dispersing progeny conidiospores or in separating the diseased host from healthy conspecifics. For example, *Entomophthora muscae*, termed the "enslaver pathogen," redirects infected yellow dung flies at death by unidentified signals to orientate on the plant in an abnormal posture to maximize spore dispersal onto healthy conspecifics (Maitland 1994). Similarly, Andersen et al. (2009) examined the behavior of the arboreal ant *Camponotus leonardi* infected with *Ophiocordyceps unilateralis*. Infected ants just prior to death descend from the tree canopy to a height of 25 cm and attach to the plant by their mandibles. This disease-induced "death grip" places the mycosed insect in a habitat optimal for spore production and dispersal. The fungal signals involved in the programming of this phenotype have yet to be discovered. It is interesting to note that the resulting ant graveyards are avoided by healthy ants (Pontoppidan et al. 2009).

2.4 Fate of Mycosed Insects

Post-mortem, various mycosed insects emit signals that enhance or suppress the transmission of disease to healthy conspecifics. For example, adult house flies killed by *E. muscae* are more attractive to healthy males than are healthy females. Initially, this circumstance was attributed to their swollen abdomens, which provide a visual cue to males (Moller 1993). However, later work by Zurek et al. (2002) suggested that chemical cues attract males; they attempt to copulate with the mycosed flies. The result is the horizontal transmission of the fungus to the male and to subsequent female partners. It is probable that similar scenarios occur with other insect mycopathogens; one can predict that fungi producing the sticky hydrophilic conidiospores that rely on direct contact with the mycosed specimens have mechanisms to attract healthy conspecifics.

Social insects, defined by their genetic homogeneity and colony lifestyle, are contained within a controlled habitat and interact continuously with nestmates. In many cases, social insects inhabit a soil environment that harbors numerous pathogens; one would expect widespread epizootics. However, these insects express highly effective behavioral defenses that neutralize infectious threats to the colony. Detection of fungal propagules induces grooming behaviors to reduce propagule contact with the colony (Staples and Milner 2000). Wasps, bees, ants, and termites also exhibit necrophoric (undertaking) behavior; they transport and bury dead insects away from the colony (Howard and Tschinkel 1976; Julian and Cahan 1999; Renucci et al. 2011; Trumbo and Robinson 1997; Wilson et al. 1958). Complementing this hygienic activity is altruistic behavior, in which unattended, sick insects move away from the colony into the foraging arena. In E. O. Wilson's early work, an ester and a fatty acid extracted from the dead insect were found to elicit the necrophoric response. This work established a precedent: compounds emitted from the dead or diseased insect serve as signals triggering this behavior (Masterman et al. 2001; Visscher 1983). The collection of volatiles emitted from mummified honey bees infected with *A. apis* induces necrophoric responses by adult bees (Swanson et al. 2009). Interestingly, the ability to detect the fungal volatiles is correlated with the degree of colony hygiene; colonies that contain bees with high olfactory sensitivity take the shortest time to remove sick and dead brood. The trigger of the necrophoric response is not always the release of volatiles from diseased insects. Recent work by Choe et al. (2009) demonstrated that the healthy Argentine ant *Linpithema humile* responds to the loss of ant-produced colony recognition cues rather than to the production of "death cues" by dead conspecifics. The loss of the normal chemical signatures stimulates necrophoric behaviors. The fate of the mummified insect varies according to the mycopathogen and its habitat; certain pathogens produce a variety of resting structures designed to allow the fungus to persist over long periods. In other cases, the mummified insect serves as a nutrition source for mycophagy and/or mycoparasitism.

3 Insect Mycophagy

There are many species of insects and other arthropods that use fungi as a source of nutrition and shelter. For the most part, fungi digesting woody plant material provide insects with a food source that is nutritionally comparable to seeds, fruit, and animal biomass. As pointed out by Martin (1979), as a component of the food chain, fungi do an excellent job in assimilating high-carbon cellulose substrate (wood) and converting it into a more balanced nutrition source. Insects also sequester fungal enzymes and use them to digest plant and fungal material and, potentially, to detoxify xenobiotics (Boddy and Jones 2008). Mycophagy can be simply the ingestion of yeast, mycelium, or sporocarps, or it may involve elegant symbiosis such as that exhibited by fungal-farming termites and ants. Key to these associations is the ability of the insects to detect and identify the chemical signatures produced by fungal substrates. In many cases, these signatures are VOCs released by fruiting bodies that function much like flower scents in their attraction of insect pollinators (Table 2). The ability of the fungi to attract insects is supported by the mimicry exhibited by the sapromyophilous plants that produce

Table 2 Selected examples of fungal chemicals that induce various insect responses

Compound(s)	Fungal source	Responsive insect(s)	Effect
Triglycerides	*Nigrospora sphaerica*	*Tribolium confusum*	Aggregation
Rugulosin	Conifer fungal endophytes	*Choristonuera fumiferana*	Repellent
Naphthalene	*Muscodor vitigenus*	*Cephus cinctus*	Repellent
3-(4-methylfuran-3-yl) propanol-1-ol	Fungal endophyte	*Eysarcoris ventralis*	Repellent
Blend of 5 alcohols, ethyl acetate and acetaldehyde	*Fusarium verticilloides*	Nitidulid sap beetles	Attractant
Isoamyl acetate, 2 phenylethyl acetate	*Pichia pinus* and *Hansenula holstii* (yeast symbionts)	*Dendroctonus frontalis*	Synergists to plant attractant
Methl(Z)-3-methyldodec-2-enoate, chokol K	Epichloe endophytes	*Botanophila* flies	Attractant
Blend of benzaldehyde, phenyl ethyl alcohol, phenylacetaldehyde, indole,	*Puccinia punctiformis (suavelons)*	*Ceutorhynchus litura*	Attractant
Benzaldehyde, phenyl ethyl alcohol, phenylacetaldehyde, phenylethyl esters	*Puccinia monoica*	Dipteran pollinators	Attractants
Blend of floral fragrances (jasmine lactone, methyl benzoate, phenylether alcohol, methyl palmitrate) and pheromone-like chemicals (6-methyl-5-hepten-2-one, γ-caprolactone)	*Puccinia arrhenatheri*	Various insect pollinators	Attractants
Fermentation volatile blend	*Monilinia vaccinii-corymbosi*	Spectrum of insects	Attractant
1-Octen-3-ol	*Trametes* fruiting bodies	*Cis boleti*	Attractant
From a complex blend both 2-methyl-1-propanol and 1-heptene	*Verticillium bulbosum*	*Onychiurus armatus*	Attractant
Secondary metabolites, gliotoxins, sterigmatocystin	*Aspergillus* spp.	*Drosophila, Folsomia candida*	Antifungivory, repellent
Plant alkaloids	*Epichloë* and *Neotyphodium* spp. (fescue grass endophytes)	*Listronotus bonariensis*	Repellent
Fruit odor made of a complex of esters, alcohols, ketone, and aldehyde	*Ceratocystis fagacearum*	Nitidulid beetles	Attractant
Oligosulfides	Phallaceae	Carrion flies	Attractant

Phenol, indole, and p-cresol	Phallaceae	Filth flies (fecal)	Attractant
Oxygenated monoterpenes	*Ophiostoma* (blue stain fungus)	Bark beetle parasitoid *Roptrocerus xylophagrum*	Synergists to plant VOCs
Blend of ethyl acetate, acetaldehyde, 2-pentanol,3-methyl butanol	*Saccharomyces cerevisiae*	Nitidulid *Carpophilus hemipterus*	Attractant
1-Octen-3-ol	*Aspergillus* sp.	Weevil parasitoid *Lariophagous distiuendus*	Repellent
Blend of acetaldehyde and other components	*Ablosternum spp.* (symbionts of *Sirex*)	Three parasitoids of *Sirex*	Attractant
Phenethyl acetate, 2-phenylethanol, and benzyl alcohol	*Ascosphaera apis* (causal agent chalkbrood)	*Apis mellifera*	Induces hygienic behavior in worker bees
ß-Glucosidase	*Fibularhizoctonia*	*Reticulitermes* species	Egg mimic

mushroom-mimicking flowers (Kaiser 2006). For example, orchids within the genus *Dracula* produce mushroom-mimicking flowers that release VOCs containing a blend of oct-1-en-3-ol, oct-1-en-3-one, octan-3-ol, and octan-3-one, compounds produced commonly by mushrooms. A second example is the angiosperm *Aristolochia arborea*, which produces fly-trapping flowers that morphologically mimic the toadstool in the genus *Marasmius* and emit a mushroom-like scent. Both flowers are highly attractive to fungus gnats that act as pollinators.

4 Insect Fungivory

The nutritional resource of insect fungivores is the patchily distributed fruiting bodies or sporocarps. Sporocarps are either perennial (bracket polypores) or ephemeral (*e.g., Agaricales*) in providing substrate to insects. The majority of insect fungivores are found in specific families of flies and beetles that have evolved the ability to assimilate fungal biomass as their major nutrient source. Typically, insects targeting the ephemeral food source have short life spans, are generalists, and display efficient search patterns to find food. Historically, the insects feeding on sporocarps were considered to be polyphagous, feeding on wide range of fruiting bodies. Jonsell and Nordlander (2004) reported that this supposition is true for those insects utilizing late-stage decaying fruiting bodies; however, primary colonizers often are monophagous. It has been suggested that the chemical composition of the fruiting bodies serves as host selection cues for the insects. One might speculate that the olfactory cues emitted from newly formed sporocarps serve as a signature used by monophagous fungivores, whereas the common odors produced by different mature/decaying sporocarps serve as generic cues attractive to polyphagous fungivores. In many cases, adult emergence of the monophagous fungivore is timed with sporocarp production.

The relationships existing between the insect fungivores and their food sporocarps have served as models for community ecology studies dealing with populations using patchily distributed resources. Indeed, extensive studies have been conducted on the insects attacking the wood-decaying bracket fungi in Northern Europe (Hanski 1989; Komonen et al. 2000). In part, these studies were stimulated by the impact of intensive forest management on biodiversity. The removal of dead wood has threatened hundreds of forest insects, including species which rely on the bracket fungi as their primary food source (Berg et al. 1994). Behavioral studies by Jonsell and Nordlander (1995) determined that fungal odors may serve as attractants to some insect species but not other fungivorous insects. Using fruiting bodies of the red-banded polypore *Fomitopsis pinicola* and the tinder polypore *Fomes fomentarius* fungi as baits in a spruce or birch habitat, the investigators trapped ~30 beetle taxa showing significant attraction to fungal odors. Ten of these taxa corresponded to species reared from field-collected fruiting bodies. The predominant taxa included the beetles *Cis glabratus* and *C. quadridens*; both were selectively attracted to their hosts. However, none of the beetles known to feed on *F. fomentarius* were attracted by the odor traps. Other beetles, including several monophagous anobiids, were attracted to exposed fruiting bodies at a rate ten times higher than to the odor traps. Jonsell and Nordlander (1995) suggested that chopped fruiting bodies used in the traps likely were producing damage-induced components that were repellent to primary colonizers. The volatiles emitted from fruiting body samples were collected, analyzed, and assayed for biological activity (Faldt et al. 1999). The volatile blend emitted from fruiting bodies varied according to the species, to the age (vegetative vs. sporulating phases), and to the condition (live vs. chopped) of fruiting body samples. Some of the major volatiles collected from *F. pinicola* included the sequisterpene β-barbatene and the 8-carbon aliphatic compounds oct-1-en-3-ol, octan-3-one, and oct-1-ene. From *F. fomentarius*, volatiles included octan-3-one, β-phellandrene, cis-furanoid linalool oxide, and β-myrcene. Oct-1-en-3-ol, alone or blended with octan-1-ol, octan-3-ol, and nonan-1-ol, was tested in bioassays for the ability to attract insects in field cages. Of the multitude of insects that were trapped in control and baited cages, only six species showed clear preference to baited field cages. The oct-1-en-3-ol, or mushroom alcohol, attracted many species of polyphagous saproxylic beetles, whereas the blend, acting as a kairomone, attracted the staphylinid, *Lordithorn lunulatus,* an insect predator commonly found in mushrooms. In addition to the beetles, *Epinotia tedella*, an herbivore of mature spruce needles, was strongly attracted to oct-1-en-3-ol baited traps. Potentially, mature needles harbor an abundance of the fungal endophyte *Lophodermium piceae*, a species reported to emit oct-1-en-3-ol.

The response of ciid beetles to fungal volatiles derived from different host and non-host fruiting bodies was examined by Gueverra et al. (2000). In olfactory tubes, walking beetles discriminate among fungal aromas in a fashion identical to that observed with intact fruiting bodies. The VOCs vary among fungal species and the subtle differences in the aroma blend determine the attractiveness of a fruiting body to a beetle species. The VOCs from the fruiting body of *Trametes gibbosa,* partitioned using gas chromatography (GC), were identified by mass spectral analysis (Thakeow et al. 2008). The resulting fractions, tested by GC-electroantennography (GC-EAG, Fig. 1), demonstrated that the 1-octen-3-ol, present as both (*R*) and (*S*) enantiomers (93:7), produces a dominant response in *Cis boleti* at the physiological level emitted by the fungus. Furthermore, it was demonstrated that female beetles are much more sensitive to the predominant (*R*) enantiomer than are males. The beetle attractant compounds, such as C_8 aliphatic alcohols and ketones produced by the bracket fungi, also are produced by fungi (*e.g., Agaricales)* characterized by ephemeral fruiting bodies, and these compounds were reported initially as oviposition attractants for the mushroom phorid fly *Megaselia halterata* (Grove and Blight 1983). Later assays by Pfeil and Mumma (1993) demonstrated that pure 1-octen-3-ol and 3-octanone acted as repellents to *M. halterata*. The spawn compost at a certain age is known to produce VOCs attractive to these flies (Pfeil and Mumma 1992); resolving the underlying chemistry would provide valuable tools for monitoring and trapping this mycophagous pest insect.

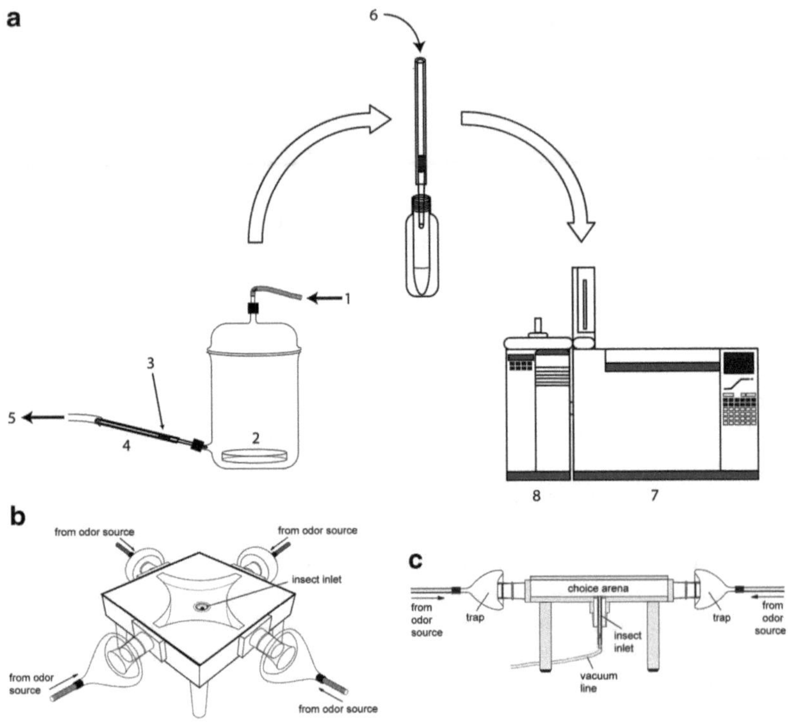

Fig. 1 Apparatus for chemical and biological analysis of volatile organic compounds. (**a**) Volatile collection system requiring: *1* clean air into the system, *2* test sample, *3* Super-Q packing material to trap volatile chemicals, *4* removable filter containing packing material, *5* vacuum for pulling air through the system, *6* solvent added to filter to remove chemicals from packing material, *7* gas chromatograph where solution with volatile chemicals is injected to separate out the individual chemicals, *8* mass spectrometer and/or FTIR to identify chemicals by characteristic chemical spectra. (**b**) A four-armed airflow olfactometer that requires a constant stream of purified air guided over each of four odor sources (or blank controls) into a choice arena, into which the test insect is introduced though a central inlet. Orientation and behavior of the insect in the four produced odor fields are recorded for a determined time period and evaluated using Observer XT (Noldus Information Technology) software. (**c**) Side view of the olfactometer. A central vacuum line directs the airflow and assures the generation of four odor fields in the arena

4.1 *Fungivory Blended with Mutualism*

The relationships between insects and fungi are complex and involve multiple trophic levels. One example of multitrophic interaction is the relationship between the fungal endophytes in the genus *Epichloë*, their plant host (perennial grasses), and anthomyiid flies in the genus *Botanophila* (previously reported as *Phorbia).* Different *Epichloë* species infect and grow as intercellular hyphae in the grasses of the subfamily Pooideae. During their vegetative growth, these fungi produce a complex of secondary compounds, including various ergot alkaloids, which are believed to provide the plant protection against herbivory (Bacon et al. 1986). In the spring, these fungi develop external spermatial stroma; their development on the leaf sheaf that covers the inflorescence prevents flowering (Kohlmeyer and Kohlmeyer 1974). This endophyte is heterothallic and cannot undergo self-fertilization. In nature, the *Botanophila* flies vector spermatia (conidia) between opposite mating types; the spermatia were reported not to be dispersed passively by wind or rain (Welch and Bultman 1993). After finding the stroma, the adult flies test its suitability for oviposition and, while doing so, ingest the spermatia. The fly, while seeking additional oviposition sites, mechanically vectors the spermatia by dragging its ovipositor over the stroma and defecating viable conidia. In certain habitats, these flies are important fungal vectors; the development of peritheca on grasses lacking fly populations is less than 20%, versus a ~70% production of stroma on grasses at sites with this insect (Bultman and White 1988). These field observations were verified by controlled exclusion experiments that demonstrated that this insect is a major vector. Bultman et al. (1998) proposed that this interaction represents one of the few examples of an obligate insect-mediated fertilization of fungi. Recently, the obligate nature of this relationship has been challenged by Górzyńska et al. (2011), who reported that non-cultivated grasses that harbor the endophyte *E. typhina* produce stroma and peritheca at sites lacking *Botanophila* flies, demonstrating that in nature additional mechanisms can provide for the transfer of spermatia between opposite mating types.

The endophytes that use the gamete-transferring *Botanophila* flies as "fungal pollinators" suffer from fungivory; these insects selectively target the stroma produced by this fungus for oviposition. Emerging larvae feed on the stromata and/or the developing peritheca. However, there are processes that dampen resource exploitation by these flies (Bultman et al. 2000). On the one hand, the endophyte responds to insect fungivory by increasing its reproductive output; on the other hand, the insect is sensitive to density, so the more eggs deposited, the higher the rate of egg and larval mortality. The result is a balanced mutualism that provides benefit to both the fungal endophyte and the insect vector.

Of interest is how these flies find suitable stromata on which to oviposit. Two volatiles, the sesquiterpene alcohol chokol-K and the methylester methyl (*Z*)-3-methyldodec-2-enoate (MME), originating from distinct biochemical pathways, play important roles. Initially, Schiestl et al. (2006) collected the VOCs emitted from *Epichloë* stromata on several host grasses and tested them using GC-EAG analysis on dissected fly heads. The single compound that elicits a positive EAG response is the alcohol chokol K. It has broad fungitoxic properties, inhibits germination of various fungal species (Steinebrunner et al. 2008a), is proposed to protect the stroma from attack by mycoparasites, and provides resistance against secondary infection by plant pathogens. It should be noted that the concentration needed to inhibit fungal germination is considered to be the highest ecologically relevant concentration (Steinebrunner et al. 2008a). Likely, these fungal metabolites are sequestered by the fungivore and protect the fly against insect mycopathogens. Chokol K was hypothesized to have originated as an antimicrobial defense mechanism that has been exploited secondarily as a

Botanophila attractant (Schiestl et al. 2006). The second compound, MME, was elucidated by Steinebrunner et al. (2008c) as a novel second VOC that was detected in the head-space of *Epichloë* species and elicited an EAG response from *Botanophila* flies. Follow-up studies examined the production of the VOCs by different *Epichloë* species to determine if blends of these two compounds are selective to different *Botanophila* fly genotypes (Steinebrunner et al. 2008b). Results demonstrated that the stromata of different *Epichloë* genotypes all produce unique blends of the chokol K, MME, and other compounds. The amounts of VOCs produced by the different *Epichloë* genotypes are influenced by plant host differences. Analysis of headspace VOCs produced under identical *in vitro* conditions demonstrated that the production of chokol K is genetically encoded in *Epichloë*, whereas plant-mediated signaling is proposed to be involved in the production of MME (Steinebrunner et al. 2008b). Both the intact stromata of *E. clarkii* and *E. typhina* and the synthetic blends designed from the headspace chemistry are highly attractive to the same fly taxon (Steinebrunner et al. 2008a, c). The chokol K and MME blends comprise a unique signal specifically recognized by *Botanophila* flies and which provides the female fly a "private channel" to find a nutrient source for her offspring.

5 Fungal Kairomones: Infochemicals

Various non-fungivorous insects use chemicals emitted from yeasts and fungi growing on plant material as cues. The presence of a fungus or yeasts on plant material provides information about food quality. For example, fungi associated with stored grains and fruits produce VOCs that serve as oviposition attractants for stored grain beetles (Starratt and Loschiavo 1970), dried fruit beetles (Phelan and Lin 1991), and several moth species (Honda et al. 1988). Often, attractive VOC blends are associated with the fungi developing on the target food substrate and reflect a combination of fungal metabolites and components released by plant cell lysis. In addition to serving as infochemicals for insect herbivores, fungal VOCs also are used by insect predators and parasitoids as host location cues.

Early work by Madden (1968) reported that the various hymenopterous parasitoids of the wood wasp *Sirex noctuilo* use the VOCs produced by the fungal symbiont *Amylostereum* sp. to find their respective egg and larval hosts. This basidiomycete, a causal agent of white rot in conifers, is harbored in the specialized cuticular structures or mycangia of the wood wasp and gains ingress into the plant via ovipositing female wood wasps. The insect develops in a cryptic habitat that supports the growth of both the insect and the fungal symbiont. This plant pathogen, fed upon by developing *Sirex* larvae, supplements insects with hydrolytic enzymes that operate in concert with gut bacteria to assimilate ingested lignocellulose (Adams et al. 2011; Kukor and Martin 1983). Interestingly, the egg parasitoid *Ibalia leucospoides* prefers the VOCs produced from egg drills at the time of egg eclosion, whereas the larval parasitoid *Rhyssa persuasoria* responds to logs at

5 months post-inoculation, suggesting that different parasitoids recognize VOC blends produced when there are suitable host targets (Madden 1968; Spradber 1974). Retesting *I. leucospoides* in various assays (Martinez et al. 2006) has demonstrated that the parasitoids respond in a dose-dependent manner to *Amylostereum areolatum* VOCs in the presence or absence of tree and potential host volatiles. The authors point out that the use of host mutualists, such as *Amylostereum*, provides the parasitoid with a reliable VOC cue that is amplified over time by an ever-increasing fungal biomass.

The parasitoids that attack bark beetles furnish another example of insect fungi used for host location. Like the wood wasps, bark beetles inhabit a cryptic habitat and harbor fungal associates that are inoculated by the beetles and develop extensively in the tree tissue. Sullivan and Berisford (2004) conducted a series of experiments that tested the attractiveness of pine tree bolts inoculated with bluestain fungi to the female pteromalid *Roptrocerus xylophagorum* and the braconid *Spathius pallidus*. In the laboratory, bolts infested with the bluestain fungus *Ophiostoma ips* or *O. minus* were more attractive to *R. xylophagorum* or *S. pallidus,* respectively, than mock-inoculated bolts. It should be noted that bark beetles *Ips grandicollis*, surface-sterilized as eggs to remove fungal associates, are as attractive to *R. xylophagorum* as are those beetles harboring *O. ips*. In addition, under field conditions the parasitoids do not discriminate between fungal-inoculated and mock-inoculated bolts, suggesting that fungal volatiles may not act as long-range host location cues. A second study with bark beetles included the natural enemies of *Ips pini* and its fungal/yeast associates *O. ips* and *Pichia scolyti* (Boone et al. 2008). Results showed that the pteromalid *Heydenia unica* is attracted to both bark-beetle-colonized logs and to *O. ips*-inoculated logs, whereas the dolichopodid predator *Medetera* is attracted to logs colonized by *P. scolyti*. The authors suggest that the fungal associates, by metabolizing plant terpenoids, produce oxygenated monoterpenes that are attractive to pteromalid parasites.

A third example in which an insect parasitoid uses fungal signals to locate insect hosts involves the frugivorous *Drosophila*. These insects feed on decaying and fermenting fruit, which support the growth of bacteria and yeasts, respectively (Atkinson and Shorrocks 1977). Using choice olfactometer assays, Vet et al. (1984) determined that the braconid larval parasitoid *Asobara tabida* is attracted to fruit at the early stages of fermentation, corresponding to the period when it would contain hosts most suitable for parasitization. The host-finding cues used by these parasitoids have not been investigated but are assumed to be yeast fermentation products. A situation opposite to the drosophilid-yeast-parasitoid interaction exists with the granary weevil-fungus-parasitoid (Steiner et al. 2007). In this scenario, infestations of grains by weevils lead to mold growth that produces localized hot spots, resulting in extensive insect mortality. In the olfactometer, both moldy wheat and moldy insect feces repel females of the larval parasitoid *Lariophagus distinguendus*. The major VOC emanating from the moldy substrates is 1-octen-3-ol, plus lesser amounts of 3-octanone and 3-octanol. At high dosages (>300 ng), the 1-octen-3-ol is repellent to *L. distinguendus*, but low dosages (30 ng) are neutral. Interestingly, 1-octen-3-ol repels *M. halterata* yet attracts mycophagous *Cis* beetles (see Sect. 4). Steiner et al. (2007) suggest that the avoidance of mold by the parasitoid enables it also to avoid suboptimal host patches.

Certain fungi can be manipulated directly by an insect parasite, affecting host biology. Such is the case with *Kodamaea ohmeri,* a yeast associate of the small hive beetle, *Aethina tumida,* an insect parasite of the honey bee *Apis mellifera* (Torto et al. 2007b). Adult beetles are believed to acquire this flower-associated yeast when they feed on pollen. The beetles, harboring the yeast in their digestive tracts, are attracted to low levels of the honey bee alarm pheromone emitted from hives. Once inside the hive, they deposit the yeast into the honeycomb. The yeast proliferates there, producing and releasing a complex of VOCs that include isopentyl acetate (IPA), 2-heptanone, and methyl benzoate; two are important constituents of the bee alarm pheromone. The volatiles attract numerous beetle adults and the resulting infestation forces adult bees to abscond, leaving the hive defenseless against the beetles that consume the wax, honey, pollen, and brood. Interestingly, *A. tumida*, in addition to developing in hives, is attracted to and can develop in fruit infested with yeasts such as *K. ohmeri* (Benda et al. 2008). The identification of yeast volatiles as attractants has led to the development of an in-hive, baited trap to detect this invasive insect (Torto et al. 2007a).

6 Manipulations of Host Plants by Fungal Pathogens

Many plant-pathogenic fungi, being sessile organisms, have developed mechanisms to manipulate host resources to optimize their transmission by insects. In certain cases, the plant pathogen modulates the plant to produce VOCs that are attractive to insect vectors. An example is the ascomycete *Ophiostoma novo-ulmi*, the causal agent of Dutch elm disease. This disease, vectored by the native elm bark beetle *Hylurgopinus rufipes*, European elm bark beetle, *Scolytus multistriatus,* and Asian banded elm bark beetle, *Scolytus schevyrewi,* has spread throughout North America and has decimated this tree species. McLeod et al. (2005) collected and analyzed VOCs emitted from healthy and infected American elms and from *in vitro* *O. novo-ulmi* cultures. They found that fungus-infected trees produce a blend of VOCs containing the monoterpene (−)-β-pinene and the sesquiterpenes (−)-α-cubebene, (−)-spiroaxa-5, 7-diene, and (+)-δ-cadinene that are highly attractive to adult male and female *H. rufipes*. The presence of all four components, at the ratio present in the natural blend, is necessary to attract the insects. The levels of sesquiterpenes are upregulated by infection, and their synthesis, as well as that of the monoterpene, is directed by the plant's biosynthetic machinery. *In vitro* grown fungi emit none of the components found in the blend produced by infected trees. The fungal signals that induce elm trees to produce the elm-beetle-attractive semiochemicals are unknown. Whether or not the disease-induced odor blend that is attractive to *H. rufipes* also attracts *S. multistriatus* and the recently introduced *S. schevyrewi* is unknown, but both species respond to host plant VOCs (Lee et al. 2010). Interestingly, the transmission of spores by the *Scolytus* species is enhanced by mycophagous, phoretic mites that harbor *O. novo-ulmi* in specialized sporothecae (Moser et al. 2010). Similarly, the *Ceratocystis* pathogen produces volatiles that are highly attractive to sap-sucking nitidulid beetles (Chang and Jensen 1974; Hinds 1972; Lin and Phelan 1992). In the case of the oak wilt disease, the fungus *C. fagacearum* forms a mycelial mat that eventually splits tree bark; at this point, it sporulates and emits a fruity volatile blend that is attractive to nitidulids (Lin and Phelan 1992). Beetles attracted to the VOCs are coated with the sticky spores and serve as passive vectors which fertilize receptive hyphae of heterologous mats. Chemical analysis revealed that *in vitro* grown *C. fagacearum* produces 16 components, including 9 esters, 5 alcohols, 1 ketone and 1 aldehyde. Many of these volatiles have been detected as food odors, and several have been identified as nitidulid attractants, suggesting that these pathogens are producing a VOC blend that mimics the insect-attractant plant cues.

A second example is the plant disease caused by certain *Monilinia* spp. that infect various fruits. These pathogens have a two-phase life cycle: the first involves the infection of vegetative shoots by ascospores (meiospores) and the second the production of conidia (mitospores) that invade the flower and cause mummification of the fruit. Several species, including *M. vaccina-corymbosi* and *M. oxycocci*, belonging to the Disjunctoriae, produce short-lived conidia that are transported to flowers by insect pollinators (Batra 1991). The *M. vaccina-corymbosi,* casual agent of mummy-berry disease in high-bush blueberries, infects plant shoots via ascospores. Vegetative growth in plant shoots results in the production of conidia in localized lesions; these lesions have ultraviolet reflectance properties similar to those of the blueberry flower calyxes, and they emit a sweet odor that attracts insect pollinators (Batra 1991; Batra and Batra 1985). When transported to flowers by insects, this conidial stage acts as a pollen mimic; it has evolved the ability to penetrate flowers via the gynoecial pathway (Ngugi and Scherm 2004). The conidia attach to the tip of the stylus and produce a germ tube that is guided by surface chemistries to grow down the stylar canal into the ovary (Ngugi et al. 2002). Vegetative growth in the fruit leads to the production of pseudosclerotia in the form of mummy berries. Other *Monilinia* spp., belonging to the Junctoriae pathogens of the Rosaceae, produce long-lived conidia on rotting fruits. A wide spectrum of insects attracted to fermentation products feed on these fruits and imbibe the conidia that are then transported to uninfected fruits. The insect attractants in these cases are products of fermentative yeasts and molds.

A somewhat more complex manipulation of host plants for pathogen transmission by insects is observed with the parasite-castrating smuts. For example, the anther smut *Microbotryum violaceum* (syn. *Ustilago violacea*) infects female dioecious campion species and induces the formation of male sex organs, thus creating hermaphroditic plants; infection of male plants does not alter the formation of the male sex organ (Uchida et al. 2003). The ovaries formed on the hermaphrodites are sterile. The anther structures produced in male plants and hermaphrodites contain teliospores rather than plant pollen. There also is pistal smut *Salmacisia buchloëana* (syn. *Tilletia buchloëana*) that infects dioecious buffalograss. This plant pathogen infects male plants and induces the formation of nonfunctional female sex organs (pistils) that harbor fungal teliospores (Chandra and Huff 2008). The male anthers, not containing teliospores, remain undeveloped. The parasitic castration induced by these pathogens results in sex organs that are filled with teliospores that are disseminated subsequently by diurnal and nocturnal insect pollinators such as wasps, syrphid flies, bumblebees, and moths (Altizer et al. 1998). In certain cases, these smut pathogens induce the infected plant to develop more flowers per inflorescence at times preceding flower production by uninfected plants (Jennersten 1988). Furthermore, infected flowers remain open for longer intervals than healthy flowers. The amount of nectar per inflorescence in infected plants is similar to that produced by healthy plants. These traits should favor insect-mediated pollination events for the diseased plant. Interestingly, different insects display different preferences when given the choice between smut-infected and healthy plants. Both the diurnal bumblebee and dipteran pollinators visit infected plants less than they do healthy plants, whereas the nocturnal lepidopteran pollinators do not discriminate between infected and healthy plants (Jennersten and Kwak 1991).

7 Mimicry

Mimicry typically involves one biological entity (the mimic) copying another entity (the model) as a tactic to increase the fitness of the mimic. In the case of insect-fungal relationships, the mimic is usually the fungus component and the model a floral component, and the fitness parameter involves insect-mediated spore transmission. An exception is the mimicry expressed by the "cuckoo fungus" inhabiting the colonies of the lower termites in the genus *Reticulitermes* (Matsuura et al. 2000). The fungus, a basidiomycete in the genus *Fibularhizoctonia*, produces sclerotia in the form of round balls (termite balls) that in size and shape mimic termite eggs (Matsuura 2006). The mimics produce the cellulose-degrading ß-glucosidase that is an important component of the termite egg-recognition pheromone (Matsuura et al. 2009). Termites tend the mimics as if they were eggs. As the termite balls age, the termites place them in middens, where they germinate and proliferate, producing a new generation of termite balls that are transported back into the nursery by worker termites. Other forms of insect mimicry by fungi have yet to be discovered; at present, the majority of them involve floral models.

7.1 Floral Mimicry

The association between insects and fungi exhibiting floral mimicry is analogous to that observed in insect pollinators that transmit pollen to the flowering plants. The plant-pathogenic fungi mimic the components of the flower-pollinator interaction in order to disperse fungal gametes. Typically, they accomplish this feat by producing distinct pseudoflowers that emit a floral fragrance and display ultraviolet reflectance properties of a true flower. The floral chemistry produced by the sexual stage of these fungi normally does not match that of the healthy host flower they infect

(Raguso and Roy 1998), thus reducing the likelihood of transferring fungal gametes to healthy flowers and limiting competition between the flower and its mimic.

Certain rust species that display floral mimicry reprogram plant development to produce growth deformities such as galls and witches' brooms. Biotropic development of the mycelia in the plant meristem alters phytohormone levels, resulting in disorganized growth at plant apices and the emergence of multiple shoot axes (Kilaru et al. 2007). For example, the rust *Puccinia arrhenatheri* that systemically infects the common Barberry *Berberis vulgaris* L. causes the plant to produce witches' brooms (Naef et al. 2002). This fungus produces sexual structures; the spermatogonia on the upper and lower leaves are bright yellow in color, have the ultraviolet reflectance of a flower, and emit a strong, sweet-smelling scent. The sugary nectar they produce contains the fungal gametes (spermatia) that require transport to an opposite mating type. Insect exclusion experiments conducted by Naef et al. (2002) demonstrated that insect visitation plays a critical role in the transfer of spermatia and in aeciospore formation. Field observations demonstrated that insects, belonging mainly to the Hymenoptera and Diptera, preferentially visit witches' broom plants. The composition of the headspace VOCs from infected and healthy leaves and flowers is complex, with each containing a unique blend of compounds. Healthy leaves emit green-leaf odors at levels 50 times less than those measured in the infected tissues and flowers. The *B. vulgaris* flowers emit a spermy odor that shares only two minor components with the complex emitted by the spermatia-infected leaves (Naef et al. 2002). The VOC blend released by the spermatia differs from the green leaf and flower blends and can be placed into three functional groups: floral fragrances known to attract insect pollinators, pheromone-like compounds, and defensive compounds. The floral fragrances associated with the spermatia-infected leaves, found in other true flowers but not *B. vulgaris* flowers, are believed to attract both diurnal and nocturnal pollinators. Several of the compounds produced by spermatia-infected leaves, including the indole and phenethyl alcohol, are also components of the fragrance produced by the sexual stages of the Canadian thistle rust, *P. puntiformis* (Connick and French 1991).

Another example of a floral mimic is the rust *P. monoica* that infects various herbaceous mustards (Roy 1993). Infection by this pathogen induces the plant to produce more leaves and leaf rosettes, to grow taller, and to grow pseudoflowers at the crowns of the elongated stems. These sweet-smelling pseudoflowers are comprised of clusters of bright yellow leaves composed mainly of spermatogonia, receptive hyphae, and the sugary spermatial fluid. They produce significantly more nectar (18- to 70-fold) than co-occurring flowers (Roy 1993). As with the spermatia-infected *B. vulgaris* leaves, the *P. monoica* pseudoflowers have a UV-reflectance that is identical to the yellow flowers of other angiosperms. Significantly, they attract a range of insect pollinators that function to transfer the fungal gametes to opposite mating types. In addition to the insect pollinators, these pseudoflowers also attract pollinator predators.

Exclusion cage studies demonstrated the importance of the insect pollinator; plants in cages with no pollinators produce no pseudoflowers with aeciospores, whereas 100% of the open pollinated pseudoflowers contain aeciospores (Roy

1993). Follow-up field studies by Roy (1994) examined the impact of pseudo-flowering plants on the pollination of co-blooming healthy buttercups. Both plant types, when placed in a mixed planting, receive higher rates of visitation than do monocultures, suggesting that pseudoflowers operate as Mullerian floral mimics. The pseudoflowers present the pollinators with three attractants, the visual (UV-reflectance) and olfactory (fragrance) signals and the reward (sugar exudate) for their pollination services. Typically, diurnal pollinators use a combination of signals to find a suitable nectar/pollen source; visual cues are used initially to locate flowers and odor is used as a short-distance cue (Dobson 1993). Roy and Raguso (1997), studying the role of visual and olfactory attractants, determined that their attractiveness varies according to the pollinator. The solitary halictid bees initially use yellow-colored visual cues, but they respond more quickly when yellow is combined with the floral fragrance. For dipteran visitations, the role of visual and odor cues varies among the different species, but odor appears more important than color as a long-distance cue. Due to the cool night temperatures at the sampling site (altitude >2,700 m), there are no visitations by nocturnal pollinators. A chemical analysis of the volatiles released by various species of *Puccinia* growing on their respective *Arabis* hosts revealed an array of aromatic compounds, including phenylacetaldehyde and 2-phenyl ethanol that, to humans, possess a pseudo-flower fragrance (Raguso and Roy 1998). These two compounds, along with indole and benzaldehyde, also are found in the headspace of *P. punctiformis* (see above, Connick and French 1991). In general, similar VOC profiles from different *Puccinia* are produced on different host plants.

Not all floral mimicry results in the production of typical fragrances associated with flowers. For instance, the stinkhorn fungi (Phallaceae) produce fruiting structures or gleba containing a mucilaginous mass of spores. These structures, depending on the stinkhorn species, may produce oligosulfides and/or a complex of phenol, indole, and p-cresol, releasing putrid scents that attract a variety of insects, including filth flies, that serve as vectors. This "sapromyiophily", observed also in various angiosperms and mosses, is an example of convergent evolution in both the plants and fungi that results in the recruitment of insects for dispersal (Johnson and Jurgens 2010). The odors produced by the carrion and fecal mimics are used by flies as ovipositional cues; however, the gleba do not support the development of insect progeny.

8 Summary

At present, only the initial aspects of communication that exist between fungi and the insects have been elucidated. Both the volatile organic compounds (VOCs) and the contact chemistries serve as communication signals. To date, insect chemical ecologists have conducted much research on insect-fungal interactions, focusing on insects' response to fungal VOCs. Insects, the larger, more mobile organisms, are the responders, whereas the sessile fungi are the producers of the volatile signals. Various interactions have been identified that benefit the insect:

fungivores use fungal VOCs to find appropriate fruiting bodies; parasitoids and predators cue on the volatiles produced by fungal associates of targeted hosts; insects utilize fungal volatiles as infochemicals to find appropriate oviposition and feeding sites; and insects directly use fungal volatiles to modify their habitats. In many cases, the fungal VOCs are a blend of compounds that at specific concentrations and in precise combinations operate as insect attractants.

On the other hand, fungi also benefit from the responses of insects to VOCs. Certain fungi produce attractants that mimic floral scents, insect pheromones, carrion, or feces that attract insects for the primary function of spore dispersal. In some cases, these attractants are complemented by the production of insect nutrients. However, little is known about the response of fungi to insect VOCs; it is unlikely that these organisms, lacking the chemoreceptors/CNS of insects, respond to VOCs with the sensitivity or fidelity displayed by insects. However, in confined habitats in which high insect densities exist, as is the case with social insects, it is possible that insect VOCs act as fumigants, regulating fungal growth and development.

What is known is that fungi respond to chemicals that come into contact with the cell wall. This fact is clearly observed in cuticular chemistries that either stimulate or suppress germ tube formation of entomopathogenic fungi. Furthermore, as-yet unidentified compounds are responsible for the synchronous switch from budding to the hyphal-growth program that occurs in the insect hemocoel. Potentially, this switch is due to a host or fungal cue that operates as a quorum-sensing agent that triggers the fungus to switch to the tissue-invasive phenotype. Fungi developing in the insects produce an array of metabolites, including enzymes, immunosuppressive and cytolytic peptides, and cell wall components, that potentially serve as pathogen-associated molecular patterns (PAMPs). The bidirectional cross-communication that occurs between those fungi that invade or that exist as insect symbionts is unknown territory. Recent large-scale sequencing projects are providing a foundation for conducting functional molecular studies required to decipher these interactions.

Acknowledgements In part this work was supported by funding provided through a specific cooperative program between P. Teal (CMAVE, USDA/ARS) and D. Boucias (University of Florida, Gainesville).

References

Adams AS, Jordan MS, Adams SM, Suen G, Goodwin LA, Davenport KW, Currie CR, Raffa KF (2011) Cellulose-degrading bacteria associated with the invasive woodwasp *Sirex noctilio*. ISME J 5:1323–1331

Altizer SM, Thrall PH, Antonovics J (1998) Vector behavior and the transmission of anther-smut infection in *Silene alba*. Am Midl Nat 139:147–163

Andersen SB, Gerritsma S, Yusah KM, Mayntz D, Hywel-Jones NL, Billen J, Boomsma JJ, Hughes DP (2009) The life of a dead ant: the expression of an adaptive extended phenotype. Am Nat 174:424–433

Atkinson W, Shorrocks B (1977) Breeding site specificity in domestic species of *Drosophila*. Oecologia 29:223–232

Bacon CW, Lyons PC, Porter JK, Robbins JD (1986) Ergot toxicity from endophyte-infected grasses – a review. Agron J 78:106–116

Barbosa P, Krischik VA, Jones CG (1991) Microbial mediation of plant-herbivore interactions. Wiley, New York/Chichester

Batra LR (1991) World species of *Monilinia* (Fungi): their ecology, biosystematics and control (Mycological Memoir). J. Cramer, Berlin

Batra LR, Batra SWT (1985) Floral mimicry induced by mummy-berry fungus exploits hosts pollinators as vectors. Science 228:1011–1013

Beauvais A, Latge JP, Vey A, Prevost MC (1989) The role of surface components of the entomopathogenic fungus *Entomophaga aulicae* in the cellular immune response of *Galleria mellonella* (Lepidoptera). J Gen Microbiol 135:489–498

Benda ND, Boucias D, Torto B, Teal P (2008) Detection and characterization of *Kodamaea ohmeri* associated with small hive beetle *Aethina tumida* infesting honey bee hives. J Apic Res 47:194–201

Berg A, Ehnstrom B, Gustafsson L, Hallingback T, Jonsell M, Weslien J (1994) Threatened plant, animal, and fungus species in Swedish forests: distribution and habitat associations. Conserv Biol 8:718–731

Boddy L, Jones TH (2008) Interactions between basidiomycota and invertebrates. Br Mycol Soc Symp Ser 28:155–179

Bogus MI, Kedra E, Bania J, Szczepanik M, Czygier M, Jablonski P, Pasztaleniec A, Samborski J, Mazgajska J, Polanowski A (2007) Different defense strategies of *Dendrolimus pini*, *Galleria mellonella*, and *Calliphora vicina* against fungal infection. J Insect Physiol 53:909–922

Bogus MI, Czygier M, Golebiowski M, Kedra E, Kucinska J, Mazgajska J, Samborski J, Wieloch W, Wloka E (2010) Effects of insect cuticular fatty acids on *in vitro* growth and pathogenicity of the entomopathogenic fungus *Conidiobolus coronatus*. Exp Parasitol 125:400–408

Boone CK, Six DL, Zheng YB, Raffa KF (2008) Parasitoids and dipteran predators exploit volatiles from microbial symbionts to locate bark beetles. Environ Entomol 37:150–161

Bot ANM, Ortius-Lechner D, Finster K, Maile R, Boomsma JJ (2002) Variable sensitivity of fungi and bacteria to compounds produced by the metapleural glands of leaf-cutting ants. Insect Soc 49:363–370

Boucias DG, Pendland JC (1987) Detection of protease inhibitors in the hemolymph of resistant *Anticarsia gemmatalis* which are inhibitory to the entomopathogenic fungus, *Nomuraea rileyi*. Experientia 43:336–339

Boucias DG, Pendland JC (1993) The galactose binding lectin from the beet armyworm, *Spodoptera exigua*: distribution and site of synthesis. Insect Biochem Mol Biol 23:233–242

Boucias DG, Pendland JC (1998) Principles of insect pathology. Kluwer Academic, Boston

Boucias DG, Pendland JC, Latge JP (1988) Nonspecific factors involved in attachment of entomopathogenic deuteromycetes to host insect cuticle. Appl Environ Microbiol 54:1795–1805

Boucias DG, Stokes C, Storey G, Pendland JC (1996) The effects of imidacloprid on the termite *Reticulitermes flavipes* and its interaction with the mycopathogen *Beauveria bassiana*. Pflanzenschutz-Nachrichten Bayer 49:103–144

Bulmer MS, Bachelet I, Raman R, Rosengaus RB, Sasisekharan R (2009) Targeting an antimicrobial effector function in insect immunity as a pest control strategy. Proc Natl Acad Sci USA 106:12652–12657

Bultman TL, White JF (1988) "Pollination" of a fungus by a fly. Oecologia 75:317–319

Bultman TL, White JF, Bowdish TI, Welch AM (1998) A new kind of mutualism between fungi and insects. Mycol Res 102:235–238

Bultman TL, Welch AM, Boning RA, Bowdish TI (2000) The cost of mutualism in a fly-fungus interaction. Oecologia 124:85–90

Chandra A, Huff DR (2008) *Salmacisia*, a new genus of Tilletiales: reclassification of *Tilletia buchloeana* causing induced hermaphroditism in buffalograss. Mycologia 100:81–93

Chang VCS, Jensen L (1974) Transmission of pineapple disease organism of sugarcane by nitidulid beetles in Hawaii. J Econ Entomol 67:190–192

Charnley AK (2003) Fungal pathogens of insects: cuticle degrading enzymes and toxins. Adv Bot Res 40:241–321

Choe DH, Millar JG, Rust MK (2009) Chemical signals associated with life inhibit necrophoresis in Argentine ants. Proc Natl Acad Sci USA 106:8251–8255

Connick WJ, French RC (1991) Volatiles emitted during the sexual stage of the Canada thistle rust fungus and by thistle flowers. J Agric Food Chem 39:185–188

Crespo R, Juarez MP, Cafferata LFR (2000) Biochemical interaction between entomopathogenous fungi and their insect-host-like hydrocarbons. Mycologia 92:528–536

Crespo R, Pedrini N, Juarez MP, Dal Bello GM (2008) Volatile organic compounds released by the entomopathogenic fungus *Beauveria bassiana*. Microbiol Res 163:148–151

Dobson HEM (1993) Floral volatiles in insect biology. In: Bernays EA (ed) Insect-plant interactions. CRC Press, Boca Raton, pp 47–81

Douglas AE (2009) The microbial dimension in insect nutritional ecology. Funct Ecol 23:38–47

Faldt J, Jonsell M, Nordlander G, Borg-Karlson AK (1999) Volatiles of bracket fungi *Fomitopsis pinicola* and *Fomes fomentarius* and their functions as insect attractants. J Chem Ecol 25:567–590

Fargues J (1984) Adhesion of the fungal spore to the insect cuticle in relation to pathogenicity. In: Roberts DW, Aist JR (eds) Infection processes of fungi, Foundation conference reports. Rockefeller, New York, pp 90–110

Febvay G, Decharme M, Kermarrec A (1984) Digestion of chitin by the labial glands of *Acromyrmex octospinosus* Reich (Hymenoptera: Formicidae). Can J Zool 62:229–234

Fransen JJ, Vanlenteren JC (1993) Host selection and survival of the parasitoid *Encarsia formosa* on greenhouse-whitefly, *Trialeurodes vaporariorum*, in the presence of hosts infected with the fungus *Aschersonia aleyrodis*. Entomol Exp Appl 69:239–249

Gao QA, Jin K, Ying SH, Zhang YJ, Xiao GH, Shang YF, Duan ZB, Hu XA, Xie XQ, Zhou G, Peng GX, Luo ZB, Huang W, Wang B, Fang WG, Wang SB, Zhong Y, Ma LJ, St Leger RJ, Zhao GP, Pei Y, Feng MG, Xia YX, Wang CS (2011) Genome sequencing and comparative transcriptomics of the model entomopathogenic fungi *Metarhizium anisopliae* and *M. acridum*. PLoS Genet 7(1):e1001264

Gibson CM, Hunter MS (2010) Extraordinarily widespread and fantastically complex: comparative biology of endosymbiotic bacterial and fungal mutualists of insects. Ecol Lett 13:223–234

Gillespie JP, Bailey AM, Cobb B, Vilcinskas A (2000) Fungi as elicitors of insect immune responses. Arch Insect Biochem Physiol 44:49–68

Golebiowski M, Malinski E, Bogus MI, Kumirska J, Stepnowski P (2008) The cuticular fatty acids of *Calliphora vicina*, *Dendrolimus pini* and *Galleria mellonella* larvae and their role in resistance to fungal infection. Insect Biochem Mol Biol 38:619–627

Gorzynska K, Lembicz M, Olszanowski Z, Leuchtmann A (2011) *Botanophila-Epichloe* interaction in a wild grass, *Puccinellia distans*, lacks dependence on the fly vector. Ann Entomol Soc Am 104:841–846

Gottar M, Gobert V, Matskevich AA, Reichhart JM, Wang CS, Butt TM, Belvin M, Hoffmann JA, Ferrandon D (2006) Dual detection of fungal infections in *Drosophila* via recognition of glucans and sensing of virulence factors. Cell 127:1425–1437

Gross J, Muller C, Vilcinskas A, Hilker M (1998) Antimicrobial activity of exocrine glandular secretions, hemolymph, and larval regurgitate of the mustard leaf beetle *Phaedon cochleariae*. J Invertebr Pathol 72:296–303

Grove JF, Blight MM (1983) The oviposition attractant for the mushroom phorid *Megaselia halterata* – the identification of volatiles present in mushroom house air. J Sci Food Agric 34:181–185

Guevara R, Rayner ADM, Reynolds SE (2000) Orientation of specialist and generalist fungivorous ciid beetles to host and non-host odours. Physiol Entomol 25:288–295

Hanski I (1989) Fungivory: fungi, insects and ecology. In: Wilding N, Collins NM, Hammond PM, Webber JF (eds) Insect-fungus interactions. Academic, London, pp 25–68

Hinds TE (1972) Insect transmission of *Ceratocystis* species associated with aspen cankers. Phytopathology 62:221–226

Holder DJ, Keyhani NO (2005) Adhesion of the entomopathogenic fungus *Beauveria* (*Cordyceps*) *bassiana* to substrata. Appl Environ Microbiol 71:5260–5266

Honda H, Ishiwatari T, Matsumoto Y (1988) Fungal volatiles as oviposition attractants for the yellow peach moth, *Conogethes punctiferalis* (Guenee) (Lepidoptera: Pyralidae). J Insect Physiol 34:205–211

Howard RW, Lord JC (2003) Cuticular lipids of the booklouse, *Liposcelis bostrychophila*: hydrocarbons, aldehydes, fatty acids, and fatty acid amides. J Chem Ecol 29:615–627

Howard DF, Tschinkel WR (1976) Aspects of necrophoric behavior in the red imported fire ant, *Solenopsis invicta*. Behaviour 56:157–180

Hung S-Y, Boucias DG (1992) Influence of *Beauveria bassiana* on the cellular defense response of the beet armyworm, *Spodoptera exigua*. J Invertebr Pathol 60:152–158

Huxham IM, Lackie AM, McCorkindale NJ (1989) Inhibitory effects of cyclodepsipeptides, destruxins, from the fungus *Metarhizium anisopliae* on cellular immunity in insects. J Insect Physiol 35:97–105

Jennersten O (1988) Insect dispersal of fungal disease: effects of *Ustilago* infection on pollinator attraction in *Viscaria vulgaris*. Oikos 51:163–170

Jennersten O, Kwak MM (1991) Competition for bumblebee visitation between *Melampyrum pratense* and *Viscaria vulgaris* with healthy and *Ustilago*-infected flowers. Oecologia 86:88–98

Jin K, Zhang YJ, Fang WG, Luo ZB, Zhou YH, Pei Y (2010) Carboxylate transporter gene *JEN1* from the entomopathogenic fungus *Beauveria bassiana* is involved in conidiation and virulence. Appl Environ Microbiol 76:254–263

Johnson SD, Jurgens A (2010) Convergent evolution of carrion and faecal scent mimicry in fly-pollinated angiosperm flowers and a stinkhorn fungus. S Afr J Bot 76:796–807

Jonsell M, Nordlander G (1995) Field attraction of Coleoptera to odours of the wood-decaying polypores *Fomitopsis pinicola* and *Fomes fomentarius*. Ann Zool Fenn 32:391–402

Jonsell M, Nordlander G (2004) Host selection patterns in insects breeding in bracket fungi. Ecol Entomol 29:697–705

Julian GE, Cahan S (1999) Undertaking specialization in the desert leaf-cutter ant *Acromyrmex versicolor*. Anim Behav 58:437–442

Kaiser R (2006) Flowers and fungi use scents to mimic each other. Science 311:806–807

Kerwin JL (1982) Biological aspects of the interaction between two entomogenous fungi, *Coelomomyces psorophorae* and *Entomophthora culicis*, and their dipteran hosts. Dissert Abstr Int B Phys Sci Eng 42:1291

Kerwin JL (1984) Fatty acid regulation of the germination of *Erynia variabilis* conidia on adults and puparia of the lesser housefly, *Fannia canicularis*. Can J Microbiol 30:158–161

Kilaru A, Bailey BA, Hasenstein KH (2007) *Moniliophthora perniciosa* produces hormones and alters endogenous auxin and salicylic acid in infected cocoa leaves. FEMS Microbiol Lett 274:238–244

Kohlmeyer J, Kohlmeyer E (1974) Distribution of *Epichloe typhina* (Ascomycetes) and its parasitic fly. Mycologia 66:77–86

Komonen A, Penttila R, Lindgren M, Hanski I (2000) Forest fragmentation truncates a food chain based on an old-growth forest bracket fungus. Oikos 90:119–126

Kukor JJ, Martin MM (1983) Acquisition of digestive enzymes by siricid woodwasps from their fungal symbiont. Science 220:1161–1163

Lamberty M, Zachary D, Lanot R, Bordereau C, Robert A, Hoffmann JA, Bulet P (2001) Insect immunity – constitutive expression of a cysteine-rich antifungal and a linear antibacterial peptide in a termite insect. J Biol Chem 276:4085–4092

Latgé J-P, Boucias DG, Fournet B (1988) Structure of the exocellular polysaccharide produced by the fungus, *Nomuraea rileyi*. Carbohydr Res 181:282–286

Lecuona R, Riba G, Cassier P, Clement JL (1991) Alterations of insect epicuticular hydrocarbons during infection with *Beauveria bassiana* or *B. brongniartii*. J Invertebr Pathol 58:10–18
Lee JC, Hamud SM, Negron JF, Witcosky JJ, Seybold SJ (2010) Semiochemical-mediated flight strategies of two invasive elm bark beetles: a potential factor in competitive displacement. Environ Entomol 39:642–652
Lin HC, Phelan PL (1992) Comparison of volatiles from beetle-transmitted *Ceratocystis fagacearum* and four non-insect-dependent fungi. J Chem Ecol 18:1623–1632
Lord JC (2001) Response of the wasp *Cephalonomia tarsalis* (Hymenoptera: Bethylidae) to *Beauveria bassiana* (Hyphomycetes: Moniliales) as free conidia or infection in its host, the sawtoothed grain beetle, *Oryzaephilus surinamensis* (Coleoptera: Silvanidae). Biol Control 21:300–304
Mackichan J, Thomsen L, Kerwin J, Latge JP, Beauvais A (1995) Unsaturated fatty acids are the active molecules of a glucan-synthase-inhibitory fraction isolated from entomophthoralean protoplasts. Microbiology 141:2757–2762
Madden JL (1968) Behavioural responses of parasites to the symbiotic fungus associated with *Sirex noctilio* F. Nature 218:189–190
Maitland DP (1994) A parasitic fungus infecting yellow dungflies manipulates host perching behavior. Proc R Soc Lond Ser B Biol Sci 258:187–193
Martin MM (1979) Biochemical Implications of insect mycophagy. Biol Rev Cambridge Philos Soc 54:1–21
Martinez AS, Fernandez-Arhex V, Corley JC (2006) Chemical information from the fungus *Amylostereum areolatum* and host-foraging behaviour in the parasitoid *Ibalia leucospoides*. Physiol Entomol 31:336–340
Masterman R, Ross R, Mesce K, Spivak M (2001) Olfactory and behavioral response thresholds to odors of diseased brood differ between hygienic and non-hygienic honey bees (*Apis mellifera* L.). J Comp Physiol A Sens Neural Behav Physiol 187:441–452
Matsuura K (2006) Termite-egg mimicry by a sclerotium-forming fungus. Proc R Soc B Biol Sci 273:1203–1209
Matsuura K, Tanaka C, Nishida T (2000) Symbiosis of a termite and a sclerotium-forming fungus: sclerotia mimic termite eggs. Ecol Res 15:405–414
Matsuura K, Yashiro T, Shimizu K, Tatsumi S, Tamura T (2009) Cuckoo fungus mimics termite eggs by producing the cellulose-digesting enzyme beta-glucosidase. Curr Biol 19:30–36
Mazet I, Pendland JC, Boucias DG (1994) Comparative analysis of phagocytosis of fungal cells by insect hemocytes versus horse neutrophils. Dev Comp Immunol 18:455–466
McLeod G, Gries R, von Reuss SH, Rahe JE, McIntosh R, Konig WA, Gries G (2005) The pathogen causing Dutch elm disease makes host trees attract insect vectors. Proc R Soc Lond Ser B Biol Sci 272:2499–2503
Meyling NV, Pell JK (2006) Detection and avoidance of an entomopathogenic fungus by a generalist insect predator. Ecol Entomol 31:162–171
Moller AP (1993) A fungus infecting domestic flies manipulates sexual behavior of its host. Behav Ecol Sociobiol 33:403–407
Moser JC, Konrad H, Blomquist SR, Kirisits T (2010) Do mites phoretic on elm bark beetles contribute to the transmission of Dutch elm disease? Naturwissenschaften 97:219–227
Mueller UG, Schultz TR, Currie CR, Adams RMM, Malloch D (2001) The origin of the attine ant-fungus mutualism. Q Rev Biol 76:169–197
Myles TG (2002) Alarm, aggregation, and defense by *Reticulitermes flavipes* in response to a naturally occurring isolate of *Metarhizium anisopliae*. Sociobiology 40:243–255
Naef A, Roy BA, Kaiser R, Honegger R (2002) Insect-mediated reproduction of systemic infections by *Puccinia arrhenatheri* on *Berberis vulgaris*. New Phytol 154:717–730
Ngugi HK, Scherm H (2004) Pollen mimicry during infection of blueberry flowers by conidia of *Monilinia vaccinii-corymbosi*. Physiol Mol Plant Pathol 64:113–123
Ngugi HK, Scherm H, Lehman JS (2002) Relationships between blueberry flower age, pollination, and conidial infection by *Monilinia vaccinii-corymbosi*. Phytopathology 92:1104–1109

Ochiai M, Ashida M (2000) A pattern-recognition protein for beta-1,3-glucan. The binding domain and the cDNA cloning of beta-1,3-glucan recognition protein from the silkworm, *Bombyx mori*. J Biol Chem 275:4995–5002

Oi DH, Pereira RM (1993) Ant behavior and microbial pathogens (Hymenoptera: Formicidae). Florida Entomol 76:63–74

Ormond EL, Thomas APM, Pell JK, Freeman SN, Roy HE (2011) Avoidance of a generalist entomopathogenic fungus by the ladybird, *Coccinella septempunctata*. FEMS Microbiol Ecol 77:229–237

Ortius-Lechner D, Maile R, Morgan ED, Boomsma JJ (2000) Metapleural gland secretion of the leaf-cutter ant *Acromyrmex octospinosus*: new compounds and their functional significance. J Chem Ecol 26:1667–1683

Pedrini N, Crespo R, Juarez MP (2007) Biochemistry of insect epicuticle degradation by entomopathogenic fungi. Comp Biochem Physiol C Toxicol Pharmacol 146:124–137

Pendland JC, Boucias DG (1992) Ultrastructural localization of carbohydrate in cell walls of the entomogenous hyphomycete *Nomuraea rileyi*. Can J Microbiol 38:377–386

Pendland JC, Boucias DG (1993) Variations in the ability of galactose and mannose specific lectins to bind to cell wall surfaces during growth of the insect pathogenic fungus *Paecilomyces farinosus*. Eur J Cell Biol 60:322–330

Pendland JC, Boucias DG (1998) Characterization of monoclonal antibodies against cell wall epitopes of the insect pathogenic fungus, *Nomuraea rileyi*: differential binding to fungal surfaces and cross-reactivity with host hemocytes and basement membrane components. Eur J Cell Biol 75:118–127

Pendland JC, Boucias DG (2000) Comparative analysis of the binding of antibodies prepared against the insect *Spodoptera exigua* and against the mycopathogen *Nomuraea rileyi*. J Invertebr Pathol 75:107–116

Pfeil RM, Mumma RO (1992) Air sampling of volatiles from *Agaricus bisporus* in a mushroom facility and from mushroom compost. Hortscience 27:416–419

Pfeil RM, Mumma RO (1993) Bioassay for evaluating attraction of the phorid fly, *Megaselia halterata*, to compost colonized by the commercial mushroom, *Agaricus bisporus*, and to 1-octen-3-ol and 3-octanone. Entomol Exp Appl 69:137–144

Phelan PL, Lin HC (1991) Chemical characterization of fruit and fungal volatiles attractive to dried-fruit beetle, *Carpophilus hemipterus* (L) (Coleoptera: Nitidulidae). J Chem Ecol 17:1253–1272

Pontoppidan MB, Himaman W, Hywel-Jones NL, Boomsma JJ, Hughes DP (2009) Graveyards on the move: the spatio-temporal distribution of dead *Ophiocordyceps*-infected ants. PLoS One 4: e4835

Poulsen M, Bot ANM, Nielsen MG, Boomsma JJ (2002) Experimental evidence for the costs and hygienic significance of the antibiotic metapleural gland secretion in leaf-cutting ants. Behav Ecol Sociobiol 52:151–157

Raguso RA, Roy BA (1998) 'Floral' scent production by *Puccinia rust* fungi that mimic flowers. Mol Ecol 7:1127–1136

Renucci M, Tirard A, Provost E (2011) Complex undertaking behavior in *Temnothorax lichtensteini* ant colonies: from corpse-burying behavior to necrophoric behavior. Insect Soc 58:9–16

Roh KB, Kim CH, Lee H, Kwon HM, Park JW, Ryu JH, Kurokawa K, Ha NC, Lee WJ, Lemaitre B, Soderhall K, Lee BL (2009) Proteolytic cascade for the activation of the insect toll pathway induced by the fungal cell wall component. J Biol Chem 284:19474–19481

Rosengaus RB, Jordan C, Lefebvre ML, Traniello JFA (1999) Pathogen alarm behavior in a termite: a new form of communication in social insects. Naturwissenschaften 86:544–548

Roy BA (1993) Floral mimicry by a plant pathogen. Nature 362:56–58

Roy BA (1994) The effects of pathogen-induced pseudoflowers and buttercups on each others insect visitation. Ecology 75:352–358

Roy BA, Raguso RA (1997) Olfactory versus visual cues in a floral mimicry system. Oecologia 109:414–426

Roy HE, Steinkraus DC, Eilenberg J, Hajek AE, Pell JK (2006) Bizarre interactions and endgames: entomopathogenic fungi and their arthropod hosts. Annu Rev Entomol 51:331–357

Roy HE, Brown PMJ, Rothery P, Ware RL, Majerus MEN (2008) Interactions between the fungal pathogen *Beauveria bassiana* and three species of coccinellid: *Harmonia axyridis*, *Coccinella septempunctata* and *Adalia bipunctata*. Biocontrol 53:265–276

Schiestl FP, Steinebrunner F, Schulz C, von Reuss S, Francke W, Weymuth C, Leuchtmann A (2006) Evolution of 'pollinator'-attracting signals in fungi. Biol Lett 2:401–404

Schwemmler W, Gassner G (1989) Insect endocytobiosis: morphology, physiology, genetics, evolution. CRC Press, Boca Raton

Silbering AF, Benton R (2010) Ionotropic and metabotropic mechanisms in chemoreception: 'chance or design'? EMBO Rep 11:173–179

Sloman IS, Reynolds SE (1993) Inhibition of ecdysteroid secretion from *Manduca* prothoracic glands in vitro by destruxins – cyclic depsipeptide toxins from the insect pathogenic fungus *Metarhizium anisopliae*. Insect Biochem Mol Biol 23:43–46

Sosa-Gomez DR, Boucias DG, Nation JL (1997) Attachment of *Metarhizium anisopliae* to the southern green stinkbug *Nezara viridula* cuticle and fungistatic effect of cuticular lipids and aldehydes. J Invertebr Pathol 69:31–39

Spradber JP (1974) Responses of *Ibalia* species (Hymenoptera: Ibaliidae) to fungal symbionts of siricid woodwasp hosts. J Entomol 48:217–222

Staples JA, Milner RJ (2000) A laboratory evaluation of the repellency of *Metarhizium anisopliae* conidia to *Coptotermes lacteus* (Isoptera: Rhinotermitidae). Sociobiology 36:133–148

Starratt AN, Loschiavo SR (1970) Chemical stimuli from fungus *Nigrospora sphaerica* that induce aggregation of confused flour beetle, *Tribolium confusum* Duval, pest. Abstr Pap Am Chem Soc 1970:21

Steinebrunner F, Schiestl FP, Leuchtmann A (2008a) Ecological role of volatiles produced by *Epichloe*: differences in antifungal toxicity. FEMS Microbiol Ecol 64:307–316

Steinebrunner F, Schiestl FP, Leuchtmann A (2008b) Variation of insect attracting odor in endophytic *Epichloe* fungi: phylogenetic constrains versus host influence. J Chem Ecol 34:772–782

Steinebrunner F, Twele R, Francke W, Leuchtmann A, Schiestl FP (2008c) Role of odour compounds in the attraction of gamete vectors in endophytic *Epichloe* fungi. New Phytol 178:401–411

Steiner S, Erdmann D, Steidle JLM, Ruther J (2007) Host habitat assessment by a parasitoid using fungal volatiles. Front Zool 4:3 (10 pp)

St Leger RJ, Joshi L, Roberts D (1998) Ambient pH is a major determinant in the expression of cuticle-degrading enzymes and hydrophobin by *Metarhizium anisopliae*. Appl Environ Microbiol 64:709–713

St Leger RJ, Nelson JO, Screen SE (1999) The entomopathogenic fungus *Metarhizium anisopliae* alters ambient pH, allowing extracellular protease production and activity. Microbiology SGM 145:2691–2699

Storey GK, Vandermeer RK, Boucias DG, McCoy CW (1991) Effect of fire ant (*Solenopsis invicta*) venom alkaloids on the in vitro germination and development of selected entomogenous fungi. J Invertebr Pathol 58:88–95

Sullivan BT, Berisford CW (2004) Semiochemicals from fungal associates of bark beetles may mediate host location behavior of parasitoids. J Chem Ecol 30:703–717

Swanson JAI, Torto B, Kells SA, Mesce KA, Tumlinson JH, Spivak M (2009) Odorants that induce hygienic behavior in honeybees: identification of volatile compounds in chalkbrood-infected honeybee larvae. J Chem Ecol 35:1108–1116

Tartar A, Shapiro AM, Scharf DW, Boucias DG (2005) Differential expression of chitin synthase (*CHS*) and glucan synthase (*FKS*) genes correlates with the formation of a modified, thinner cell wall in *in vivo*-produced *Beauveria bassiana* cells. Mycopathologia 160:303–314

Thakeow P, Angeli S, Weissbecker B, Schutz S (2008) Antennal and behavioral responses of *Cis boleti* to fungal odor of *Trametes gibbosa*. Chem Senses 33:379–387

Torto B, Arbogast RT, Van Engelsdorp D, Willms S, Purcell D, Boucias D, Tumlinson JH, Teal PEA (2007a) Trapping of *Aethina tumida* Murray (Coleoptera: Nitidulidae) from *Apis mellifera* L. (Hymenoptera: Apidae) colonies with an in-hive baited trap. Environ Entomol 36:1018–1024

Torto B, Boucias DG, Arbogast RT, Tumlinson JH, Teal PEA (2007b) Multitrophic interaction facilitates parasite-host relationship between an invasive beetle and the honey bee. Proc Natl Acad Sci USA 104:8374–8378

Trumbo ST, Robinson GE (1997) Learning and task interference by corpse-removal specialists in honey bee colonies. Ethology 103:966–975

Tsitsigiannis DI, Keller NP (2007) Oxylipins as developmental and host-fungal communication signals. Trends Microbiol 15:109–118

Uchida W, Matsunaga S, Sugiyama R, Kazama Y, Kawano S (2003) Morphological development of anthers induced by the dimorphic smut fungus *Microbotryum violaceum* in female flowers of the dioecious plant *Silene latifolia*. Planta 218:240–248

Veal DA, Trimble JE, Beattie AJ (1992) Antimicrobial properties of secretions from the metapleural glands of Myrmecia gulosa (the Australian bull ant). J Appl Bacteriol 72:188–194

Vega FE, Blackwell M (2005) Insect-fungal associations: ecology and evolution. Oxford University Press, Oxford

Vet LEM, Janse C, Vanachterberg C, Vanalphen JJM (1984) Microhabitat location and niche segregation in two sibling species of drosophilid parasitoids: *Asobara tabida* (Nees) and *A. rufescens* (Foerster) (Braconidae: Alysiinae). Oecologia 61:182–188

Vey A (1985) Efficiency of the encapsulation reaction in insects: effect on the structure and the viability of the encapsulated fungus. Dev Comp Immunol 9:174–174

Vilcinskas A, Matha V, Götz P (1997) Inhibition of phagocytic activity of plasmatocytes isolated from *Galleria mellonella* by entomopathogenous fungi and their secondary metabolites. J Insect Physiol 43:475–483

Visscher PK (1983) The honey bee way of death: necrophoric behavior in *Apis mellifera* colonies. Anim Behav 31:1070–1076

Wang C, Leger RJS (2006) A collagenous protective coat enables *Metarhizium anisopliae* to evade insect immune responses. Proc Natl Acad Sci USA 103:6647–6652

Wang CS, Hu G, St Leger RJ (2005) Differential gene expression by *Metarhizium anisopliae* growing in root exudate and host (*Manduca sexta*) cuticle or hemolymph reveals mechanisms of physiological adaptation. Fungal Genet Biol 42:704–718

Welch AM, Bultman TL (1993) Natural release of *Epichloe typhina* ascospores and its temporal relationship to fly parasitism. Mycologia 85:756–763

Wheeler Q, Blackwell M (1984) Fungus-insect relationships: perspectives in ecology and evolution. Columbia University Press, New York

Wilson EO, Durlach NI, Roth LM (1958) Chemical releasers of necrophoric behavior in ants. Psyche 65:108–114

Yanagawa A, Yokohari F, Shimizu S (2009) The role of antennae in removing entomopathogenic fungi from cuticle of the termite, *Coptotermes formosanus*. J Insect Sci 9:9

Yoshida S, Yamashita M, Yonehara S, Eguchi M (1990) Properties of fungal protease inhibitors from the integument and hemolymph of the silkworm and effect of an inhibitor on the fungal growth. Comp Biochem Physiol B Biochem Mol Biol 95:559–564

Zhang YJ, Zhang JQ, Jiang XD, Wang GJ, Luo ZB, Fan YH, Wu ZQ, Pei Y (2010) Requirement of a mitogen-activated protein kinase for appressorium formation and penetration of insect cuticle by the entomopathogenic fungus *Beauveria bassiana*. Appl Environ Microbiol 76:2262–2270

Zurek L, Watson DW, Krasnoff SB, Schal C (2002) Effect of the entomopathogenic fungus, *Entomophthora muscae* (Zygomycetes: Entomophthoraceae), on sex pheromone and other cuticular hydrocarbons of the house fly, *Musca domestica*. J Invertebr Pathol 80:171–176

Index

G. Witzany (ed.), *Biocommunication of Fungi*, DOI 10.1007/978-94-007-4264-2,

MIX
Papier aus verantwortungsvollen Quellen
Paper from responsible sources
FSC® C105338

If you have any concerns about our products,
you can contact us on
ProductSafety@springernature.com

In case Publisher is established outside the EU,
the EU authorized representative is:
Springer Nature Customer Service Center GmbH
Europaplatz 3, 69115 Heidelberg, Germany

Printed by Libri Plureos GmbH
in Hamburg, Germany